Electron Crystallography

NATO Science Series

A Series presenting the results of scientific meetings supported under the NATO Science Programme.

The Series is published by IOS Press, Amsterdam, and Springer in conjunction with the NATO Public Diplomacy Division

Sub-Series

I. **Life and Behavioural Sciences**	IOS Press
II. **Mathematics, Physics and Chemistry**	Springer
III. **Computer and Systems Science**	IOS Press
IV. **Earth and Environmental Sciences**	Springer

The NATO Science Series continues the series of books published formerly as the NATO ASI Series.

The NATO Science Programme offers support for collaboration in civil science between scientists of countries of the Euro-Atlantic Partnership Council. The types of scientific meeting generally supported are "Advanced Study Institutes" and "Advanced Research Workshops", and the NATO Science Series collects together the results of these meetings. The meetings are co-organized by scientists from NATO countries and scientists from NATO's Partner countries – countries of the CIS and Central and Eastern Europe.

Advanced Study Institutes are high-level tutorial courses offering in-depth study of latest advances in a field.
Advanced Research Workshops are expert meetings aimed at critical assessment of a field, and identification of directions for future action.

As a consequence of the restructuring of the NATO Science Programme in 1999, the NATO Science Series was re-organized to the four sub-series noted above. Please consult the following web sites for information on previous volumes published in the Series.

http://www.nato.int/science
http://www.springer.com
http://www.iospress.nl

Electron Crystallography
Novel Approaches for Structure Determination of Nanosized Materials

edited by

Thomas E. Weirich

Central Facility for Electron Microscopy,
RWTH Aachen University,
Germany

János L. Lábár

Research Institute for Technical Physics and Materials Science,
Budapest, Hungary

and

Xiaodong Zou

Structural Chemistry,
Stockholm University, Sweden

 **Springer**

Published in cooperation with NATO Public Diplomacy Division

Proceedings of the NATO Advanced Study Institute on
Electron Crystallography: Novel Approaches for Structure Determination
of Nanosized Materials
Erice, Italy
10–14 June 2004

A C.I.P. Catalogue record for this book is available from the Library of Congress.

ISBN-10 1-4020-3919-0 (PB)
ISBN-13 978-1-4020-3919-5 (PB)
ISBN-10 1-4020-3918-2 (HB)
ISBN-13 978-1-4020-3918-8 (HB)
ISBN-10 1-4020-3920-4 (e-book)
ISBN-13 978-1-4020-3920-1 (e-book)

Published by Springer,
P.O. Box 17, 3300 AA Dordrecht, The Netherlands.

www.springer.com

Printed on acid-free paper

Table of Contents

D. Applications

List of (attending) contributors to this volume

A. Avilov
Shubnikov Institute of Crystallography
RAS, 59 Leninsky prospekt
Moscow 117333, Russia

G. Cox
BASF Aktiengesellschaft
GKP/A - M300
Polymer Phyiscs
Ludwigshafen 67056, Germany

M. Deanko
Dpt. Metal Phys. Inst. Phys. SAS
Dubravska cesta 9
Bratislava 84511, Slovakia

L. Demchenko
National Technical University of Ukraine
37 Peremogy prospect, Kyiv 3056, Ukraine

I. Djerdj
Department of Physics
Faculty of Science
Bijenička c. 32, POB 331
Zagreb 10002, Croatia

C. J. Gilmore
Department of Chemistry
Joseph Black Building
University of Glasgow
Glasgow G14 8QQ, Scotland

T. Gorelik
Institut für Physikalische Chemie
Johannes Gutenberg-Universität Mainz
Welderweg 11
Mainz 55099, Germany

I. Hargittai
Institute of General and Analytical Chemistry
Budapest University of Technology and Economics
POBox 91, Budapest 1521, Hungary

M. Hargittai
Structural Chemistry Research Group
of the Hungarian Academy of Sciences
at Eötvös University,
P. O. Box 32, Budapest 1518, Hungary

S. Hovmöller
Structural Chemistry
Arrhenius Laboratory
Stockholm University
Stockholm 106 91, Sweden

O.V. Hryhoryeva
National Technical University of Ukraine
37, Peremohy Ave., Kyiv 3056, Ukraine

J. Jansen
National Centre for HREM
Delft University of Technology
Rotterdamseweg 137
Delft 2628 AL, The Netherlands

H. Jiang
Nanotechnology, VTT
Technical Research Center of Finland,
P. O. Box 1602,
Espoo 02044, Finland

D.M. Kepaptsoglou
Laboratory of Physical Metallurgy
School of Mining and Metallurgical Engineering
National Technical University of Athens
9, Heroon Polytechniou str, Zografou Campus
Athens 157 80, Greece

C. Kisielowski
National Center for Electron Microscopy
Lawrence Berkeley National Laboratory
Berkeley CA 94720, USA

V. Klechkovskaya
Lab. of Electron Diffraction
Shubnikov Institute of Crystallography
RAS, 59 Leninsky prospekt
Moscow 117333, Russia

U. Kolb
Institut für Physikalische Chemie
Johannes Gutenberg-Universität Mainz
Welderweg 11
Mainz 55099, Germany

C. Kübel
FEI Company
Building AAE
Achsteweg Noord 5
Eindhoven 5651 GG, The Netherlands

J.L. Labar
Research Institute for Technical Physics and Materials Science
Thin Film Physics Department
Konkoly Thege ut 29-33, Budapest 1121, Hungary

G. Lepeshov
Shubnikov Institute of Crystallography
RAS, 59 Leninsky prospekt
Moscow 117333, Russia

J. Mayer
Gemeinschaftslabor für Elektronenmikroskopie
Rheinisch-Westfälische Technische Hochschule Aachen
Ahornstrasse 55
Aachen 52074, Germany

E. Montanari
Dip. di Chimica GIAF
Università di Parma
Parco Area delle Scienze 17A
Parma 43100, Italy

J.P. Morniroli
Laboratoire de Métallurgie Physique et Génie des Matériaux
UMR CNRS 8517
Université des Sciences et Technologies de Lille
Villeneuve d'Ascq Cédex 59655, France

M. Nickolsky
IGEM-RAS
Staromonetny per. 35
Moskow 119017, Russia

S. Nicolopoulos
Lab de Electrochimie
Université Libre de Bruxelles
Dpt. Science des Materiaux
Bruxelles 1050, Belgium

D. Nihtianova
Central Laboratory of Mineralogy and Crystallography
Acad. Georgi Bonchev Str., bl. 107,
Sofia 1113, Bulgaria

P. Oleynikov
Structural Chemistry
Arrhenius Laboratory
Stockholm University
Stockholm 106 91, Sweden

O.V. Pobydaylo
Institute for Metal Physics
36 Vernadsky St.
Kiev 3680, Ukraine

T. Shumilova
Institute of Geology
Pervomayskaya st 54
Syktyavkar 167982, Russia

J.C.H. Spence
Department of Physics and Astronomy
Arizona State University
P. O. Box 1504 Tempe AZ 85287, USA

P. Švec
Dpt. Metal Phys. Inst. Phys. SAS
Dubravska cesta 9
Bratislava 84511, Slovakia

C.Y. Tang
Beijing National Laboratory for Condensed Matter Physics
Institute of Physics, Chinese Academy of Sciences
Beijing 100080, China

O. Terasaki
Structural Chemistry
Arrhenius Laboratory
Stockholm University
Stockholm 106 91, Sweden

A. Tonejc
Department of Physics
Faculty of Science
Bijenička c. 32, POB 331
Zagreb 10002, Croatia

M. Tsuji
Laboratory of Polymer Condensed States
Division of States and Structures III
Institute for Chemical Research, Kyoto University
Uji, Kyoto-fu 611-0011, Japan

D. Wang
Fritz-Haber-Institut der Max-Planck-Gesellschaft
Faradayweg 4-6
Berlin 14195, Germany

Y.M. Wang
Beijing National Laboratory for Condensed Matter Physics
Institute of Physics, Chinese Academy of Sciences
Beijing 100080, China

T.E. Weirich
Gemeinschaftslabor für Elektronenmikroskopie
Rheinisch-Westfälische Technische Hochschule Aachen
Ahornstrasse 55
Aachen 52074, Germany

J.M. Zuo
Materials Science and Engineering and Materials Research Laboratory
University of Illinois
Urbana-Champaign
1304 W. Green Street
Urbana IL 61801, USA

X.D. Zou
Structural Chemistry
Arrhenius Laboratory
Stockholm University
Stockholm 106 91, Sweden

Preface

The discovery of electron diffraction in the laboratories of C.J. Davisson and G.P. Thomson in 1927 certainly ranks as one of the monumental discoveries in the history of science. Soon after this invention, it was realised that electrons - like X-rays - are diffracted by the atoms in matter and thus can be used for structure determination. Because the attainable scattered intensities for electrons are about 10^6 times larger than for X-rays, electron diffraction allows us to study the interior structure of matter down to a few cubic-nanometers in size. This is far beyond the capabilities of X-ray crystallography or even a modern synchrotron. However, with increasing understanding of the electron scattering process, in particular by development of n-beam scattering theory during the 1950s of the past century, it became general knowledge that it could normally *not* be assumed that the kinematical or single-scattering approximation holds even for very thin crystals. For this reason nearly all activities in this direction declined within a few years from that time, even though the groups of Z.G. Pinsker, B.K. Vainshtein and B.B. Zvyagin in Russia had proved the feasibility of this approach since about 1940. While X-ray diffraction turned over the following years into a powerful quasi-automatic method for structure determination, structure analysis by electron diffraction was not seriously developed further for a period of more than twenty years. It took until 1976 to perform the first structure determination by direct phasing methods from electron diffraction data of an organic compound by D.L. Dorset & H. Hauptman, and another eight years for the first *ab-initio* structure determination of a heavy-metal oxide from high-resolution electron microscopy images by S. Hovmöller and co-workers. Despite these achievements, the earlier skepticism against structural results from electrons continued, and remained a common mindset until a few years ago.

Nevertheless, during the last decade we have witnessed several exciting achievements in this field. This includes structural and charge density studies on organic molecules and protein structures, complicated inorganic and metallic materials in the amorphous, nano-, meso- and quasi-crystalline state and also development of new software, tailor-made for the special needs of electron crystallography. Moreover, these developments have been accompanied by a now available new generation of computer controlled electron microscopes equipped with high-coherent field-emission sources, cryo-specimen holders, ultra-fast CCD cameras, imaging plates, energy filters and even correctors for electron optical distortions. Thus, a fast and semi-automatic data acquisition from small sample areas, similar to what we today know from imaging plates diffraction systems in X-ray crystallography, can be envisioned for the very near future. This progress clearly shows that the contribution of electron crystallography is quite unique, as it enables us to reveal the intimate structure of samples with high accuracy, but on much smaller samples than have ever been investigated by X-ray diffraction. As a tribute to these tremendous recent achievements, this NATO Advanced Study Institute was devoted to the novel approaches of electron crystallography for structure determination of nano-sized materials. The course took place 9 – 20 June 2004 at the renowned Ettore Majorana Centre for Scientific Culture in Erice, Sicily and brought together 87 participants from 32 different countries.

The present volume contains the lectures given during this ASI and covers almost all theoretical and practical aspects of advanced transmission electron microscopy techniques and crystallographic methods that are relevant for determining structures of organic and inorganic materials. Moreover a number of extended abstracts on the presented posters during this ASI have been added to this volume.

Supplementary material (videos, handouts) of the course is available at `http://erice2004.docking.org/`

We are grateful for a number of sponsors of this ASI. First and foremost is the support from NATO (ASI 979903) acknowledged. The International Association for the Promotion of Co-operation with Scientists from the New Independent States (NIS) of the Former Soviet Union (INTAS) gave generous financial support as did the International Union of Crystallography, FEI Company (Kassel, Germany), Carl Zeiss NTS (Oberkochen, Germany), Ditabis (Pforzheim, Germany) and Gatan (Munich, Germany). We are also extremely grateful to Prof. Paola Spadon for her constant aid toward making this ASI possible. Special thanks are also given to Prof. Lodovico Riva di Sanseverino for his organisational skills, constant attention to all details, his valuable advice based on more than 30 years experience with these courses, and for competent managing of the budget of this ASI. Last but not least, we thank John Irwin for his rapid troubleshooting of all IT problems, and for setting up the web site for the supplementary material.

Thomas E. Weirich, Janos L. Labar, Xiaodong Zou
Aachen, Budapest, Stockholm, Spring 2005

Introduction

What is Electron Crystallography?

Xiaodong Zou
Structural Chemistry, Arrehnius Laboratory, Stockholm University, SE-106 91 Stockholm, Sweden

Abstract: Electron crystallography is the quantitative use of different information by electron scattering to study perfect crystal structures as well as defects and interfaces. The development of electron crystallography is summarized. Electron crystallography is compared with X-ray crystallography. Examples of recent developments in electron crystallography are presented.

Key words: Electron diffraction, crystallographic image processing, 3D reconstruction, Structure factors, ultrafast electron crystallography, aberration corrected electron microscope, oversampling phasing method

1. ELECTRON CRYSTALLOGRAPHY – AN INTRODUCTION

Most properties of matter depend on its structure. Many different techniques have been developed over the centuries for structure analysis of matter. Crystallography is the most important technique for obtaining the complete atomic structure of matter in a crystalline phase. Among the three main radiation sources used in crystallography, X-ray is the dominant one. X-ray crystallography was founded in 1912, and is still today the most important technique for studying atomic structures of crystalline materials. Electron diffraction (ED) of single crystals was discovered fifteen years later and thereafter the wave property of electrons was exploited in the invention of the electron microscope. Since then, electron microscopes have been used in many fields as a tool for exploring and visualizing the microscopic world in all its beauty. Between the first electron microscopy (EM) images by

3

T.E. Weirich et al. (eds.), Electron Crystallography, 3–16.
© 2006 *Springer. Printed in the Netherlands.*

Ruska 1933 and today's high resolution electron microscopy (HREM) images with atomic resolution lies seventy years of untiring engineering. More recently, the unprecedented power of computers has made it possible to analyze quantitatively, and even further improve, the HREM images. The amalgamation of quantitative electron diffraction and high (atomic) resolution electron microscopy with image processing has created a new powerful tool for structure analysis - electron crystallography. Modern analytical transmission electron microscopes now combine structure analysis by ED and HREM with chemical analysis by energy dispersive spectroscopy (EDS) and electron energy loss spectroscopy (EELS). This makes electron microscopy a powerful and complete analytical tool for studying materials. As a result, the concept of electron crystallography is further extended to the quantitative use of different information by electron scattering to study perfect crystal structures as well as defects and interfaces.

2. DEVELOPMENT OF ELECTRON CRYSTALLOGRAPHY

2.1 Electron diffraction analysis started early in Moscow

Structure analysis of crystals by electron diffraction (ED) was first pursued in 1937-1938, by a group of crystallographers in the former Soviet Union, led by Pinsker and Vainshtein. For twenty years, they used their own designed electron diffraction cameras with relatively low accelerating voltages (<100 kV) to collect electron diffraction data of a large number of materials, such as basic salts, metal nitrides and carbides, semiconducting alloys and clay minerals, as well as some organic crystals. Most of the materials were oblique textured and gave electron diffraction of texture patterns (Fig. 1). Using this electron diffraction data, they developed electron diffraction into a complete and independent structure analysis method. They addressed various problems which then could not be solved by X-ray diffraction, such as location of hydrogen atoms in crystals. These studies led to great expectations for the future of electron diffraction. Vainshtein wrote in his book (1956): *"There is no doubt now that electron diffraction may be used for the complete analysis of crystals whose structure is unknown"*. This pioneering work was summarized in a review article by Vainshtein, Zvyagin and Avilov (1992).

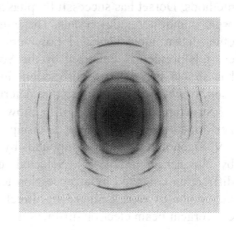

Figure 1 An oblique texture pattern of lizardite.

The method of structure analysis developed by the Soviet group was based on the kinematical approximation that ED intensity is directly related (proportional) to the square of structure factor amplitudes. The same method had also been applied by Cowley in Melbourne for solving a few structures. In 1957 Cowley and Moodie introduced the *n*-beam dynamical diffraction theory to the scattering of electrons by atoms and crystals. This theory provided the basis of multi-slice calculations which enabled the simulation of dynamical intensities of electron diffraction patterns, and later electron microscope images. The theory showed that if dynamical scattering is significant, intensities of electron diffraction are usually not related to structure factors in a simple way. Since that day, the fear of dynamical effects has hampered efforts to analyze structures by electron diffraction.

Experimental electron diffraction amplitudes, just as X-ray diffraction, do not contain the phase information of the structure factor which is needed for solving unknown structures. Various methods, such as the Patterson method and different trial and error techniques, were used for finding phase information in X-ray crystallography. These methods were used by the Soviet group also for phasing electron diffraction data. In 1953, Hauptmann and Karle introduced the so-called direct methods, for solving the phase problem. The direct methods, combined with the development of computers, accelerated very much the development of X-ray crystallography. In 1976, Dorset and Hauptmann in Buffalo for the first time applied the direct methods to electron diffraction data. This pushed structure analysis by electron diffraction a significant step forward. Using the direct phasing methods, Dorset has successfully phased electron diffraction data of various organic and inorganic crystals. These include aromatic molecules, lipids and other linear molecules and polymers, as well as some of the inorganic crystals previously collected by the Soviet group (Dorset 1995). His work has shown that electron diffraction, just as X-ray diffraction data, can be used for *ab initio* crystal structure determination.

Structure determination of unknown crystals by electron diffraction was performed by several research groups, on Al-Fe alloys by Gjønnes *et al.* (1998), on metal-cluster compounds by Weirich *et al.* (2000) and on zeolites by Wagner *et al.* (1999). Selected area electron diffraction or electron diffraction collected by a precession technique were used and the structure factor phases were deduced by direct methods, Patterson method or from convergent beam electron diffraction.

2.2 Electron microscopy can be combined with image processing

Quantitative electron microscopy (EM) for structure analysis can be traced back to the work by Aaron Klug and co-workers at the MRC laboratory of Molecular Biology in Cambridge in the 1960-ies. Klug started to study viruses by electron microscopy, soon after the double helix of DNA and the first protein structures were solved in that laboratory. His group developed optical diffraction and computerized image processing and first demonstrated 3D reconstruction by electron microscopy (DeRosier and Klug, 1968). This 3D reconstruction method is based on the fact that both phase and amplitude information are present in electron microscopy images and can be extracted from the Fourier transform of images by digitized image processing. They discussed various ways of determining the crystallographic structure factor phases which are needed for solving crystal structures and point out for the first time that these phases can be extracted directly from the Fourier transform of an EM image.

The 3D reconstruction method introduced by DeRosier and Klug created a revolution in structural molecular biology. Hundreds of macromolecular structures, including membrane proteins and viruses, were determined by this method. This method was further developed by Henderson and Unwin, and they determined first the 3D structure of a membrane protein, bacteriorhodopsin, to atomic resolution (Henderson and Unwin, 1975; Hendersson *et al.*, 1990). Since then several other membrane structures, including PhoE porin and light harvesting complex, have also been solved to near atomic resolution (3-4 Å) (Jap, Walian and Gehring, 1991; Kühlbrandt, Wang and Fujiyoshi, 1994). Due to his development of crystallographic electron microscopy and his structural elucidation of biologically important nucleic acid-protein complexes, Aron Klug was awarded the Nobel Prize in Chemistry in 1982, the only Novel Prize on electron crystallography.

The resolution of EM images of crystals of biological macromolecules is limited by large unit cells, poor crystal ordering and radiation sensitivity. Most inorganic crystals give much higher resolution than biological samples do, due to smaller unit cells, better ordering and less radiation sensitivity. It is relatively easy to get high resolution electron microscopy images and diffraction patterns from them. The first EM images near atomic resolution (4Å) showing details within a unit cell were obtained from a thin crystal of $Ti_2Nb_{10}O_{29}$ by Ijima (1971; 1978), then working at Arizona State University. It was found that the contrast of HREM images changed with optical conditions and crystal thickness. In order to interpret the contrast changes, O'Keefe in Melbourne (1973) developed the multi-slice calculation method to simulate series of images of $Ti_2Nb_{10}O_{29}$ under different defocus and thickness condition and successfully interpreted the contrast changes in the experimental images. The success of image simulation led many microscopists to study the effects of the contrast transfer function and specimen thickness on images. Soon a consensus was reached that experimental HREM images can never be directly interpreted and used for solving atomic structures, but have to be confirmed by image simulation.

There are several disadvantages with the image simulation method. A nearly correct structure model is needed beforehand. This is often not available, especially for relatively complicated structures. Images are compared visually and no quantitative figure of merit is used for judging how well images and simulations agree.

The crystallographic electron microscopy developed by Klug *et al.* has also great impacts on inorganic crystals. Klug (1978-79) applied the crystallographic image processing method to a through-focus series of HREM images of a $GeNb_9O_{25}$ crystal and corrected for the contrast changes caused by the contrast transfer function. They showed that the correct structure projection could be retrieved from every single image in the series. This was the first attempt to interpret images of inorganic crystals without simulation.

The crystallographic image processing method was then used by Hovmöller's group in Stockholm for solving crystal structures of a number of niobium oxides (Hovmöller *et al.* 1984). In the projected potential maps obtained by crystallographic image processing from experimental HREM images at 2.5Å resolution, all heavy atoms could be resolved. Atomic positions could be determined with an accuracy of about 0.1 Å. In the mean time, Li Fan-hua and Fan Hai-fu's group in Beijing developed the image processing method, combined with image deconvolution and phase extension by maximum entropy and direct methods, to solve the crystal structure of $K_2Nb_{14}O_{36}$ (Hu, Li and Fan, 1992). They also applied the direct methods for solving unconventional crystal structures such as quasicrystals (Xiang *et al.*, 1990) and incommensurately modulated structures (Mo *et al.*, 1992).

Since 1990, more and more structures have been solved from HREM images and electron diffraction and more and more scientists have become interested in structure analysis by electron crystallography (Hovmöller 1992; Zou, 1995). This pushed electron crystallography a big step forward. The term "electron crystallography" was first introduced and soon been accepted.

2.3 Combining HREM images and electron diffraction

In HREM images of inorganic crystals, phase information of structure factors is preserved. However, because of the effects of the contrast transfer function (CTF), the quality of the amplitudes is not very high and the resolution is relatively low. Electron diffraction is not affected by the CTF and extends to much higher resolution (often better than 1Å), but on the other hand no phase information is available. Thus, the best way of determining structures by electron crystallography is to combine HREM images with electron diffraction data. This was applied by Unwin and Henderson (1975) to determine and then compensate for the CTF in the study of the purple membrane.

Electron diffraction has also been used for extending the resolution of images, using the direct methods from X-ray crystallography (Fan, *et al.*, 1985; Hu *et al.*, 1992). Another method, based on the maximum entropy-likelihood technique was introduced by Bricogne and Gilmore (1990). The maximum entropy-likelihood method has been used for phase extension of electron diffraction data. Two successful applications are the extension of HREM image resolution of the organic structure perchlorocoronene from 3.2 Å to 1 Å (Dong et al, 1992) and the 2D projection of the membrane protein bacteriorhodopsin (Gilmore and Shankland, 1993).

Several other techniques, such as electron holography (Lichte, 1986) and convergent beam electron diffraction (CBED) have also been developed for structure analysis. CBED can provide information not only on the lattice parameters and the symmetry of crystals, but also accurate structure-factor amplitudes and phases (Høier *et al.*, 1993). Accurate structure factor determination by CBED can provide information on the location of valence electrons. However, it is more favourable for thick crystals (> 500 Å) with small unit cells (< 10 Å). Structure analysis by CBED has been summarized in two review articles (Spence, 1993; Tanaka, 1994).

3. ELECTRON CRYSTALLOGRAPHY HAVE SOME ADVANTAGES OVER X-RAY CRYSTALLOGRAPHY

X-ray crystallography is still the best technique for complete and accurate determination of crystal structures. However, electron crystallography has several important advantages over X-ray crystallography:

- Electrons interact with matter much stronger than X-rays. Thus much smaller crystals than those needed for single crystal X-ray diffraction can be analyzed by electron crystallography. For single crystal X-ray diffraction, crystals should have a size larger than $100 \times 100 \times 100$ μm^3 (or $10 \times 10 \times 10$ μm^3 on a synchrotron). For electron diffraction, crystals can be as small as $10 \times 10 \times 10$ nm^3. Even smaller crystals, down to some 10-20 unit cells, can be studied by HREM imaging. Electron crystallography has a promising future for structure analysis of crystals too small for X-ray diffraction analysis, such as grain boundaries, metastable phases etc.

- Electrons can be focused by magnetic lenses to form an image. High resolution electron microscopy images of crystals can be obtained, while X-ray imaging is not possible. The phase information which is lost when registering diffraction patterns is preserved in HREM images.

- The mechanism by which electrons interact with crystals is different from that of X-rays. X-rays detect electron density distribution in crystals, while electrons detect electrostatic potential distribution in crystals. Electron crystallography may be used for studying some special problems related to potential distribution such as the oxidation states of atoms in the crystal.

- Almost all crystals suitable for X-ray powder diffraction can be studied by electron diffraction. Several of the most demanding problems with powder diffraction are overcome by electron diffraction. There is no problem of overlapping reflections in electron diffraction and all diffraction spots can be unambiguously indexed. There is no problem of under-determination (less data than unknown parameters) for electron diffraction since 10-100 times more reflections than parameters can be obtained by ED, whereas in X-ray powder diffraction the over-determination is close to one.

- HREM images can be used for studying defects, local structure and chemical variations in crystals, as well as interfaces.

In conclusion, electron crystallography can be used to extend the range of samples amenable to structure analysis beyond those which can be studied by single crystal X-ray diffraction. Moreover, HREM images can supply some initial low resolution phases for X-ray diffraction which may aid in phase determination in X-ray crystallography.

4.   STRUCTURE REFINEMENT

A complete crystal structure determination (after collecting the data and finding the unit cell and the symmetry) can be divided in two distinct steps: solving the structure and refining it. Solving the structure means finding a rough model of at least the most important (= heaviest) atoms. For solving a structure, any method, including guessing, may be used.

The structure model arrived at from for example HREM images needs to be verified and further improved. This can be done by refinement of the structure model by least-squares methods against data from selected area electron diffraction or convergent beam electron diffraction. The accuracy of atomic coordinates obtained from a 2D projection by HREM to 2 Å resolution is about 0.2 Å. This limited accuracy is due only to the sparse amount of experimental data, with typically no more than one reflection per atomic coordinate. Electron diffraction data extends to much higher resolution, typically 0.8 to 0.5 Å, and so contains about ten times more reflections than the corresponding HREM image. These data have the further advantage compared to the image data that they are not distorted by the contrast transfer function (CTF) of the lens. The fact that SAED patterns do not contain phase information is no disadvantage for the refinement, because only the amplitudes are used at this stage of a structure determination.

Structure refinement based on kinematical scattering was already applied by the Russian scientist 60 years ago. Weirich *et al.* (1996) first solved the structure of an unknown $Ti_{11}Se_4$ by HREM combined with crystallographic image processing. Then they used intensities extracted from selected area electron diffraction patterns of a very thin crystal and refined the structure to a precision of 0.02 Å for all the atoms. Wagner and Terasaki *et al.* (1999) determined the 3D structure of a new zeolite from selected area electron diffraction, based on kinematical approach.

The disadvantage selected area electron diffraction is that they contain a higher proportion of multiply scattered electrons than HREM images, mainly due to the factor that the SAED patterns are usually taken from a larger and thicker region than those thinnest parts of HREM images which are cut out and used for image processing.

Structure refinement based on dynamical scattering was developed by Zandbergen and Jansen (Zandbergen *et al.*, 1997; Jansen *et al.*, 1998), known as the MSLS software. Electron diffraction from crystal regions with relatively homogenous thicknesses was used. Both the crystal orientation, crystal thickness and the atomic coordinates could be refined simultaneously.

5. ELECTRON CRYSTALLOGRAPHY – THE FUTURE OF CRYSTALLOGRAPHY

Electron crystallography is still a young research field. The continuous development of modern electron microscopy techniques challenges scientists to explore the possibilities and push the limit of electron crystallography, and to tackle more difficult problems. Here I will present some of the developments in electron crystallography during the past few years and hope that more and more young scientists will join the development of electron crystallography and apply this unique technique to challenging problems in material sciences.

5.1 Extending image resolution by through-focus exit wave reconstruction

Modern electron microscopes with field emission electron sources provide brighter and more coherent electrons. Images with information of crystal structures up to 1 Å can be achieved. A through-focus exit wave reconstruction method was developed by Coene *et al.* (1992; 1996) to retrieve the complete exit wave function of electrons at the exit surface of the crystal. This method can be applied to thicker crystals which can not be treated as weak-phase object. It is especially useful for studying defects and interfaces (Zandbergen *et al.*, 1999).

5.2 Electron diffraction imaging on nanostructured materials

Most of the phasing methods on electron diffraction, such as the direct methods and Patterson method, are only valid for crystalline materials with large number of repeating unit cell. When electron diffraction patterns are taken a finite sample (with the size only a few times of the unit cell dimensions) using a coherent electron beam, they can be sampled at a spacing finer than the inverse of the size of the sample (=Nyquist frequency). In such a case, there will be a region of zeros (outside the sample) in the inverse Fourier transform of the diffraction pattern. The finer the sampling frequency, the larger the region of zeros. When the region of zeros is larger than the sample, the phase at each point of the whole electron diffraction pattern (both on and between the diffraction spots) could be directly retrieved using the iterative phase retrieval algorithm. This approach was applied to simulated electron diffraction patterns and they could show that 3D structure of a zeolite nanocrystal could be determined *ab initio* at a resolution of 1 Å from 29 electron diffraction patterns (Miao *et al.*, 2000). This oversampling phasing method was demonstrated by Zuo *et al.* (2003) to imaging the 3D structure of a double carbon nanotube.

5.3   Lens-aberration corrected electron microscopy

The point resolution of an electron microscope is limited by the spherical aberration of the objective lens. Haider *et al.* (1995; 1998) developed a corrector to be implemented into a standard transmission electron microscope to correct the spherical aberration. They showed that the point resolution could be improved from 2.4 Å to 1.4 Å.

With the help of the the spherical aberration corrector, Jia *et al.* (2003) could adjust the spherical aberration coefficient of the objective lens to a negative value and imaging directly oxygen atoms in perovskite ceramics $SrTiO_3$ and high-Tc superconductor $YBa_2Cu_3O_7$ by HREM. They could see oxygen deficency and ordering directly from the HREM images. They could even measure the oxygen content at a twin boundary in $SrTiO_3$ (Jia *et al.*, 2004).

5.4   3D reconstruction from HREM images on inorganic crystals

HREM images are only projections of the 3D structure. Most of the structure determination from HREM images of inorganic compounds are on crystals with at least one short unit cell axis (<5 Å). The structures can be solved from a single projection with a short axis, where at least the heavy atoms do not overlap. Although the resolution of electron microscopes can be improved by the spherical aberration correction, for complex structures with large unit cell axes, atoms may still overlap in HREM images even at a resolution of better than 1Å. The only way to solve the overlapping problem is to collect several images from different directions and combine these images into a 3D structure. Nowadays electron microscopes can provide images with resolutions beyond 1.5 Å for inorganic crystals. This resolution is sufficient to resolve interatomic spacing, which means that it should be possible to resolve all atoms.

In 1992 Wenk *et al.* in Berkeley for the first time combined 2D HREM images into a 3D reconstruction for solving the structure of an inorganic crystal. They combined, using image processing, HREM images (to a resolution of 1.38 Å) taken in five different orientations of the silicate mineral staurolite and constructed a 3D electron potential map. In this map all atoms (Fe, Al, Si and oxygen) were clearly resolved. This work showed that 3D electron crystallography has a great potential also in structure determination of inorganic crystals - perhaps even more promising for inorganic structures than for organic and biological structures, because of the higher resolution.

We have reconstructed the 3D structure of a complex quasicrystal approximant v-AlCrFe ($P6_3/m$, $a = 40.687$ and $c = 12.546$ Å) (Zou *et al.*, 2004). Due to the huge unit cell, it was necessary to combine crystallographic data from 13 projections to resolve the atoms. Electron microscopy images containing both amplitude and phase information were combined with amplitudes from electron diffraction patterns. 124 of the 129 unique atoms (1176 in the unit cell) were found in the remarkably clean calculated potential maps. This investigation demonstrates that inorganic crystals of any complexity can be solved by electron crystallography.

The 3D reconstruction method was also applied to solve the 3D structures of a series mesoporous materials (Keneda *et al.* 2002).

5.5   Ultrafast electron crystallography

Most of chemical reactions and phase transitions occur in a time scale of femto- to pico- seconds. Electron crystallography provides the only technique to determine the entire structure changes within such a time scale, thanks for the strong interaction between electrons and matter. Recently femtosecond ultrafast electron crystallography was developed by Zewail's group (Ihee *et al.*, 2001; Ruan *et al.*, 2004) and Siwick *et al.* (2003). Zewail was awarded the Nobel Prize in 1999 for his studies of the transition states of chemical reactions using femtosecond spectroscopy. The ultrafast electron diffraction technique employs properly timed sequences of ultrafast pulses - a femtosecond laser pulse - to initiate the reaction and then followed by an ultrashort electron pulses to probe the ensuing structural change in the sample. The resulting electron diffraction patterns are then recorded on a CCD camera. This sequence of pulses is repeated, timing the electron pulse to arrive before or after the laser pulse; in effect, a series of snapshots of the evolving molecular structure are taken. Each time-resolved diffraction pattern can then, in principle, be inverted to reveal the three-dimensional molecular structure that gave rise to the pattern at that specific time delay. However, in practice, a key challenge lies in recovering the molecular structural information that is embedded in the as-acquired diffraction images. Another challenge is, to obtain EM images of the sample and retrieve the structure information from them.

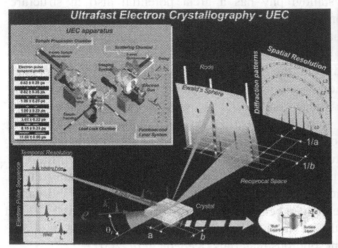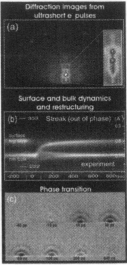

Figure 2 Ultra-fast Electron Crystallography for studying transient structures of
molecules, surfaces and phase transitions, from (http://www.its.caltech.edu/~femto/).

   Electron crystallography provides unique possibilities of studying
nanostructured materials. The aims of this course are to show theoretically
that electron crystallography can be used for crystal structure analysis; to
describe different methods of solving crystal structures by electron
crystallography and to demonstrate how these methods are used practically.

References

Bricogne, G. and Gilmore, C.J., *Acta Cryst.* **A46** (1990) 284.

Coene, W.M.J., Janssen, A.J.E.M., Op de Beeck, M., Van Dyck, D. *Phys. Rev. Lett.* **69** (1992)
     3743-3746.

Coene, W.M.J., Thust, A., Op de Beeck, M., Van Dyck, D. *Ultramicroscopy* **64** (1996) 109-
     135.

Cowley, J.M. and Moodie, A.F. *Acta Cryst.* **10** (1957) 609-619.

DeRosier, D.J. and Klug, A., *Nature (London)* **217** (1968) 130

Dong, W., Baird, T., Fryer, J.R., Gilmore, C.J., McNicol, D.D., Bricogne, G., Smith, D.J.,
     O'Keefe, M.A. and Hovmöller, S., *Nature (London)* **355** (1992) 605.

Dorset, D.L. and Hauptman, H.A., *Ultramicroscopy* **1** (1976) 195.

Dorset, D.L., *Structural Electron Crystallography*, Plenum Press, New York, 1995.

Fan, H.F., Zhong, Z.Y., Zheng, C.D. and Li, F.H., *Acta Cryst.* **A41** (1985) 163.

Gilmore, C.J. and Shankland, K., *Ultramicroscopy* **49** (1993) 132.

Gjønnes, K., Cheng, Y., Berg, B.S. and Hansen, V., *Acta Cryst.* **A54** (1998)102-119.

Hauptman H. and J. Karle, *Solution of the phase problem I. The centrosymmetric crystal.* ACA monograph **No. 3**. Ann Arbor, (Edwards Brothers, Inc., 1953).

Hendersson, R., Baldwin, J.M., Downing, K.H., Lepault, J. and Zemlin, F., *J. Mol. Biol.* **213** (1990) 899

Henderson, R. and Unwin, P.N.T., *Nature (London)* **257** (1975) 28

Hovmöller, S., A. Sjögren, G. Farrants, M. Sundberg, and B.-O. Marinder, *Nature* **311** (1984) 238.

Hovmöller, S., *Ultramicroscopy* **41** (1992) 121.

Hu, J.J., Li, F.H. and Fan, H.F., *Ultramicroscopy* **41** (1992) 387.

Høier, R., Bakken, L.N., Marthinsen, K. and Holmestad, R., *Ultramicroscopy* **49** (1993) 159.

Ihee, H., Lobastov, V.A., Gomez, U.M., Goodson, B.M., Srinivasan, R., Ruan, C.Y. and Zewail, A.H., *Science* **291** (2001) 458-462.

Iijima, S., *J. Appl. Phys.* **42** (1971) 5891

Ijima, S., O'Keefe, M.A. and Buseck, P., *Nature (London)* **274** (1978) 322

Jansen, J., D. Tang, H.W. Zandbergen and H. Schenk, Acta Cryst. A**54**, 91(1998).

Jap, B.K., Walian, P.J. and Gehring, K., *Nature (London)* **350** (1991) 167

Jia, C.L., Lentzen, M. and Urban, K., *Science* **299** (2003) 870-873.

Jia, C.L. and Urban, K., *Science* **303** (2004) 2001-2004.

Kaneda M., Tsubakiyarna, T., Carlsson, A., Sakamoto, Y., Ohsuna, T., Terasaki, O., Joo, S.H. and Ryoo, R., *J. Phys. Chem.* ***B*** **106** (2002) 1256-1266.

Klug, A. *Chimica Scripta* **14**, 245 -256 (1978-79).

Kühlbrandt, W., Wang, D.N. and Fujiyoshi, Y., *Nature (London)* **367** (1994) 614

Lichte, H., *Adv. Opt. El. Microsc.* **12** (1991) 25.

Miao, J., Sayre, D., *Acta Cryst.* **A56** (2000) 596.

Miao, J., Ohsuna, T., Terasaki, O., Hodgson, K.O. and O'Keefe, M.A., *Phys. Rev. Lett.* **89** (2002) 155502-1 –4.

Mo, Y.D., Cheng, T.Z., Fan, H.F., Li, J.Q., Sha, B.D., Zheng, C.D., Li, F.H. and Zhao, Z.X., *Supercond. Sci. Technol.* **5** (1992) 69

O'Keefe, M.A., *Acta Cryst.* **A29** (1973) 389

Pinsker, Z.G., *Diffraktsiya elektronov* (Izd-vo Akad. Nauk SSSR., Moscow 1949) [English transl,: *Electron diffraction* (Butterworths, London 1952)].

Ruan, C.Y., Lobastov, V.A., Vigliotti, F., Chen, S. and Zewail, A.H., *Science* **304** (2004) 80.

Siwick, B.J., Dwyer, J.R., Jordan, R.E. and Dwayne Miller, R.J., Science 302 (2003) 1382-1385. solid-liqiud transition

Spence, J.C.H., *Acta Cryst.* **A49** (1993) 231.

Tanaka, M., *Acta Cryst.* **A50** (1994) 261.

Unwin, P.N.T. and R. Henderson, *J. Mol. Biol.* **94** (1975) 425.

Vainshtein, B.K. *Strukturnaya elektronographiya*, (Izd-vo Akad. Nauk SSSR., Moscow 1956) [English transl,: *Structure Analysis by Electron Diffraction*, (Pergamon Press Ltd., Oxford, 1964)].

Vainshtein, B.K., B.B. Zvyagin and A.S. Avilov, in *Electron Diffraction Techniques*, Vol. 1, ed. by J.M. Cowley, (Oxford Univ. Press, Oxford, 1992) p.216.

Wagner, P., Terasaki, O., Ritsch, S., Nery, J.G., Zones, S.I., Davis, M.E. & Hiraga, K. (1999). *J. Phys. Chem.* B103, 8245-8250.

Weirich, T.E., R. Ramlau, A. Simon, S. Hovmöller and X.D. Zou, *Nature* **382** (1996) 144.

Weirich, T.E., Zou, X.D., Ramlau, R., Simon, A., Cascarano, G.L., Giacovazzo, C. and Hovmöller, S., *Acta Cryst.* **A56**, (2000) 29 - 35.

Wenk, H.-R., K.H. Downing, M. Hu and M.A. O'Keefe, *Acta Cryst.* **A48**, (1992) 700.

Xiang, S.B., Li, F.H. and Fan, H.F., *Acta Cryst.* **A46** (1990) 473.

Zandbergen, H.W., Andersen, S.J. and Jansen, J., *Science* 277 (1997) 1221 –1224.

Zandbergen, H.W., Bokel, R., Connolly, E. Jansen, J. *Micron* **30** (1999) 395-416.
Zou, X.D., *Chemical Communication* No. **5**, Stockholm University (1995).
Zou, X.D., Z.M. Mo, S. Hovmöller, X.Z. Li & Kuo, K.H. *Acta Cryst.***A59** (2003) 526.
Zuo, J.M., Vartanyants, I., Gao, M., Zhang, R. And Nagahara, L.A., *Science* 300 (2003) 1419.

EXPLOITING SUB-ÅNGSTROM ABILITIES:
What Advantages Do Different Techniques Offer?

C. Kisielowski

National Center for Electron Microscopy, Ernest Orlando Lawrence Berkeley National Laboratory, One Cyclotron Road, Bldg. 72, Berkeley CA 94720, USA

Abstract: This contribution reviews the current status of sub Ångstrom electron microscopy. High Resolution Transmission Electron Microscopy (HRTEM) and Scanning Transmission Electron Microscopy (STEM) are considered and compared for imaging applications. While both techniques provide comparable sub Ångstrom resolution around 0.8Å noise levels and chemical discrimination are dissimilar, which result in complementary characteristics for the detection of different elements. In particular, the ability to detect single atoms benefits greatly from the ongoing resolution enhancement and corrections of lens aberrations. As a result it is feasible aiming at single atom analyses in three dimensions.

Key words: HRTEM, STEM, resolution, precision, sensitivity, single atom detection.

1. INTRODUCTION

Traditionally the performance of HRTEM is judged in terms of its ability to resolve two adjacent atom columns. Resolution is ruled by a few basic principles[1]: A position dependent image intensity g($\mathbf{r}$) is described as a convolution of the specimen function f($\mathbf{r}$) with a point spread function h($\mathbf{r}$). It is convenient to express this convolution in real space as a product in reciprocal space:

(1) $G(\mathbf{u}) = H(\mathbf{u})F(\mathbf{u})$

H($\mathbf{u}$) is the Fourier Transform of h($\mathbf{r}$) and is called the contrast transfer function (CTF). $\mathbf{u}$ is a reciprocal–lattice vector that can be expressed by image Fourier coefficients. The CTF is the product of an aperture function A($\mathbf{u}$), a wave attenuation function E($\mathbf{u}$) and a lens aberration function B($\mathbf{u}$) = exp(iχ($\mathbf{u}$)). Typically, a mathematical description of the lens aberration function to lowest orders builds on the Weak Phase Approximation and yields the expression:

(2) $\chi(u) = C_s \lambda^3 u^4 \pi/2 + \pi f \lambda u^2$

17

T.E. Weirich et al. (eds.), Electron Crystallography, 17–29.

Equation (2) describes the effect of defocus f and spherical aberration Cs on the object function and λ is the electron wavelength. Higher order aberrations can be considered[2]. An attenuation of the electron wave is taken into account by a spatial damping cnvclope

(3) $E_s(u) = \exp[-(\pi\alpha/\lambda)^2 (C_s\lambda^3 u^3 + f\lambda u)^2]$,

and a temporal one that is produced by mechanical and electrical instabilities and ultimatively limits the resolution.

(4) $E_t(u) = \exp[-1/2\ \pi^2\lambda^2\ \Gamma^2 u^4]$ with $\Gamma = C_c[4(\Delta I/I)^2 + (\Delta E/eV)^2 + (\Delta V/V)^2]^{1/2}$

In equations (3) and (4), α is the semi angle of the illuminating convergence cone, C_c is the coefficient of chromatic aberration, ΔE is the thermal energy spread of the electrons of charge e and $\Delta I/I$ and $\Delta V/V$ are the relative fluctuations of lens current and high voltage, respectively.

The Rayleigh resolution criterion[3] can be satisfied within noise limits if spatial frequencies are recorded to the Scherzer point resolution[4]:

(5) $\rho_s = 0.66\ C_s^{1/4}\lambda^{3/4}$

Field emission technology opened the possibility to transfer misphased information beyond the Scherzer point resolution to the information limit that is imposed by the temporal damping envelope:

(6) $\rho_i = (\pi C_c \lambda/2^{1/2}\cdot 4(\Delta I/I)^2 + (\Delta E/eV)^2 + (\Delta V/V)^2)^{1/4}$

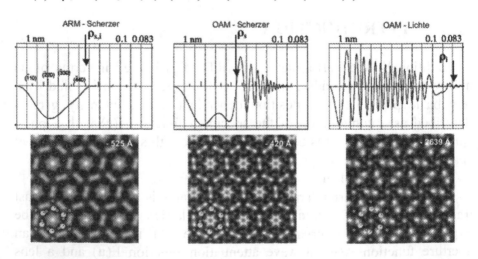

Figure 1. CTF and simulated lattice images of $Si_3N_4[0001]$ for selected microscopes and focus values. Structural models are superimposed.

Figure 1 shows CTF's for NCEM's Atomic Resolution Microscope (ARM) and One Angstrom Microscope (OAM) with corresponding simulated lattice images of $Si_3N_4[0001]$. CTF calculations and image simulations are widely available through software packages[5,6,7]. It is seen that the Scherzer resolution is close to the information limit if thermal electron emitters are utilized. In contrast, the use of field emission technology separates the Scherzer resolution limit greatly from the information limit. The rapid CTF oscillation in the spatial frequency range between both values misphases information and introduces large image delocalization. The specific case of Lichte defocus[8] is of relevance here because lattice images recorded at Lichte defocus exhibit the smallest image delocalization:

$$(7) \quad f_L = -0.75 \, C_s \, \lambda^2 \, u_{max}^2$$

Resolution extension by use of other defocus values is discussed. H. Hashimoto and H. Endoh suggested the use of an aberration-free focus[9]. However, prior knowledge of the relevant image Fourier coefficients is required that must suitably match the CTF oscillations but such components cannot be predicted if non- periodic structures are recorded. M.A. O'Keefe[10] suggested the use of an alpha-null defocus to further optimize information transfer at sub Ångstrom values for exit wave reconstruction. While information gain has not been demonstrated, significant drawbacks are a huge image delocalization and sampling limitations. Therefore, Lichte defocus remains the most common defocus setting for the recording of focal series for exit wave reconstructions.

It is stressed that lattice images are beam interferograms with *no* direct relation to the underlying crystal structure. In general, image simulations are required to link the image intensities to atom column positions. Only if thin samples are imaged at Scherzer defocus with electrons of limited coherence length (thermal emitter) the situation simplifies and contrast minima may be related to column positions. But even in this case, image simulations are required to determine the sample thickness by matching lattice images to calculated defocus/thickness map. Obviously, the introduction of field emission technology has wiped out a last option for simple image interpretation as seen from the OAM-Scherzer image of Figure 1 where silicon columns are located on intensity maxima and nitrogen columns on minima. It is currently debated if the utilization of specific apertures can restore one simple structure-image relationship.

For a long time HRTEM resolution was limited by a Scherzer point resolution of about 1.8 Å[11] and was the only "bulk-like" imaging technique with atomic resolution. Over the last decade electron microscopy made unprecedented progress that allowed breaking the 1 Ångstrom barrier by HRTEM and STEM.

2.   SUB ÅNGSTROM RESOLUTION

In a high voltage approach with Scherzer imaging, F. Phillipp et al. demonstrated a resolution of about 1 Å utilizing electrons accelerated to 1.25 MeV[12] and recorded diamond [004] image Fourier components at 0.89 Å[13]. H. Ichinose resolved C-C and C-Si dumbbell distances in SiC[14]. Gold (337) lattice fringes (0.498 Å) were recorded with a 1.25 MeV field emission microscope. The authors suggested an information limit of 0.6 Å[15].

STEM matured as a reliable imaging tool with atomic resolution[16]. Images are formed by detection of electrons scattered to high angles (High Angle Annular Dark Field STEM). The image formation process is incoherent and provides large contrast discrimination for elements of different atom number Z (Z-contrast imaging). Nellist and Pennyook reported sub Ångstrom resolution in 1998 detecting the silicon (444) image Fourier coefficient at 0.78 Å[17]. More recently, Batson et al. highlighted 0.75Å resolution obtained in a STEM operated at 120 kV[18] by processing the contrast of single gold atoms. Later, the argument was relaxed to 0.78 ± 0.15 Å resolution[19]. It remained unexplained why such a resolution could not be achieved in images of periodic crystal structures.

A mid-voltage (100-300kV) approach that aimed for 0.8Å resolution was pursued in the Brite Euram Project No. 3322[20]. The project built on combining improved FEG performance, better lens technology and stabilities with holographic image reconstruction that extends resolution to the microscope's information limit[21,22,23]. One Ångstrom resolution was approached but could not be reached. The Berkeley One Ångstrom Project initiated at NCEM joined the Brite Euram collaboration in a later stage and aimed at a mandatory information limit of 1.2 Å with the expectation that 1 Ångstrom performance may be achieved on basis of a CM300 FEG/UT named OAM. Image reconstruction by off-axis electron holography[21], extended resolution electron microscopy (EREM)[24] and electron exit wave reconstruction[22, 23] were considered as supporting software approaches. A resolution close to 1 Ångstrom was achieved by off axis electron holography[25]. A successful implementation of EREM remains to be demonstrated although a related scheme has recently been implemented[26]. The exit wave reconstruction process by W. M. J. Coene and A. Thust was successfully implemented and sub Ångstrom resolution was published in 2000 for the first time by resolving the 0.89Å dumbbell distance in diamond [110] [27]. Obviously, the result exceeded NCEM's performance expectations. The achievement was described in detail in 2001 together with confirming results that highlight the resolution of aluminum and oxygen columns in sapphire (0.85Å) and the resolution of gallium and nitrogen columns in GaN at 1.13 Å[28]. Unlike lattice images the electron exit wave can depict the crystal structure directly with an outstanding signal to noise ratio. The effect of point spread is removed from the image in the reconstruction process and a hardware correction of 2- & 3-fold astigmatism and coma boosts the image quality[29]. The original 0.89Å result from diamond [110] [27] gained popularity in successive publications[30, 31].

The reliability of the exit wave reconstruction process was independently confirmed by making available to the scientific community a reconstructed focal series from the OAM[32]. L.J. Allen et al. demonstrated striking agreement for the identical data set[33] with an iterative method for exit wave function reconstruction.

The OAM Project sets new standards for HRTEM performance and makes sub Ångstrom capabilities available to a large user community through operation of the OAM at the NCEM[28, 30]. It is noted that a resolution of 0.9 Å is readily accessible on a daily basis. However, great efforts are required to obtain a larger resolution.

3. RESOLUTION VERIFICATION

The recording of Young's fringes from amorphous layers is the most conservative test for resolution in HRTEM. Results depend on the choice of test materials since heavy elements scatter electrons stronger to large angles. Young's fringes recorded by the OAM extend beyond 1 Ångstrom[30]. A verification of sub Ångstrom capabilities, however, requires further considerations since the spatial frequency spectrum of amorphous materials is structured and not "white", which can lead to an underestimation of information transfer.

The 0.91 Å spacing of (222) image Fourier coefficients from molybdenum [$\bar{1}$12] are directly visible in lattice images from the OAM recorded successively over the course of a day[28]. Their detection proves sub Ångstrom information transfer since they are of lowest order and cannot be excited by non-linear beam interference's. Similarly, an intensity enhancement of the diamond (004) image Fourier coefficients could be achieved by sweeping the edge of the *spherical* damping envelope across 0.89 Å by increasing the underfocus of the objective lens[28]. Therefore, the information limit imposed by the *temporal* envelope of the OAM must lie beyond 0.89 Å. Measurements of Fourier periods[34] $f_p = 2 / \lambda u^2$ of image Fourier coefficients from focal series of lattice images can in principle verify the presence of a largest spatial frequency that marks the information limit. The lowest order Si (220) Fourier coefficient in Si [100] lattice images yields a Fourier period that agrees with theory[28]. However, oscillations of higher order coefficients exhibit large distortions that suggest the presence of non-linear beam interferences. Nevertheless, an information limit of 0.78 Å was reported for the OAM by referring to the detection of high order Si (444) Fourier periods[35]. This discrepancy is not understood since the appropriate data are not available. Calculations of information limits usually do not consider error bars[30].

A direct removal of image Fourier coefficients from a reconstructed electron exit wave seems a most suitable way for resolution verification[28]. Here, Figure 2 shows for the first time the contribution of Si (444) image Fourier coefficients to the separation of the 0.78Å dumbbell distance in Si [112] in the phase of an electron exit wave. The focal series was recorded at Lichte defocus with the OAM and reconstructed in 2001.

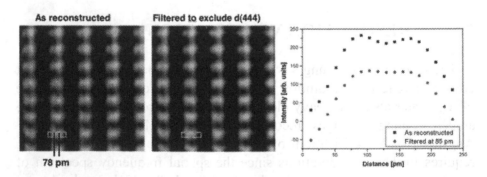

Figure 2. Proof of 0.78Å resolution in the phase of the exit wave from silicon [112] by removing (444) image Fourier coefficients.

In STEM applications the highest recorded image Fourier coefficient is usually taken as a direct proof for resolution[17]. Contributions of this component to the separation of atom columns are often not considered or compared with noise levels. Further, unisotropic signal transfer is disregarded. P. Batson considers intensity measurements on single atoms for probing resolution[18, 19].

In general, definitions for TEM resolution are more rigorous than for STEM and it is desirable to find common ground for an objective comparison between STEM and HRTEM resolution. Ultimately, one value of resolution lies in the separation of atom columns at non-periodic lattice sites as shown in Figure 3[35].

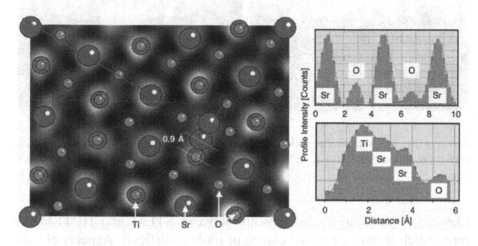

Figure 3. The phase of the electron exit wave depicts the structure of one unit cell of a Σ13 boundary in $SrTiO_3$. Ti/O, Sr, and O columns are resolved and intensity profiles across selected directions are shown. Intensity on O-columns is generated by ~ 10 atoms. Sub Ångstrom resolution is achieved.

4. STEM OR HRTEM?

It is sometimes argued that - by reciprocity - STEM and HRTEM can perform similar tasks since a STEM can be thought of as a HRTEM with detector and electron source being exchanged if the detector is a point detector on the optical axis[37]. Indeed, bright field - *and* annular dark field imaging can be executed in either a STEM or a HRTEM as shown for the latter in Figure 4[38].

However, the theorem of reciprocity is a wave optical argument that does not consider intensities where we easily find the differences. For example, if one thinks of a STEM as an inverted HRTEM one would not detect any intensity in an image since it is an inherent property of a point detector to collect no intensity. On the other extreme side, the ability to form an intense and focused probe is a valuable ability that boosts local spectroscopy[39]. Obviously, the best choice of tools cannot be a matter of exclusion but must relate to the problem at hand that needs solving[40].

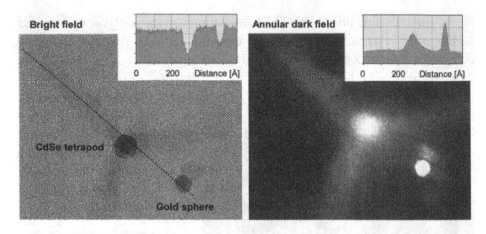

Figure 4. Bright - and annular dark field image of a CdSe tetrapod and an adjacent gold particle recorded in a TEM. The line profiles demonstrate Z-contrast in the dark field image.

Currently, quantitative comparisons between STEM and HRTEM are rare, which is why a rational choice of tools is difficult. Aspects of such a comparison evolve and are outlined for imaging applications next.

5.   RESOLUTION AND PRECISION

Resolution gain is an achievement that enables new science since closely spaced atom columns can be resolved in many more projections (Figure 5) [41] and at interfaces such as shown in Figure 3 where a relaxation driven Sr column splitting is observed at 0.9Å. In particular, investigations of materials containing light elements with short bond length such as SiC, BN, diamond, GaN, or sapphire will benefit from sub Ångstrom resolution[28].

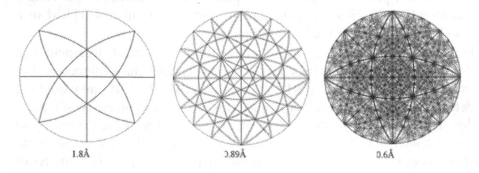

Figure 5. Stereographc representation of the number of accessible lattice orientations for atomic resolution analysis of diamond at different resolution limits.[41]

By now, resolution approaches physical limits given by the width of the 1s Bloch wave state that acts as a resolution limiting cross section[42]. It is a priori not clear if a further increase of resolution is scientifically beneficial even though some information gain may still be possible for heavier elements with scattering cross sections around 0.5 Å. However, there are other equally important aspects that require considerations.

D. Van Dyck et al.[43] pointed out that resolution ρ and precision ξ are linked by the electron dose N: $\xi = \rho / N^{1/2}$. Graphically speaking, one depicts resolution as the width of a distribution function and precision as its center of gravity. Images on CCD's are typically recorded with counts that range from 500 to 20000 events per pixel. Therefore, a microscope of 0.8Å resolution can principally pinpoint an atom column position to a precision of 0.04 − 0.006 Å! In the past, HRTEM was indeed utilized to measure small displacement fields using a suitable approaches[44, 45, 46]. C.L. Jia and A. Thust[47] showed that a grain boundary expansion of 19 pm can be measured to a precision of ±4 pm from a reconstructed electron exit wave. Image distortions and residual lens aberrations are limiting factors. The physical relevance of displacement measurements is significant: It was estimated that 1 picometer of bond length change corresponds typically to 50 meV of energy change[49]. Thereby, electron microscopy is linked to total energy calculations that nowadays can be used to formulate a structure atom-by-atom[48]. Using this estimate, one would for example expect that a column displacement by 10 pm could move a defect energy level in silicon half way across its band gap and it becomes quite obvious that electron microscopy moves into a position that can help refine theoretical calculations. A attempt to match theory and experiment with identical numbers of atoms is currently pursued for the case of a 30^0 partial dislocation in GaAs[50, 51] where individual column positions can be measured to a precision of up to ±2.4 pm as shown in Figure 6. J.R. Jinschek et al. recently demonstrated that local displacement measurements are now sensitive enough to detect the displacement field caused by single indium atoms in a GaN matrix[52].

Clearly, precision must be a small fraction of resolution and any structural model must fit column position within a few picometer as shown in Figures 3 and 6. Deviations from this rule indicate the need to refine structural models, the presence of systematic errors, or an over interpretations of data. Unfortunately, this basic rule is sometimes disregarded by ignoring mismatches of up to 1 Å = 100 pm[53, 54]. Systematic errors often relate to the presence of scanning noise, sample tilt, or to unfavorable specimen geometry for an exit wave reconstruction

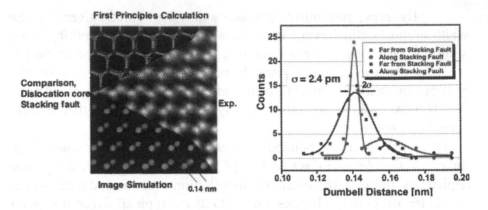

Figure 6. Merging theory and experiment with a 30^0 partial dislocation in GaAs. Dumbbell distances are measured with 2.4 pm of precision.

process that may be a property of the sample itself if it is spherical or may be introduced during sample preparation by surface roughening. It was estimated that one nanometer of defocus change shifts intensity maxima in an exit wave image by about 4 pm[55].

6.   SENSITIVITY

It is most important to realize that resolution enhancement has directly boosted the sensitivity for the detection of single atoms, in particular, since the development was accompanied by correcting lens aberrations for the electron source[56] and the objective lens[57, 58]. An approach to quantify this progress was developed by analyzing amplitude and phase of electron exit waves in an Argand plot (Figure 7). Since amplitude alterations depend on the achievable contrast in the lattice images of an electron microscope under consideration, they yield circles of different diameter that can be characterized by a largest phase value. Experiments were performed with wedge shaped gold samples in the first extinction distance[1]. The result suggests an improvement factor of almost 10 between a traditional microscope (1.9 Å Scherzer resolution, thermal emitter) and the OAM[59].

In addition the signal to noise ratio for the detection of single gold atoms in a STEM was determined[59] and Figure 7 compares the results with data from literature[18,60] and theoretical expectations. Clearly, single atom sensitivity is principally obtainable for most elements of the periodic system. Further, it is seen that the strength of HRTEM lies in the detection of light atoms[28, 47, 61] as demonstrated in Figure 3 where the

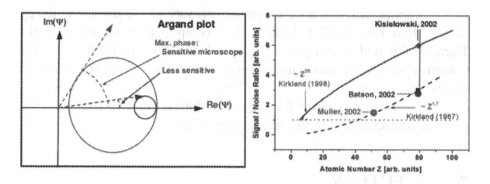

Figure 7. Left: Schematized Argand plot for microscopes of different sensitivity. Right: Sensitivity estimation for STEM (bottom curve) and HRTEM (top curve).

intensity on the oxygen columns is generated by only 10 ± 1 oxygen atoms. On the other hand STEM discriminates better between heavy elements in a light matrix[60] but at a smaller signal to noise ratio that imposes a detection limit around $Z=40$. Practically, an exploitation of such abilities is limited by several factors including the radiation damage caused by the 300 keV electron used in the OAM.

Applications that would benefit from the described achievements are commonly bottlenecked by a lack of general procedures to interpret intensity data. Recently, it was suggested to interpret STEM data by extraction of single atom scattering cross sections[37, 59]. For HRTEM, Figure 8 highlights a suitable method for intensity quantification from reconstructed electron exit waves.

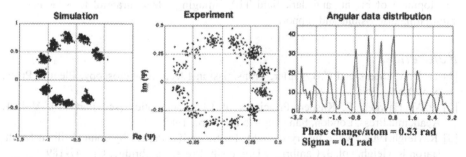

Figure 8. Channeling plot for gold [110] revealing the discrete nature of atom columns.

The method extends previous work of W. Sinkler, L.D. Marks [62] and allows for a comparison with the 1s-state model[63] that approximates well an Argand plot in the first extinction distance while providing a promising path to go from an exit wave to the crystal structure[64]. J.R. Jinschek et al.[52] demonstrated by simulations and experiments on gold samples that a channeling plot (Figure 8) reveals the discrete nature of atom columns, provides an appropriate measure of the phase change per single atom (Au: 0.53 rad/atom) and is applicable to any material. Thereby, the method is a general approach to access the third dimension. Chemical sensitivity was demonstrated by the detection of single indium atoms in a GaN matrix and is limited by sample preparation. A conservative error estimate for phase resolution yields 0.1 rad but 0.02 rad seems reachable with the current technology.

7.   FUTURE

In conclusion, aberration corrected electron microscopy with sub Ångstrom resolution looks at a bright future. It has plowed the way for single atom detection, which enables a wealth of novel scientific investigations and may finally remove the problem of crystal projection into a single image plane by application of electron tomography[65]. In general, an expansion of electron microscopy is of crucial importance to a further development of nano-technology[66]. In the US, the construction of an aberration corrected deep sub Ångstrom electron microscope that operates at voltages as low as 100 kV is pursued within the TEAM Project[65].

Acknowledgements

The project was sponsored by the Director, Office of Science, Office of Basic Energy Sciences, of the U.S. Department of Energy under Contract No. DE-AC03-76SF00098. The author acknowledges the large commitment of J.R. Jinschek and S. Bals to the development of bright- and dark field TEM imaging. He is grateful to R. Kilaas for stimulating discussions and support.

References

[1] D. B. Williams and C.B. Carter, Transmission Electron Microscopy, Plenum Press, New York, 1996
[2] P.W. Hawkes, E. Kasper, Principles of Electron Optics (see Part IV), Vol. 1, Academic Press, London, San Diego, New York, 1989
[3] Rayleigh,1902. Wave theory of light. In: Scientific Papers by John William Strutt, Baron Rayleigh, vol. 3. Cambridge University Press, Cambridge, UK, 47-189
[4] O. Scherzer, Z. Physik 101 (1936) 593
[5] R. Kilaas, 45th Ann. Proc. EMSA 1987, 66
[6] P. Stadelmann, Ultramicroscopy 21 (1979) 131
[7] M. V. Sidorov http://clik.to/ctfexplorer, 2004
[8] [H. Lichte, Ultramicroscopy 38 (1991) 13
[9] H. Hashimoto, H. Endoh "Electron Diffraction 1927 – 1977" P. Dobson, J. Pendry, C. Humphreys eds., IoP, Bristol, 1978
[10] M. A. O'Keefe, Microscopy and Microanalysis 7 (2001) 916
[11] M. A. O'Keefe, Ultramicroscopy 47 (1992) 282

[12] F. Phillipp, R. Hoeschen, M. Osaki, G. Moebus, M. Ruehle, Ultramicroscopy 56 (1994) 1

[13] F. Phillipp in Proc. of the Asian Science Seminar on New Directions in: Transmission Electron Microscopy and Nano-Characterization of Materials, eds. C. Kinoshita, Y. Tomokiyo, and S. Matsumura (Kyushu University Press, Fukuoka, ISBN4-87378-558-8) (1998) 75

[14] H. Ichinose, Science and Technology of Advanced Materials 1 (2000) 11

[15] T. Kawasaki, T. Yoshida, T. Madsuda, N. Osakabe, A. Tonomura, I. Matsui, K. Kitazawa, Appl. Phys. Lett. 76 (2000) 1342

[16] S.J. Pennycook, D.E. Jessen, Phys. Rev. Let. 64 (1990) 938

[17] P. D. Nellist, S. J. Pennycook, Phys. Rev. Let. 81 (1998) 4156

[18] P. E. Batson, N. Dellby, O. L. Krivanek, Nature 418 (2002) 617

[19] P.E. Batson, Ultramicroscopy, 96 (2003) 239

[20] D. Van Dyck, H. Lichte, K.D. van der Mast, Ultramicroscopy 64 (1996) 1

[21] H. Lichte, P. Kessler, F. Lenz, W.-D. Rau Ultramicroscopy, 52 (1993) 575

[22] W. M. J. Coene, A. Thust, M. Op de Beeck, D. Van Dyck, Ultramicroscopy, 64 (1996) 109

[23] A. Thust, W. M. J. Coene, M. Op de Beeck and D. Van Dyck, Ultramicroscopy, 64 (1996) 211

[24] M.A. O'Keefe in "Microstructure of Materials", K. Krishnan ed.; San Francisco Press Inc. 1993, 121

[25] Y.C. Wang, M.A. O'Keefe, M. Pan, E.C. Nelson, C. Kisielowski, Proc. ICEM 1998, Electron Microscopy, 1998, 573

[26] Y. Takai1, Y. Kimura, T.Ikuta, R. Shimizu, Y.Sato, S. Isakozawa, M. Ichihashi, Journal of Electron Microscopy 48 (1999) 879

[27] C. Kisielowski, E.C. Nelson, C. Song, R. Kilaas, A. Thust, Microscopy and Microanalysis 6, 2000, 16

[28] C. Kisielowski, C.J.D. Hetherington, Y.C. Wang, R. Kilaas, M.A. O'Keefe, A. Thust, Ultramicroscopy 89 (2001) 243

[29] Y.C. Wang, A. Fitzgerald, E.C. Nelson, C. Song, M.A. O'Keefe, and C. Kisielowski, Microscopy and Microanalysis 5, 1999, 822

[30] M.A. O'Keefe, C.J.D. Hetherington, Y.C. Wang, E.C. Nelson, J.H. Turner, C. Kisielowski, J.-O. Malm, R. Mueller, J. Ringnalda, M. Pam, A. Thust, Ultramicroscopy 89 (2001) 215

[31] M. A. O.Keefe, Y. Shao-Horn, Microscopy and Microanalysis, 10 (2004) 86

[32] A. Ziegler, C. Kisielowski, R.O. Ritchie, Acta Mater. 50 (2002) 565

[33] L.J. Allen, W. McBride, N.L. O'Leary, M.P. Oxley, Ultramicroscopy, 2004, in press

[34] S. Iijima, M.A. O'Keefe, J. Microsc. 117 (1979) 347

[35] M. A. O'Keefe, E.C. Nelson, E.C. Wang, A. Thust, Phil. Mag, B 81 (2001) 1861

[36] J. Ayache, C. Kisielowski, R. Kilaas, G. Passerieux , S. Lartigue-Korinek, to be published

[37] S. Bals, B. Kabius, M. Haider, V. Radmilovic, C. Kisielowski, Solid State Comm., 2004, in press

[38] L. Raimer, Transmission Electron Microscopy, Springer, Berlin 1993, p. 123

[39] I. Arslan, N. Browning, Phys. Rev. Let. 2004, in press

[40] A. C. Diebold· B. Foran, C. Kisielowski· D. Muller, S. Pennycook· E. Principe· S. Stemmer, Microsc. Microanal. 9 (2003) 493

[41] Adapted from: NSF PanelReport on *Atomic Resolution Microscopy*, NSF (1993) and TEAM project 2003

[42] S. Van Aert, A. J. den Dekker, D. Van Dyck, A. van den Bos, Ultramicroscopy 90 (2002) 273

[43] D. van Dyck, High resolution electron microscopy, Advances in Imaging and Electron Physics 123 (2002) 105

[44] H. Seitz H, M. Seibt, F.H. Baumann, K. Ahlborn, W. Schroter, Physica Status Solidi A, 150 (1995) 625

[45] C. Kisielowski, O. Schmidt, Microscopy and Microanalyses, 4 (1998) 614

[46] M.J. Hytch, J-L. Putaux. J-M. Penisson, Nature 423 (2003) 270

[47] C.L. Jia, A. Thust, Phys. Rev. Lett. 82 (1999) 5052

[48] e.g. Workshop on Complex System, Ernest Orlando Lawrence Berkeley National Laboratory, PUB-826, 1999

[49] C.Kisielowski, E.Principe, B.Freitag, D.Hubert, Physica B 308 (2001) 1090

[50] S.P. Beckman, X. Xu, P. Specht, E.R. Weber, C. Kisielowski, D.C. Chrzan, Journal of Physics, Condensed Matter, 14 (2002) 12673

[51] X. Xu, S.P. Beckman, P. Specht, D.C. Chrzan, E.R. Weber, C. Kisielowski, Microscopy and Microanalysis, 9, Supplement 2 (2003) 498CD

[52] J.R. Jinschek, C. Kisielowski, D. Van Dyck, P. Geuens, Proceedings SPIE International Symposium on Optical Science and Technolgy, 3.-8. August 2003, San Diego/CA, Convention Center, Vol. 5187

[53] N. Shibata1, S.J. Pennycook, T. R. Gosnell, G. S. Painter3, W.A. Shelton, P. F. Becher, Nature, 2004, in press

[54] Y. Shao-Horn, L. Croguennec. C. Delmas, E.C. Nelson, M. A. O'Keefe, Nature Mater. 2 (2003) 464

[55] J.R, Jinschek, C. Kisielowski, T. Radetic, U. Dahmen, M. Lentzen, K. Urban, Mat. Res. Soc. Symp. 727 (2002) R1.3

[56] O.L. Krivanek, N. Dellby, A.J. Spence, R.A. Camps, L.M. Brown, Electron Microscopy and Analysis 1997. Proc. Electron Microscopy and Analysis Group Conference. IoP, Bristol, UK, 35-40 (1997)

[57] M. Haider, *et al.*, Nature 392 (1998) 768

[58] H. Rose, Nucl. Inst. Methods, 187 (1981) 187

[59] C. Kisielowski, J.R. Jinschek, 4th Symposium on non-stoichiometrc III-V compounds, Physik Mikrostruktuierter Halbleiter 27 (P. Specht, T.R. Wetherford, P. Kiesel, T. Marek, S. Malzer, eds.) Erlangen-Nuernberg, 2002, 137

[60] P.M. Voyles, D.A. Muller, J.L. Grazul, P.H. Citrin and H.J.L. Grossmann, Nature 416 (2002) 826

[61] C. L. Jia, M. Lentzen, K. Urban, Science 299 (2003) 870

[62] W. Sinkler, L.D. Marks, *Ultramicroscopy* 75 (1999) 251

[63] P. Geuens and D. Van Dyck, Ultramicroscopy 93 (2002) 179

[64] D. Van Dyck, P. Geuens, J. R. Jinschek, C. Kisielowski, Proceedings Ninth Frontiers Meeting of Electron Microscopy in Materials, Berkeley, 2003, in press

[65] http://ncem.lbl.gov/team3.htm

[66] http://foundry.lbl.gov/index.html

THE WAY FORWARD FOR ELECTRON CRYSTALLOGRAPHY

JONH C.H.SPENCE

Dept of Physics and Astronomy
Arizona State University, Tempe
Az. USA. 85287-1504. Spence@asu.edu.

Key words: electron diffraction, structure determination, low-dose CBED, protein structure

1.   INTRODUCTION

In an age of nanoscience, it might be thought that electron microdiffraction would be the technique of choice for analyzing microstructures, particularly when these are thin enough to allow use of single scattering theory. However, despite sustained effort over many decades, only three area of high energy electron diffraction (THEED) can be considered to be truly quantitative for this purpose - ultra-high vacuum transmission diffraction from reconstructed surfaces, cryo-em of two-dimensional protein crystals in biology, and quantitative convergent beam diffraction from crystals of known structure. The first two of these are based on single-scattering theory and the last on dynamical theory. It should not surprise us that, because of quantification, these have been the most successful applications of THEED, especially the work in biology. This provides a powerful incentive to improve quantification in other areas, such as efforts to solve inorganic and non-biological organic crystal structures by combinations of high-resolution imaging (HREM) and THEED, the analysis of diffuse scattering from defects, coherent nanodiffraction, large-angle convergent beam diffraction (LACBED), reflection high energy diffraction (RHEED), and fast gas electron diffraction. (A fourth highly successful area, that of space-group determination by CBED, does not depend on quantification at present). We have recently seen some remarkable successes in the semiquantitative structure analysis of mesoporous materials and zeolites by combining

31

T.E. Weirich et al. (eds.), Electron Crystallography, 31–42.
© 2006 *Springer. Printed in the Netherlands.*

HREM and TED, allowing structures to be solved which could not by any other method [1]. Here many additional constrains are used in the analysis, such as the form of local structural units such as octahedral. The matching of HREM and STEM images to dynamical simulations is rapidly moving from a semi-quantitative method to accurate quantification, but difficulties remain, such as the anomalous background ("Stobbs factor") for HREM and the role of thermal diffuse scattering for dark-field STEM. The highly successful use of electron microscope bright-field imaging in materials science will be considered only briefly here, since our workshop emphasizes diffraction, rather than imaging. (It is unclear, for example, whether aberration corrected tomographic imaging at about one Angstrom resolution, if possible, might provide a better method of nanocrystal structure determination than methods based on electron microdiffraction).

Seen from the perspective of X-ray diffraction, electron diffraction has the reputation of being based on data of low quality, in the sense that the diffraction intensities are unreproducible and depend very sensitively on experimental parameters (such as thickness and local orientation) which are poorly known. These are all factors which result from the strong interaction of fast electrons with matter and consequent multiple scattering. Only in the single scattering limit (or the large-thickness Blackman approximation) are the intensities, like X-ray intensities, "independent" of thickness (in the sense that all beams have the same thickness dependence, which may be normalized), and it is significant that the three most successful methods listed above depend either on this condition or on dynamical conditions which are very well characterized. Unlike X-ray diffraction, our intensities are normally not angle-integrated, our rocking-curve widths are a much larger fraction of the Bragg angle, our Ewald sphere is much larger relative to lattice spacings (so that it is normal to have many reflections simultaneously at the Bragg condition), our extinction distances are much shorter, our electrons are diffracted by the total ground state electrostatic potential rather than the electronic charge density, and multiple scattering is the normal condition. Finally, again by comparison with X-ray diffraction, electrons have the great advantage of allowing the use of lenses (which can now provide sub-Angstrom resolution) and, for field-emission electron microdiffraction, of providing the strongest continuous signal from the smallest volume of matter in all of physics. A unique property of electrons is their sensitivity to the charge state of ions in a crystal [2], which is the basis of the QCBED method mentioned above. This provides information unobtainable by X-ray diffraction, because the electron probe is smaller than one mosaic block (so that extinction errors are eliminated) and because of the strong dependence of electron atomic scattering factors on ionicity. Finally ,we must not forget that the bulk of electron diffraction work is phase identification, aimed at simply answering the question "what is it that we have in the microscope ?". This is best answered by a combination of EDX for approximate stoichiometry and element determination, CBED for space group determination, and ring patterns for d-spacing identification, using, for example, the "NIST Standard Reference Data Base 15" , a kind of powder diffraction file for electron diffraction, available on the web. (It is an intriguing speculation that a d-spacing sequence might be used to identify a crystal by entering it into a search engine such as Google !). While cell parameters can be found only approximately from spot patterns, a method of more accurate determination based on the HOLZ lines which run through the central disk has been described by Zuo [3].

2.   BLANK DISC CBED, AND STRUCTURE-INDEPENDENT TESTS FOR KINEMATIC SCATTERING

Three modes are commonly used for THEED - selected area (SAD), Koehler mode (KM) and CBED. In the first of these the source is conjugate to the detector, in the second the illumination aperture is conjugate to the sample, and in CBED the source is conjugate to the sample. These modes are related, and it must be emphasized that a CBED pattern is nothing but a set of point (SAD) patterns laid beside each other, each for a slightly different orientation. Each plane-wave component of the incident cone of illumination acts independently, generating its own point pattern, which may be "picked out" of a CBED pattern by selecting one pixel from each disk. We can understand this by imagining that the illumination aperture in CBED is closed down to a small opening which defines one incident beam direction. In that case, for a perfectly crystalline slab, the intensity within the CBED disks will not change as the illumination cone narrows, and the area illuminated grows. Because the sample is perfectly crystalline, no new information is introduced as the probe size grows. It follows that the simplest way to eliminate the sample bending and thickness variation problem is to use CBED, so that patterns are obtained from nanoscale areas. If an SAD "point" pattern is required, it can be obtained by selecting one pixel from each CBED disk. (This method will not work in biology, where, for reasons of radiation damage, the periodic redundancy of a crystal is needed to produce a statistically significant pattern). Thus CBED solves the problem of poor data quality.

For the purpose of solving crystal structures, this leaves us with the problem of ensuring that one has single-scattering conditions. The failure of this condition has been the subject of hundreds of papers - a good analysis for inorganic crystals can be found in [4] for $Nb_{12}O_{29}$, where a thickness of less than about 6nm was found to be needed to ensure single scattering. The following is a list of tests for single scattering which do not require a knowledge of the crystal structure:

1. Since dynamical electron diffraction patterns do not obey Friedel's law ($I(g) = I(-g)$ for all crystals) , whereas kinematic ones do, a crystal which is known a-priori to be non-centrosymmetric in some projection, but which produces a symmetric diffraction pattern, must be diffracting under single-scattering conditions. If it was thick enough to scatter dynamically, the pattern would lack the inversion symmetry which the crystal also lacks, and so reflect the true symmetry of the crystal.

2. The angular width $\Delta\theta$ between minima of a CBED rocking curve is d/t for single scattering, or approximately d/ξ_g if dynamical. Here t is thickness and ξ_g the two-beam extinction distance. If CBED disks of constant intensity are found in thin areas of a light-element crystal, so that $\Delta\theta$ is large, we must either have t small or ξ_g small. Since ξ_g is in fact large near the Bragg condition for light elements, it is reasonable to conclude that we have single scattering conditions.

3. The essential requirement for single scattering is that the (000) beam be much stronger than any other, and this can be checked directly from CBED patterns taken from nanometer regions near a sample edge.

4. The observation of space-group forbidden reflections is not a good test for single scattering, since , as shown by Gjonnes and Moodie, they remain forbidden under dynamical conditions at certain diffraction conditions such as the Bragg condition [2]. In SAD patterns from large areas of slightly bent crystal, however, it may be a useful test, since these cover a range of orientations.

5. The ability to match an HREM image against simulations for all thicknesses along a wedge-shaped sample is the most complete test possible. For certain nanocrystals such as MgO cubes viewed along [110], the thickness will be know precisely from geometry. This test, however, requires knowledge of the crystal structure.

The use of blank-disc CBED patterns for solving crystal structures by electron diffraction (in conjunction with direct methods for the phase problem) would seem to have many advantages:

1. Since the central beam is spread out over a disk, its intensity will lie within the dynamic range of the same detector used to detect the Bragg beams. Hence it can be recorded and used for normalization, so that absolute intensity measurements may be made, and the unsatisfactory Wilson plot method is not needed.

2. Data is collected from such a small region that variations in orientation and thickness are avoided.

3. Blank discs give an automatic, structure-independent indication of single-scattering conditions.

4. Collection of diffraction data involves less radiation damage than imaging.

5. In our work, radiation damage has been further reduced by collecting data at either liquid helium or liquid nitrogen temperature. One seeks the smallest probe size which will deliver the critical dose at a given resolution to enough unit cells to produce a statistically significant diffraction pattern, according to the Rose equation [5]. This probe size then minimizes bending and thickness variation artifacts. For the most sensitive biological materials it will be found that such a large area (and small beam convergence) are needed that the benefits of the blank disk method are lost.

In recent work [6] , we have evaluated the low-dose blank-disc CBED method for organic films of various types, including perylene, coronene, tetracontane ($C_{40}H_{82}$) , copper pthalocyanine and anthracene. Figure 1 shows such a pattern (inner reflections only) from tetracontane along [001] taken from an 8nm diameter area of film, one molecule (10nm) thick, at 18 K and 120 kV on the Zeiss Leo 912 using an in-column energy filter for elastic filtering. Patterns were recorded on a calibrated CCD camera. The cell dimensions are a=0.742nm, b=0.469nm, c=10.503nm (orthorhombic).

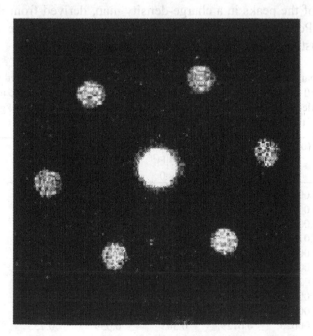

Figure 1. Low-dose CBED diffraction pattern from Tetracontane
with beam direction along [001] recorded at 18 K. from 8nm area [6].

3.   KINEMATIC APPROXIMATIONS TESTED FOR ORGANIC MONOLAYERS

Table 1 compares intensities measured from tetracontane with estimates based on several variants of the kinematic theory. This is the only theory which is simply invertible to Fourier coefficients V_g of crystal potential, which are the electron structure factors. (The inversion of the two-beam expression applied to these intensities is discussed in our paper [6]). Four variants of the kinematic theory are tested - zero and finite excitation error, and an extinction correction model which takes intensity proportional to $|V_g|$ and one which takes it proportional to $|V_g|^2$. We find that the lowest R-factor (26%) is obtained using the $|V_g|^2$ model and including the correct excitation errors. This analysis takes no account of secondary scattering [7], which can be important for long-chain polymers, and may account for the remaining discrepancies between theory and experiment. Based on these structure factor magnitudes, direct methods will give the correct solution to the phase problem, allowing the structure to be solved. If three dimensional data could be collected, our simulations show that the flipping algorithm would provide an excellent solution and potential map, which does not require any knowledge of the type of atoms present. These may then be determined from the heights of the peaks in a charge-density map, derived from the potential map using Poisson's equation. To do this, the value of the mean inner potential must be known [8].

Table 1. Comparison of structure factor modulii $|V_g|$ for tetracontane based on the known structure for hk0 reflections in the [001] zone axis. Four different kinematic approximations are used for reflections whose observed intensity exceeds three times the background. The R-factor indicates the fit of different approximations.

(hkl)	I_o(obs.)	Vg (model)	$|Vg|1$ $Ig{\sim}|F|^2$ $Sg=0$	$|Vg|2$ $Ig{\sim}|F|$ $Sg=0$	$|Vg|3$ $Ig{\sim}|F|^2$ $Sg{\neq}0$	$|Vg|4$ $Ig{\sim}|F|$ $Sg{\neq}0$
010	0.00012	0.0694	0.214	0.023	0.211	0.023
020	0.00703	1.8531	1.616	0.923	1.663	0.924
130	0.00066	0.2657	0.497	0.089	0.639	0.089
110	0.01903	1.1490	2.657	2.498	2.627	2.498
230	0.00036	0.5252	0.368	0.050	0.517	0.051
220	0.00159	0.5372	0.769	0.210	0.833	0.211
200	0.02147	2.7833	2.822	2.815	2.799	2.815
320	0.00044	0.4353	0.403	0.058	0.480	0.059
310	0.00215	0.9197	0.894	0.290	0.945	0.291
430	0.00025	0.6129	0.303	0.039	0.735	0.038
410	0.00020	0.5631	0.272	0.031	0.325	0.031
400	0.00126	0.3835	0.683	0.165	0.781	0.163
R			0.305	0.846	0.258	0.845

4.   THE PHASE PROBLEM AND THE OSZLANYI-SUTO FLIPPING ALGORITHM

While many solutions to the phase problem exist (such as MAD, direct methods, Patterson methods, heavy atom techniques, oversampling, etc) we have found that the recently developed "charge flipping" method [9] is the simplest and works best. It requires three-dimensional atomic-resolution data. The calculation iterates between real and reciprocal space, reversing the sign of all charge density values below a certain threshold. It is thus a form of density modification. Structure factor magnitudes are repeatedly replaced with their measured values. We have shown that this algorithm is equivalent to Fienup's output-output algorithm, and applied it to unknown structures using X-ray data [10]. The method does not use atomic scattering factors or require knowledge of the species present. For two-dimensional data which extends to atomic resolution several direct methods software packages have been developed (e.g SIR 97) which work well with electron diffraction data. The essence of the direct-methods algorithm is this: the sum of the phases of any three reflections taken around a closed loop in reciprocal space is independent of origin, and is probably zero, for both centrosymmetric and non-centrosymmetric crystals. Since one may choose several phases at will to fix the origin, it therefore becomes possible to iteratively determine all phases, and this is done using the tangent formula. (The number of origin-fixing phases depends on the space group and dimensionality - for tables, see [11]). In practice it is found that one needs about 8 strong reflections per inequivalent unknown atom-position to use the programs. Note that the values of the phases are more important in fixing atom positions than the structure factor magnitudes - errors in the magnitudes due to multiple scattering will mainly broaden the Cochran distribution, which gives the probability of a zero sum for the triplet.

5.   OVERSAMPLING FOR CRYO-EM OF ORGANIC MONOLAYERS

For a non-periodic object which scatters continuously, several groups have now shown that the phase problem may be solved by sampling the scattered intensity at intervals of $\square\square/2$ ([12] and references therein). Here $\square\square$ is the angle at which Bragg scattering would be generated if the object were periodically extended. This sampling at half the equivalent Bragg-angle makes the number of unknown phases just equal to the number of Fourier equations (one for each detector pixel), and is consistent with Shannon's sampling theorem. (Because the sampling of the *intensity* is optimal in the

Shannon sense, rather than oversampled, it is unfortunate that the method has come to be known as oversampling. Bragg sampling is optimal for complex amplitudes). In practice, the resulting non-linear Fourier equations for the diffracted intensity are best solved using Fienup's hybrid input-output algorithm, or its more recent variants. This algorithm can be understood using the method of projection onto non-convex sets. Useful introductory background to this subject can be found in a recent essential text [13].

Because the oversampling method of solving the phase problem requires measurements midway between Bragg directions, it has always been assumed that it is not useful for crystals. However organic monolayers (such as purple membrane) are nonperiodic in the direction normal to the layer, so that oversampled data may be collected along the diffraction rods which occur in reciprocal space. Then the phase problem can be solved along each rod independently using the Fienup hybrid input-output (HIO) algorithm. If a few high resolution images are then used to provide phases on a plane parallel to the slab, the phase relationship between different rods will be established. In this way, the need to take high resolution images at very high tilts is eliminated, and three-dimensional tomographic reconstruction of organic monolayer crystals becomes possible using diffraction data collected over a large range of angles, but images collected over only a small range around the zone axis. We have tested this method using data from the protein data base for Lysozyme protein, and find that it works well. Tilts of less than 30 degrees can be used for imaging, to obtain near-atomic resolution three-dimensional reconstructions of the charge-density [14]. The aim is to reduce the time and labor involved for tomographic cryo-electron microscopy of monolayer crystals.

6.   ELECTRON DIFFRACTION FROM LASER-ALIGNED PROTEIN BEAMS - "SERIAL CRYSTALLOGRAPHY"

For proteins, which consist almost entirely of C, O, H, and N, embedded in vitreous ice, it has been found that multiple electron scattering effects are minimal, so that a useful reconstruction of protein structures is possible using either single-particle imaging or a combination of diffraction and imaging from two-dimensional monolayer crystals. However many of the most important proteins, such as the membrane proteins involved in drug delivery, cannot be crystallized, while single particle imaging is a lengthy and tedious process, currently limited to about 1nm resolution. (It is estimated that 70% of drug molecules interact with a membrane protein, while only about 83 of these have been solved. In total, the 33,000 human genes code for about one million proteins, of which about 66,000 have been solved). In most cases the amino-acid sequence for a protein will be known, and these can be distinguished in images at about 0.3nm resolution. The secondary structure (e.g. alpha helicies) becomes visible at about 0.7nm resolution. Few protein crystals diffract to atomic resolution.

To address the problem of solving proteins which cannot be crystallized, we have devised a method for electron diffraction from a beam of hydrated proteins [15]. The proteins originate in aqueous solution, which passes through a micron-sized nozzle into vacuum. Evaporative cooling produces a single-file stream of sub-micron vitreous ice balls, most of which contain one protein. (Cryoprotectant in the water ensures the formation of vitreous rather than crystalline ice). The flight time of the iceballs is arranged to allow most of the ice to sublime, and they pass into a volume of orthogonal crossed laser and high-energy electron beams , as shown in figure 2. The non-resonant laser alignment in three dimensions of small molecules has been experimentally demonstrated [16], and depends on the anisotropic electronic polarizability. Unlike schemes for femtosecond pulsed X-ray diffraction from proteins (which diffract before they are destroyed), our plan uses continuous electron, laser and protein beams. The CCD camera shutter remains open until a statistically significant diffraction pattern accumulates, which is then read out. The laser polarization is then rotated to a new orientation, and the process repeated. Many proteins, travelling at about 30 m/sec, may fall within the electron beam at one time, however the electron beam coherence is adjusted to be about the same size as the molecule, so that no interference occurs between different molecules, and no speckle is seen. The exposure time for an LaB6 electron source is estimated to be about 80 seconds per orientation. The HIO algorithm is used to solve the phase problem. For large misalignment, the resulting patterns resemble powder diffraction patterns.

A classical analysis shows that the degree of alignment is proportional to temperature, and inversely to the laser power and difference in polarizability along orthogonal directions of the molecule. In recent work [17] our simulations show that with a continuous laser power of 10^7 Watt/cm^2 at 4K, resulting in an RMS alignment of 5 degrees, the HIO algorithm reconstructs a lysozyme charge density in three-dimensions at 0.6nm resolution, sufficient to observe secondary structure. Apparatus for these experiments has been constructed at ASU.

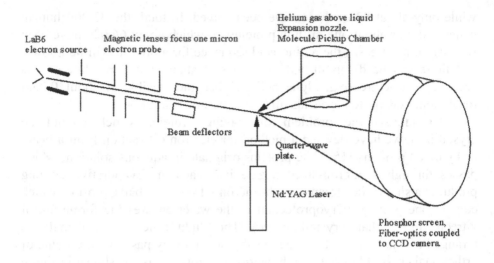

Figure 2. Diffraction camera for single-molecule electron diffraction. A Lanthanum hexaboride electron source is used. The laser and associated optics is rotated after each data readout for a new molecular beam orientation. Organic molecules are picked up within liquid helium droplets to form a molecular beam traversing the electron beam.

7.   THE WAY FORWARD

There is an urgent need in the field of nanoscience for a rapid method for solving the structures of inorganic crystallites too small to be solved by X-ray diffraction. Some success has already been achieved in this direction, as demonstrated in this book. Direct methods for phasing work best in three-dimensions, and the scientific community has come to expect three-dimensional information. Many years after it first became possible in biological electron microscopy, tomographic high resolution imaging has recently been demonstrated at 1nm resolution in materials science [18]. Advances of this imaging technique to atomic resolution using aberration corrected machines   may obviate the need for improved electron crystallography, however this imaging will be extremely difficult experimentally. For diffraction methods, the urgent priority is for a means of collecting high quality kinematic electron diffraction data from sub-micron regions in three-dimensions. This might be done under automated computer control in combination with the new aberration correctors and piezo-electric stages. Although some manufacturers now provide some software for diffraction data collection and analysis, there is little market demand to improve it. Aberration correction allows large beam tilts under electronic control, while electronic probe deflection and pattern recognition software may be able maintain the probe on the same sample area after tilting. On older machines, one used the shadow-image in out-of-focus diffraction spots. Using two hands to keep the particle within the shadow image in the central CBED disk, and two feet to control the tilts, it was possible to orient a nanocrystal without loosing it. Surely it is possible to improve on this procedure now with computer automation, and this is where instrumental development is most urgently needed. LACBED, being a low-dose method, may also be useful for organic films, as Mornirolli has shown at this workshop, however for more radiation sensitive materials the Rose equation sets fundamental limits to the use of focused probes, and either large areas containing many identical molecules must be used to reduce damage, or perhaps the "serial crystallography" method proposed above will be successful. For less sensitive materials, where a probe can be used, I have listed above some structure-independent tests for kinematic scattering, and suggested the use of "blank-disk" CBED as one approach toward this goal.

Acknowledgements
This work involves contributions from many people, including Dr. Jinsong Wu, Dr. U. Weierstall and Professors B. Doak and K. Schmidt. It was supported by ARO award DAAD190010500 and NSF award XXX.X.

References
1. S. Che, Z. Liu, T.Ohsuna, K. Sakmoto, O. Terasaki T. Tatsumi. Nature 429, p.281 (2004). A list of useful books, software, conferences and research groups in electron crystallography is given on the IUCr web page link http://www.public.asu.edu/~jspence/ElectrnDiffn.html
2. Electron Microdiffraction. J. Spence and J. Zuo. Plenum (1992).
3. J. Zuo. Ultramic. 41, p.211 (1992)
4. D. Lynch , A. Moodie, and M. O'Keeffe, Acta Cryst. A31, p.300.(1975)
5. High resolution electron microscopy. J.C.H.Spence. Oxford University Press. 2003. 3rd Edition. New York.
6. J. Wu and J.Spence. Micros and Microan. 9, p.429 (2003)
7. Structural electron crystallography. D. Dorset (1995). Plenum. New York.
8. O'Keeffe, M.A. Acta Cryst. A50, p.33 (1993)
9. Oszlanyi, G. and Suto, A. Acta Cryst. A60, p. 134 (2004).
10. J.Wu, U. Weierstall, J. Spence and C. Koch. Optics Express (2004) In press.
11. Structure determination by X-ray crystallography. M. Ladd and R. Palmer. Plenum (1994). Chapter by Rogers.

12. S. Marchesini, H.He, H. Chapman, S. Hau-Riege, A. Noy, M. Howells, U. Weierstall, J. Spence. Phys Rev. B68, p.14010 (2004).

13. Image Recovery. H. Stark. (1987) Academic Press. New York.

14. J. Spence, U. Weierstall, T. Fricke, R. Glaeser, K.Downing. J. Struct. Biol. 144, p.144 (2003).

15. J. Spence and B. Doak. Phys Rev Letts. 92, p. 198102. (2004)

16. H. Stapelfeldt, Rev Mod Phys. 75, p.543 (2003).

17. J. Spence, K.Schmidt, B. Doak, U. Weierstall, G. Hembree, P.Fromme. Acta Cryst. A (2004). Submitted.

18. P. Midgely, M.Weyland, J.Thomas, F.Johnson. Chem Comms. 2001, p.907 (2001).

SYMMETRY AND STRUCTURE

MAGDOLNA HARGITTAI, ISTVÁN HARGITTAI

*Structural Chemistry Research Group of the Hungarian Academy of Sciences at Eötvös University, P. O. Box 32, H-1518 Budapest Hungary

**Institute of General and Analytical Chemistry, Budapest University of Technology and Economics and Structural Chemistry Research Group of the Hungarian Academy of Sciences at Eötvös University, P. O. Box 91, H-1521 Budapest, Hungary

Abstract: Symmetry considerations are all-important in studying structures. This compilation provides examples in diverse areas of structural science.[1] The examples include molecular symmetry, buckminsterfullerene, the Jahn-Teller effect, chirality and dissymmetry, the double helix and other biological macromolecules, molecular packing in crystals, and quasicrystals

Key words: Point groups, space groups, molecular symmetry, buckminsterfullerene, Jahn-Teller effect, chirality, dissymmetry, molecular packing, quasicrystals

1.   INTRODUCTION

The science of structures involves symmetry at a multitude of levels.[2] The symmetries of molecules are described by point groups and those of crystals by three-dimensional space groups (Figure 1).[1] One-dimensional and two-dimensional space groups are applicable to other extended structures, such as polymeric molecules and layers. In particular, helical symmetries characterize life's most important molecules. Dissymmetry appears to be a condition of existence, in general. Crystallography was the cradle of symmetry considerations in science and has remained intimately involved with it. In spite of its long history, symmetry has remained an innovative research tool in crystallography that greatly facilitates present-day research and understanding of the most frontier topics in the science of structures. In this introductory presentation, we shall concentrate on a few examples that

43

T.E. Weirich et al. (eds.), Electron Crystallography, 43–58.
© 2006 Springer. Printed in the Netherlands.

illustrate diverse applications of the symmetry concept in our studies. These examples include molecular symmetry, the importance of symmetry in the discovery of buckminsterfullerene, the Jahn-Teller effect, chirality, the significance of C_2 symmetry in the discovery of the double helical structure of DNA, the symmetry of packing in three-dimensional crystals, and the quasicrystals.

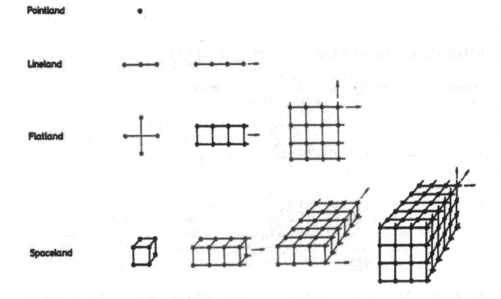

Figure 1. Point groups and space groups.[1]

2.   **MOLECULAR SYMMETRY**[1]

It is the arrangement and symmetry of the ensemble of the atomic nuclei in the molecule that is considered to be the geometry and the symmetry of the molecule. The molecules are finite structures with at least one singular point in their symmetry description and, accordingly, point groups are applicable to them. There is no inherent limitation on the available symmetries for molecules whereas severe restrictions apply to the symmetries of crystals, at least in classical crystallography.

Molecules are never motionless, even at the ground vibrational state they perform vibrations. It is only in the minimum position of the potential energy surface where a molecule would be motionless and this is a hypothetical state. Although it does not exist, it is a well-defined reference structure and is called the equilibrium structure (which is also the result of quantum chemical computations). During molecular vibrations, there are typically about 10^{12} to 10^{14} vibrations per second and the amplitudes of such vibrations are typically several percent of the internuclear distances.

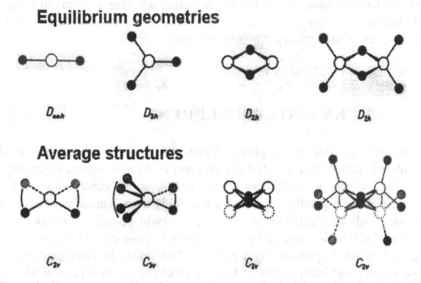

Figure 2. Consequences of low-frequency, large amplitude deformation vibrations on the geometry of some metal halides.[1]

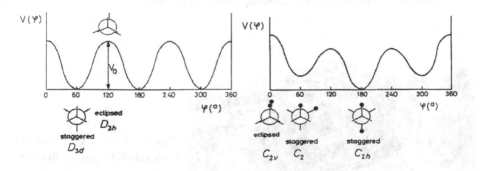

Figure 3. Different conformations with different symmetries in ethane, $H_3C–CH_3$ (left) and 1,2-dichloroethane, $ClH_2C–CH_2Cl$ (right).[1]

The symmetries of molecules usually change during the vibrations, mostly their symmetries become lower (Figure 2), sometimes they don't change, and in some special cases they may become higher. In most of our discussion, the effects of motion on molecular symmetry will be ignored for sake of simplicity. It is of interest to follow the variations of symmetry during such chemically meaningful processes as internal rotation (Figure 3), ligand substitution, or the preparation of the molecule for entering a chemical reaction. During the past half-century, considerations of orbital symmetry have entered our notion in predicting whether a chemical reaction is allowed or forbidden. Thus symmetry considerations have become part of the tools even of the synthetic organic chemist.

3.   BUCKMINSTERFULLERENE

Symmetry considerations played a crucial role in the discovery[3] of the C_{60} molecule called buckminsterfullerene and in the subsequent discovery of the whole class of fullerenes that constitute a substantial part of nanomaterials.[4] Originally, Eiji Osawa was imagining symmetrical cage-like molecules with an electronic structure that would possess aromaticity. He came up with the C_{60} molecules of icosahedral symmetry. However, he did not give it much importance and published his ideas in Japanese only.[5] A few years later and independent of Osawa, Bochvar and Gal'pern in Moscow performed a series of quantum chemical calculations on cage-like molecules that might be considered to be the host part of inclusion complexes. They arrived at the stable C_{60} molecule, again, without giving it too much importance.[6] The first experimental observation of C_{60} was or could have been in the mass spectrum of laser-evaporated graphite.[7]

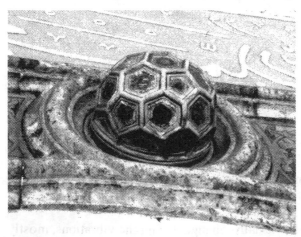

Figure 4. A truncated icosahedral decoration above a door in the Topkapi Sarai in Istambul (photograph and © by I. Hargittai).

However, it was only one of the products in the vapor and the experimenters at Exxon did not notice its presence. About a year later, another group performed similar experiments but they varied the experimental conditions and at some point they saw a huge single peak in the mass spectrum at C_{60} with a smaller one at C_{70}. They searched for a stable geometry, a closed, symmetrical one and arrived at the suggestion of the truncated icosahedral shape for C_{60}.[8] Figure 4 illustrates the buckminsterfullerene structure by an old decoration in the Topkapi Sarai in Istanbul, Turkey. For a few years, the discovery had mostly theoretical significance, and the suggested structure was also doubted by some, until in 1990 measurable quantities were produced.[9] The ensuing spectroscopic investigations produced an NMR spectrum with a single peak and an optical spectrum with four transitions showing up in the infrared, both providing strong supporting evidence for the truncated icosahedral shape and symmetry. Curl, Kroto, and Smalley received the 1996 Nobel Prize in Chemistry "for their discovery of fullerenes."

4.   JAHN-TELLER EFFECT

The Jahn-Teller effect is one of the subtle effects of structural chemistry and it is inherently related to symmetry. According to its original formulation, a nonlinear symmetrical molecule with a partially filled set of degenerate orbitals will be unstable with respect to distortion and thus it will distort to a lower-symmetry geometry and thereby removes the electronic degeneracy.[10] The Jahn-Teller effect represents an exception to the Born-Oppenheimer approximation since it involves the coupling of the electronic and nuclear motions in the molecule.

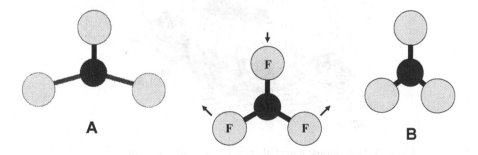

Figure 5. Jahn-Teller distortion of trigonal planar, D_{3h}-symmetry metal trihalides, such as MnF_3, AuF_3, and $AuCl_3$. Structure **A** is the ground-state and structure **B** is the transition state.

Due to this mixing a Jahn-Teller molecule is expected to be rather fluxional and this often makes it difficult to detect the effect. It is usually observed in crystals where due to the so-called Jahn-Teller cooperativity a static distortion may occur in the crystal. The most typical examples are the tetragonally distorted octahedral structures, such as MnF_3 or $KCuF_3$.

For gas-phase molecules the unambiguous experimental detection of the Jahn-Teller effect may have difficulties. Manganese trifluoride is a lucky exception, for which it was possible to prove the presence of static Jahn-Teller distortion by gas-phase electron diffraction as well as by computations.[11] The trigonal planar D_{3h}-symmetry shape is not stable and it distorts to a C_{2v} symmetry structure, with two longer and one shorter bond length and one very large and two smaller bond angles (see Figure 5). AuF_3[12] and $AuCl_3$[13] have similar structures in the gas phase and are thus also obvious examples for Jahn-Teller type distortions.

These molecules have a so-called Mexican-hat type potential energy surface as shown for $AuCl_3$ in Figure 6. Here the undistorted, D_{3h}-symmetry structure is at the center with high-energy. The three equivalent ground-state structures (**A**) can be found at the three minimum positions around the brim of the hat with the three transition-state structures (**B**) with somewhat higher energy between them.

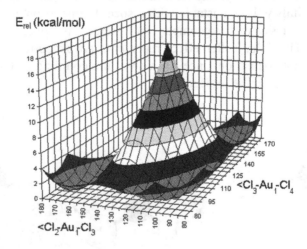

Figure 6. Mexican-hat potential energy surface of $AuCl_3$, after Ref. 14.

5.   CHIRALITY AND DISSYMMETRY

Louis Pasteur[15] (Figure 7) was the first to suggest that molecules can be chiral. In 1848, he recrystallized a salt of tartaric acid and obtained two kinds of small crystals that were mirror images of each other. The discovery followed the important discovery of optical activity by Jean Baptiste Biot. The success of the chirality concept culminated in Pierre Curie's statement, "dissymetry creates the phenomenon," that is, a phenomenon is expected to exist and can be observed only if certain symmetry elements are absent from its system.

Figure 7. Louis Pasteur's bust in front of the Pasteur Institute in Paris
(photograph and © by I. Hargittai).

There are many conspicuous examples of different actions by enantiomeric isomers of the molecules of various drugs. Suffice it to mention thalidomide, which was known as Contergan in Europe and with which many tragedies were connected before it was withdrawn from the market. Since 1992, the United States Food and Drug Administration and the European Committee for Proprietary Medicinal Products have required manufacturers to research and characterize each enantiomorph of a potential drug.[16]

Figure 8. Vladimir Prelog's *Ex Libris* by Hans Erni.

An interesting example of chirality can be observed on the drawing of Vladimir Prelog's *Ex Libris* (see Figure 8), displaying a curious positioning of hands. Prelog held several chiral memorabilia in his office,[17] among them the original of Hans Erni's drawing, which Prelog used as illustration of his Nobel lecture.[18] It is a peculiar feature of this drawing that the two hands of the youth appear as if they were turned around, inverted, if so to speak. However, he had also a different version of the drawing in which the two hands can be imagined as if the two arms were in a different position than in the familiar version. In the familiar version they are being crossed and they are being kept parallel in the modification.

6.   THE DOUBLE HELIX

In 1953, James Watson and Francis Crick[19] (Figure 9) suggested a structure for deoxyribose nucleic acid (DNA). The suggestion had important novel features. One was that it had two helical chains, each coiling around the same axis but having opposite direction. The two helices going in opposite direction, and thus complementing each other, is a simple consequence of the twofold symmetry of the whole double

Figure 9. James Watson in 2001 (photograph and © by I. Hargittai) and Francis Crick in 2004 (photograph and © by M. Hargittai).

helix with the axis of twofold rotation being perpendicular to the axis of the double helix. The other novel feature was the manner in which the two chains are held together by the purine and pyrimidine bases. "They are joined in pairs, as a single base from one chain being hydrogen-bonded to a single base from the other chain, so that the two lie side by side with identical z-coordinates. One of the pair must be purine and the other a pyrimidine for bonding to occur." A little later it is remarked that "... if the sequence of bases on one chain is given, then the sequence on the other chain is automatically determined." Thus symmetry and complementariness appear most beautifully in this model.

Concerning the helical structure of biological macromolecules in general, they relied on Linus Pauling's discovery[3] of the alpha-helix structure of proteins; and concerning the helical structure of DNA, they relied on Rosalind Franklin's X-ray diffraction pattern of DNA.

In the early 1930s W. T. Astbury and his coworkers[20] observed that the stretched, moist hair showed a drastic change in its X-ray diffraction pattern, compared with the dry, unstretched hair. This was interpreted as two forms of the polypeptide chain. One was the extended form, β-keratin, eventually called the β-pleated sheet. The other was the coiled form, α-keratin, eventually called the α-helix.

Pauling[3] decided to determine the structure, that is, the atomic arrangement, of alpha-keratin. The diffraction data indicated that the structural unit would repeat in 5.1 Å along the axis of the hair. Accordingly, two amino acid residues were expected to repeat in the alpha-keratin structure. Using his accumulated knowledge of the structures of small molecules, Pauling could predict the geometry of the peptide group. Alas, at this time he was unable to find a satisfactory solution that would comfort both his structural chemistry criteria and the 5.1 Å repeat distance along the hair axis. He returned to the puzzle of the peptide chain in 1948.

At this time, he realized the utility of a symmetry approach in his quest for the protein structure. He decided to disregard the fact that there may be up to 20 different kinds of amino acids in the chain. Rather, he assumed that they are structurally equivalent with respect to the folding of the polypeptide chain. This was a crucial simplification, and it also freed Pauling's thinking to turn in a different direction. He remembered a theorem that states that the most general operation that converts an asymmetric object into an equivalent asymmetric object (such as an L-amino acid into another molecule of the same L-amino acid) is a rotation-translation. Such a rotation around an axis combined with a translation along the axis and repetition of this operation produces a helix. Accordingly, he rotated the peptide chain around the two single bonds to the alpha carbon atoms. The amount of rotation was the same from one peptide group to the next. He kept the peptide groups planar and searched for a structure with the proper dimensions in which each NH group forms a hydrogen bond with a carbonyl group. One of these structures was the alpha helix, with 4.6 residues per turn. The corresponding repeat distance was 5.4 Å, rather than 5.1 Å.[21] This discrepancy still bothered Pauling although he felt confident that his structure was correct. Eventually, Cochran, Crick, and Vand[22] worked out the theory of diffraction of the polypeptide helix assuming that its structure is based on Pauling and Corey's alpha-helix.

7.   PACKING IN MOLECULAR CRYSTALS

The importance of symmetry in structure does not mean that the highest symmetry is the most advantageous.[1] Lucretius proclaimed about two millenia ago in his *De Rerum Natura*, that "Things whose textures have a mutual correspondence, that cavities fit solids, the cavities of the first the solids of the second, the cavities of the second the solids of the first, form the closest union."[23] In modern science Johannes Kepler recognized that the origin of the shape and symmetry of snowflakes is the internal arrangement of the building elements of water.[24] He did this observation in 1611 and we may consider it as the start of scientific crystallography.[2] Lord Kelvin's (William Thomson's) mostly forgotten geometry was a return to Lucretius' fundamental observation.[25]

As Lord Kelvin was building up the arrangement of molecular shapes he examined two fundamental variations. In one, the molecules are all oriented in the same way, while in the other, the rows of molecules are alternately oriented in two different ways. Kelvin considered the puzzle of the boundary of each molecule as a purely geometrical problem. This is the point where his successors introduced considerations for intermolecular interactions, and, ultimately, Aleksandr I. Kitaigorodskii (Figure 10) "dressed the molecules in the fur-coat of van der Waals domains." Lord Kelvin was using nearly rectilinear shapes for partitioning the plane but he did not let his molecules quite touch one another. Otherwise, he created a modern representation of molecular packing in the plane, including the recognition of complementariness in packing.

Then he came to extending the division of continuous two-dimensional space into the third dimension. He restricted his examinations to polyhedra and found one of the five space-filling parallelohedra, which were discovered by E. S. Fedorov as capable of filling the space in parallel orientation without gaps or overlaps. Fedorov was one of the three scientists who determined the number (230) of three-dimensional space groups. The other two were Arthur Schoenflies and the amateur William Barlow.

Fedorov also derived the 17 two-dimensional plane groups but their best-known presentation is by George Pólya who illustrated them with patterns that completely fill the surface without gaps or overlaps.[26] Today we would call them Escher-like patterns.[2]

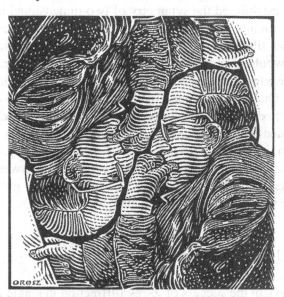

Figure 10. "Complementary Kitaigorodskii." Drawing by and © István Orosz (reproduced by permission).

An important contribution appeared in 1940, from the structural chemist Linus Pauling and the physicist turned biologist Max Delbrück,[27] dealing with the nature of intermolecular forces in biological processes. They suggested precedence for interaction between complementary parts, rather than the importance of interaction between identical parts. They argued that the intermolecular interactions of van der Waals attraction and repulsion, electrostatic interactions, hydrogen-bond formation, etc., give stability to a system of two molecules with complementary, rather than identical structures in juxtaposition. Accordingly, complementariness should be given primary consideration in discussing intermolecular interactions.

Considerations of complementarity in molecular packing culminated in the works of Kitaigorodskii.[28] His most important contribution was the prediction that three-dimensional space groups of lower symmetry should be much more frequent than those of higher symmetry among crystal structures. This was a prediction at a time when few crystal structures had been determined experimentally.

Kitaigorodskii's realization of the complementary packing of molecules was not intuition; he arrived at this principle by empirical investigation.

Today his findings appear simple, almost self-evident, a sure sign of a truly fundamental contribution.

When Kitaigorodskii finally came to the idea of using identical but arbitrary shapes, he started by probing into the best possible arrangements in the plane. He established the symmetry of two-dimensional layers that allow a coordination number of six at an arbitrary tilt angle of the molecules with respect to the tilt axes of the layer unit cell. He found that such an arrangement would always be among those that have the densest packing. In the general case for molecules of *arbitrary* shape, there are only two kinds of such layers. One has inversion centers and is associated with a non-orthogonal lattice. The other has a rectangular net, from which the associated lattice is formed by translations, plus a second-order screw axis parallel to the translation. The next task was to select the space groups for which such layers are possible. This was of great interest since it answered the question as to why there is a high occurrence of a few space groups among the crystals while many of the 230 groups hardly ever occur.

8.   QUASICRYSTALS[3]

To some extent the success of X-ray diffraction in single crystal structure determination has hindered research in areas of less ordered materials. However, some of the best scientists have paid a lot of attention to these both in materials science and in biological structures. There was a curious absence of integer number residues in the α-helix structure, in the unit cell along the fiber direction, which was a sign of formal crystallography breaking down. Bernal commented upon this in the following way: "We clung to the rules of crystallography, constancy of angles and so forth, the limitation of symmetry rotations of two-, three, four-, and six-fold, which gave us the 230 space groups, as long as we could. Bragg hung onto them, and I'm not sure whether Perutz didn't too, up to a point, and it needed Pauling to break with them with his irrational helix."[29] Bernal had great influence in extending traditional crystallography into the science of structures. A sure sign of the expansion has been the gaining importance of fivefold symmetry in it. It is remarkable that two outstanding discoveries of the mid-1980s in materials, the fullerenes and the icosahedral quasicrystals (Figure 11), are both related to fivefold symmetry. Prior to the quasicrystal discovery, for centuries excellent minds, including Johannes Kepler and Albrecht Dürer, have tried to employ regular pentagons for covering the extended surface with a pattern of repetitive fivefold symmetry without gaps or overlaps. In the early 1970, Roger Penrose came up with such a pattern.[30] Alan Mackay extended this pattern into the third dimension, and has urged experimentalists to be on the lookout for such solids in their experiments.[31] Independent of Mackay, and not even being aware of his warning, Dan Shechtman did make such an observation.[32] He used metal alloys of various compositions in rapid solidification and anticipated that this rapid solidification of the alloys would produce the predicted structures.

Figure 11. Icosahedral quasicrystals in a quenched Al/Mn sample. Sample by Ágnes Csanády (Budapest) and scanning electron micrograph by Hans-Ude Nissen (Zürich). Reproduced by permission.

Shechtman's experimental observations could not be published for some time, but when they finally were, an avalanche of theoretical and modeling papers appeared. The story of the quasicrystal discovery illustrates a development when many different threads of far-away origins come together for a unique moment of great importance, only to diverge again in many different directions. It is noteworthy though that the observation of incommensurately modulated structures had already challenged the periodicity paradigm.[33,34] It was, however, salvaged, by bringing these disturbing experiments into line, as if following a prescription by Kuhn in *The Structure of Scientific Revolutions* (see Ref. 35).

The discovery of quasicrystals has led to a paradigm change in crystallography, expressed even in a new definition of what is a crystal by one of the IUCr's commissions: "any substance is a crystal if it has a diffraction pattern with Bragg spots."

Mackay has called attention to the rather careless original definition of crystallinity, which needlessly excluded substances such as what we call today quasicrystals. In this sense the discovery was a kind of legalistic discovery. This happens when the human classificatory system is more restrictive than the laws of nature and discoveries appear to break the laws that had been artificially constructed in the first place. Pejorative words, such as deviation, imperfect, distortion, deformation, disordered, etc., may be a consequence of such human imperfection, rather than nature's. This also applies to the various degrading and upgrading adjectives of symmetry in pseudosymmetry, subsymmetry, supersymmetry, and suchlike. Molecules and atoms do not follow human-made rules of symmetry in their arrangements; rather, our symmetry rules reflect our observations.

Acknowledgements

Our research is being supported by the Hungarian National Research Fund (OTKA T037978 and OTKA T046183).

References

1 Hargittai, I., Hargittai, M., Symmetry through the Eyes of a Chemist. Second Edition. Plenum, New York, 1995.
2 Hargittai, I., "Symmetry in Crystallography." Acta Crystallogr. 1998, A54, 697.
3 Hargittai, I., Hargittai, M., In Our Own Image: Personal Symmetry in Discovery. Plenum/Kluwer Academic, New York, 2000.
4 Hargittai, I., "Lessons of a Discovery: Fullerenes and Other Clusters in Chemistry." In Martin, T. P., Ed., Large Clusters of Atoms and Molecules. NATO ASI Series, Series E: Applied Sciences, Vol. 313. Kluwer Academic, Dordrecht, Boston, London, 1996.
5 Osawa, E., "Superaromaticity." [in Japanese], Kagaku 1970, 25, 854.
6 Bochvar, D. A., Gal'pern, E. G., "Hypothetical Systems: Carbododecahedron, s-Icosahedron, and Carbo-s-Icosahedron." [in Russian], Dokl. Akad. Nauk SSSR 1973, 209, 610.
7 Rohlfing, E. A., Cox, D. M., Kaldor, A., "Production and Characterization of Supersonic Carbon Cluster Beams," J. Chem. Phys. 1984, 81, 3322.
8 Kroto, H. W., Heath, J. R., O'Brien, S. C., Curl, R. F., Smalley, R. E., "C_{60}: Buckminsterfullerene." Nature 1985, 318, 162.
9 Krätschmer, W., Lamb, L. D., Fostiropoulos, K., Huffman, D. R., "Solid C_{60}: A New Form of Carbon." Nature 1990, 347, 354.
10 Jahn, H. A.; Teller, E. Proc. R. Soc. London, Ser. A 1937, 161, 220.

11 Hargittai, M., Réffy, B., Kolonits, M., Marsden, C. J., Heully, J.-L., "The Structure of the Free MnF_3 Molecule – A Beautiful Example of the Jahn-Teller Effect." J. Am. Chem. Soc. 1997, 119, 9042.

12 Réffy, B., Kolonits, M., Schulz, A., Klapötke, T. M., Hargittai, M., "Intriguing Gold Trifluoride – Molecular Structure of Monomers and Dimers: An Electron Diffraction and Quantum Chemical Study." J. Am. Chem. Soc. 2000, 122, 3127.

13 Hargittai, M., Schulz, A., Réffy, B., Kolonits, M. "Molecular Structure, Bonding and Jahn-Teller Effect in Gold Chlorides: Quntum Chemical Study of $AuCl_3$, Au_2Cl_6, $AuCl_4^-$, $AuCl$, and Au_2Cl_2 and Electron Diffraction Study of Au_2Cl_6." J. Am. Chem. Soc. 2001, 123, 1449.

14 Schulz, A., Hargittai, M. "Structural Variations and Bonding in Gold Halides. A Quantum Chemical Study of Monomeric and Dimeric Gold Monohalide and Gold Trihalide Molecules, AuX, Au_2X_2, AuX_3, and Au_2X_6 (X = F, Cl, Br, I)." Chem. Eur. J. 2001, 7, 3657.

15 Hargittai, M., Hargittai, I., "Eternal Dissymmetry." Mendeleev Communications 2003, 3, 91.

16 Richards, A., McCague, R., Chem. Ind. 1997, June 2, 422.

17 Hargittai, I., Candid Science: Conversations with Famous Chemists. Hargittai, M., Ed., Imperial College Press, London, 2000, pp. 139-147.

18 Vladimir Prelog, "Chirality in Chemistry." in Nobel Lectures: Chemistry 1971-1980. World Scientific, Singapore, 1993, pp. 200-216.

19 Watson, J. D.; Crick, F. H. C., "Molecular Structure of Nucleic Acids." Nature 1953, 171, 737.

20 Astbury, W. T.; Street, A., "X-ray Studies of the Structure of Hair, Wool and Related Fibres. I. General." Trans. R. Soc. London 1931, A230, 75; Astbury, W. T.; Woods, H. J., "II. The Molecular Structure and Elastic Properties of Hair Keratin." ibid. 1934, A232, 333; Astbury, W. T.; Sisson, W. A., "III. The Configuration of the Keratin Molecule and its Orientation in the Biological Cell," Proc. R. Soc. London 1935, A150, 533.

21 Eventually, Linus Pauling and Francis Crick, independently, explained this discrepancy by a slight additional coiling of the helices. Because of the non-integer screw, a shift by slight coiling facilitates their best packing. According to Crick, this is a nice example of symmetry breaking by a weak interaction. Crick, F., What Mad Pursuit: A Personal View of Scientific Discovery. Basic Books, New York, 1988, p. 59.

22 Cochran, W.; Crick, F. H. C.; Vand, V., "The Structure of Synthetic Polypeptides. I. The Transform of Atoms on a Helix." Acta Crystallogr. 1952, 5, 581.

23 Lucretius, De Rerum Natura. [English translation: The Nature of Things], W.W. Norton & Co., New York, 1977.

24 Kepler, J., De Nive Sexangula. [English translation: The Six-Cornered Snowflake], Clarendon Press, Oxford, England, 1966.

25 Kelvin, Lord, Baltimore Lectures on Molecular Dynamics and the Wave Theory of Light. C. J. Clay and Sons, London, 1904, Appendix H, pp. 602-642.

26 Pólya, G. "Über die Analogie der Kristallsymmetrie in der Ebene." Z. Kristall. 1924, 60, 278.

27 Pauling, L., Delbrück, M., "The Nature of the Intermolecular Forces Operative in Biological Processes." Science 1940, 92, 77.

28 Kitaigorodsky, A. I., Molecular Crystals and Molecules. Academic Press, New York, 1973 [Russian original: Kitaigorodskii, A. I., Molekulyarnie Kristalli, Nauka, Moscow, 1971].

29 See in Olby, R., The Path to the Double Helix: The Discovery of DNA. Dover Publications, New York, 1994, p. 289.

30 Gardner, M., "Extraordinary Nonperiodic Tiling that Enriches the Theory of Tiles." Sci. Am. 1977, 236, 110.
31 Mackay, L. A., "Crystallography and the Penrose Pattern." Physica 1982, 114A, 609.
32 Shechtman, D., Blech, I., Gratias, D., Cahn, J. W., "Metallic Phase with Long Range Orientational Order and No Translational Symmetry." Phys. Rev. Lett. 1984, 53, 1951.
33 De Wolff, P. M., van Aalst, W., "The Four-Dimensional Space Group of γ-Na2Co3." Acta Crystallogr. 1972, 28A S111.
34 Janner, A., Janssen, T., "Superspace Groups." Physica A 1979, 99, 47.
35 Cahn, J. W. "Epilogue." In Proceedings of the 5th International Conference on "Quasicrystals." Avignon, 22-26 May 1995, Janot, C., Mosseri, R., Eds., World Scientific, Singapore, 1955, pp. 806-810.

Experimental techniques

INTRODUCTION TO ELECTRON DIFFRACTION

Jean-Paul Morniroli

Laboratoire de Métallurgie Physique et Génie des Matériaux, UMR CNRS 8517, USTL & ENSCL, Bâtiment C6, Cité Scientifique, 59650 Villeneuve d'Ascq cedex, France

Abstract: The main types of electron diffraction techniques are described : Selected-Area Electron Diffraction (SAED), Microdiffraction, Convergent-Beam Electron Diffraction (CBED), Large-Angle Convergent-Beam Electron Diffraction (LACBED) and electron precession. They produce spot, ring, disk or line patterns at microscopic or nanoscopic scales in correlation with the image of the diffracted area. An overview of the main applications is given.

Key words: Electron diffraction, CBED, LACBED

1. INTRODUCTION

The diffraction of visible light (with a wavelength λ in the range 350 to 700 nm) by small holes, slits or lattices is a well-known phenomenon which results from destructive and constructive interferences of coherent waves. For a lattice, diffraction may occur when the wavelength is lower than the lattice repeat distances.

Crystals are made of atoms, ions or groups of atoms which repeat along the three dimensions to form a 3D lattice. Since the repeat distances in these crystal lattices are about 0.2 nm, diffraction by visible light is impossible.

X rays are electromagnetic waves like visible light but they have a much shorter wavelength in the same range than the crystal repeat distances so that they can be diffracted. The first X-ray diffraction experiments on crystals were performed by Max von Laue in 1912. Electrons are charged particles

T.E. Weirich et al. (eds.), Electron Crystallography, 61–72.

but in agreement with the wave-particle duality proposed by Louis de Broglie in 1924, a wave is associated when they are accelerated. In 1927 Davisson and Germer obtained the first electron diffraction pattern from a nickel crystal. Later, diffraction was also performed with accelerated neutrons.

Two aspects can be considered when dealing with diffraction patterns: their geometrical aspect and the intensity of the diffracted beams.

The geometrical aspect concerns the position of the diffracted beams on a pattern; it only depends on the direct lattice of the crystal through the Bragg law $\lambda=2d_{hkl}\sin\theta_B$ - d_{hkl} being the interplanar distance of the diffracted (hkl) lattice planes and θ_B the Bragg angle. In other words, it only depends on the lattice parameters of the crystal a, b, c, α, β and γ.

The intensity of the diffracted beams is connected, among other parameters, with the nature and the positions of the atoms in the unit cell through the structure factor.

Despite very strong similarities with X-ray and neutron diffraction, electron diffraction displays some specificities, which to a large extend, are due to the interactions of the incident electrons with the specimen and to the electron wavelength. Thus, the electron cross sections are 10^4 to 10^5 times larger than the ones for X rays and neutrons meaning that the interactions of electrons with a crystal are very strong. As a result, the diffracted beams can have a large intensity and interact. This behavior is interpreted with the complex dynamical theory while X-ray and neutron diffractions are more simply described with the kinematical theory. These strong interactions have some advantages: it is possible to obtain electron diffraction patterns from very tiny diffracted volumes (in the range $1nm^3$ to $1\mu m^3$) even after very short exposure times (a few seconds). They also have some disadvantages. The main one concerns the specimens; they must be thin enough to be crossed by electrons and this is a serious limitation. Moreover, large relaxations of the diffraction conditions are observed with thin crystals meaning that diffraction occurs in a more or less large angular domain around the Bragg angles θ_B and not only at the precise Bragg angles as it is the case with X rays and neutrons.

The wavelength of the wave associated with accelerated electrons is much shorter than the neutron or X-ray wavelength (about a hundred times). As a result, the Bragg angles are very small (a few tenths of degrees) and the diffraction pattern is concentrated around the transmitted beam.

For a long time, Selected-Area Electron Diffraction (SAED) performed with a parallel incident beam and a selected-area aperture was the only experimental method available. During the three last decades, new diffraction techniques based on a convergent electron incident beam (CBED: Convergent-Beam Electron Diffraction, LACBED: Large-Angle

Convergent-Beam electron Diffraction and Microdiffraction) become available on analytical transmission electron microscopes. Most of the electron diffraction techniques use a stationary incident beam, but some specific methods like the precession method take advantage of a moving incident beam.

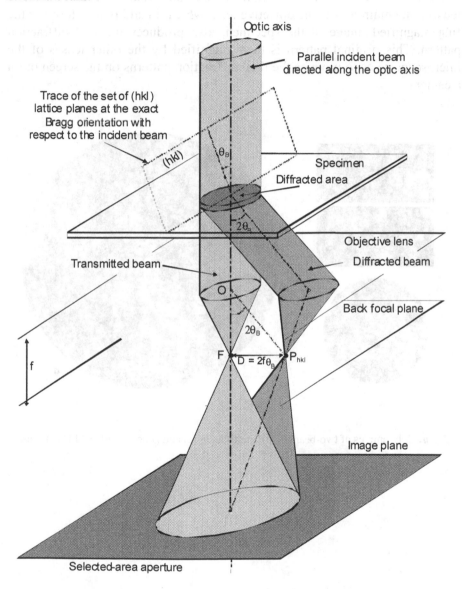

Figure 1. Selected-Area Electron Diffraction (SAED). Two-beam conditions.

2. FORMATION OF THE DIFFRACTION PATTERN WITH A STATIONARY INCIDENT BEAM

Electron diffraction patterns are usually produced with transmission electron microscopes. These instruments are composed of several magnetic lenses. The main lens is the objective lens, which, in addition to forming the first magnified image of the specimen, also produces the first diffraction pattern. This original pattern is then magnified by the other lenses of the microscope so as to produce the final diffraction patterns on the screen or on a camera.

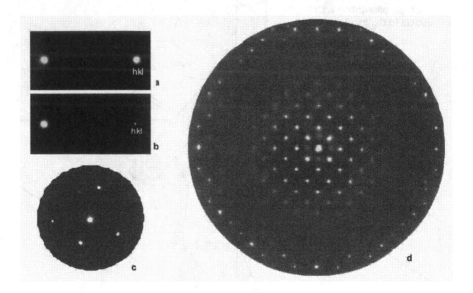

Figure 2. Examples of two-beam (a, b), multi-beam (c) and zone axis (d) SAED patterns.

Diffraction with a parallel incident beam

Consider figure 1. It shows:
- a parallel incident beam directed along the optical axis,
- a thin specimen. Suppose it is a single-crystal cut and oriented so that only one of its sets of (hkl) lattice planes is exactly at the Bragg orientation with respect to the incident beam (the angle between the (hkl) planes and the optic axis is exactly the Bragg angle θ_B). These experimental conditions correspond to the "two-beam" conditions, which are often used to obtain electron micrographs of crystal defects. At the exit face of the specimen, we observe a transmitted beam and a diffracted beam deviated by an angle $2\theta_B$. These two beams focus in the back focal plane of the objective lens to give two spots (a transmitted and a diffracted spots). Therefore, the diffraction pattern is located in the back focal plane of the objective lens and it is made of two spots (figure 2a).

What happens if the set of (hkl) lattice planes is not exactly at the Bragg orientation? As shown on figure 2b, the position of the two spots is hardly affected but the intensity of the diffracted beam is strongly modified. This behavior can be explained by means of the Ewald sphere construction.

Usually many set of lattice planes can simultaneously be exactly or close to the Bragg orientation and give a "multi-beam" pattern made of several diffracted beams as shown in the example on figure 2c. A special type of "multi-beam" pattern concerns Zone-Axis Patterns (ZAP). This type of pattern is observed when a high symmetry [uvw] direction of the crystal is parallel to the incident beam. In this case, the spots on the pattern are arranged along Laue zones (Figure 2d).

The spots located around the transmitted beam form the Zero-Order Laue Zones (ZOLZ) and they come from (hkl) lattice planes which are vertical in the microscope and in zones with the [uvw] zone axis. They are connected with the (uvw)* layer of the reciprocal lattice containing the origin O* (layer with n=0) and they give 2D information since only one reciprocal layer is involved. The other spots are located on concentric areas and belong to the High-Order Laue Zones (HOLZ). These reflections come from slightly tilted planes and are connected with other parallel and equidistant (uvw)* reciprocal layers (with n=1, 2, 3…). A pattern displaying HOLZs (at least the First-Order Laue Zone: FOLZ) is very useful since it contains 3D information coming simultaneously from different (uvw)* layers. On the experimental pattern on figure 2d we can only see the FOLZ. In order to reduce the size of the diffracted area, a selected-area aperture located in the image plane of the objective lens is placed on the electron beam (Figure 1). If the diffracted area contains many crystals (grains) with various orientations, then, the pattern is made of rings.

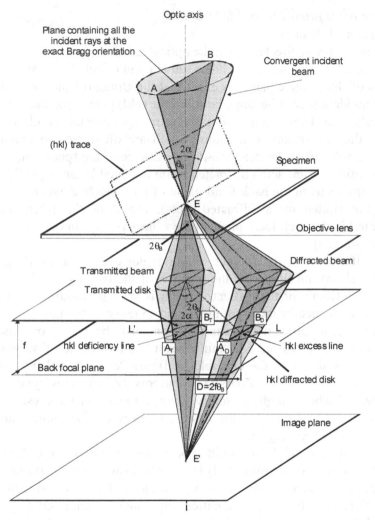

Figure 3. Diffraction with a convergent beam. Two-beam conditions.

Diffraction with a convergent beam

On figure 3, we consider the same experimental conditions as the ones on figure 1 except the incident beam that is now convergent with a semi-angle α of about 0.5°. This beam is exactly focused on the specimen with a spotsize in the range 1 to 100 nm. For the sake of clarity, the convergence angle is strongly exaggerated on the figure and the size of the illuminated area is considered to be a point. As shown on this figure, the transmitted beam and the hkl diffracted beam cut the back focal plane along two disks. The diffraction pattern is now made of two disks instead of two spots. The diameter of these two disks depends on the convergence angle.

What is the intensity distribution inside these two disks? The incident convergent beam can be considered to be composed of a set of incoherent rays having all the orientations within the incident cone. All the rays, which are in the vertical surface ABE are exactly at the Bragg orientation with respect to the (hkl) lattice planes. In the back focal plane, they give a bright line called excess hkl line and a dark deficiency hkl line (Figure 4a). Both lines run through the centers of the disks. In this case, the diffracted disk is called a Dark-Field (DF) pattern and the transmitted disk is the Bright-Field (BF) disk.

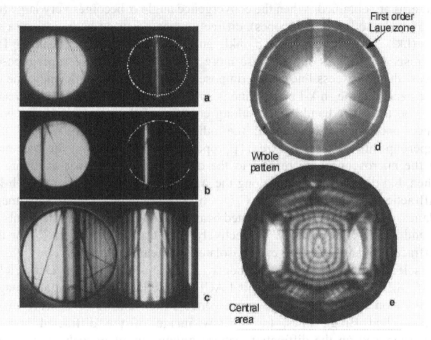

Figure 4. Two-beam (a, b, c) and zone axis (d, e) CBED patterns. Figure e is an enlargement of the central area of figure d.

If the set of (hkl) lattice planes is not exactly in Bragg orientation, then the two excess and deficiency lines are shifted inside the disks as shown on figure 4b. Depending on their intensity, two types of line are observed. If the intensity is weak, then a quasi-kinematical behavior occurs and the lines are sharp and look like the ones on figures 4a and b. Lines with a strong intensity have a dynamical behavior and they display a set of black and white fringes (figure 4c).

For a Zone-Axis Pattern, the disks are also arranged along Laue zones (Figures 4d and e). The ZOLZ reflections have usually a strong intensity and their individual fringe sets interact to produce a more or less complex fringe system visible in the central disk and giving 2D information (Figure 4e). The excess lines belonging to the HOLZs are usually weak and they form HOLZ rings. Their corresponding deficiency lines present in the central disk appear as very sharp lines. These sharp lines are very useful since they carry 3D information. They are used to identify the Bright-Field (BF) symmetry (see the chapter on symmetry determination).

Diffraction with a large-angle convergent beam

The diameter of the disks on a CBED pattern depends on the convergence angle α. When this angle is very small, Microdiffraction patterns are obtained. When the convergence angle α becomes very large (up to 5° with modern microscopes), diffraction can occur on both sides of a set of (hkl) planes giving two hkl and -h-k-l diffracted beams. The corresponding disks superimpose more or less and in these superimposed areas, the hkl excess line is superimposed with the -h-k-l deficiency line on one hand and the -h-k-l excess line is superimposed with the hkl deficiency line on the other hand. The resulting contrast of the patterns is very poor since one line is bright and the other one is dark. To remove this superimposition, Tanaka *et. al.* [1] proposed to raise or lower the specimen in the microscope with respect to the object plane as shown on figure 5. Then, hkl diffraction occurs along the $A_E B_E$ line of the specimen and -h-k-l diffraction along another line $C_E D_E$. In the conjugate object and image planes, the transmitted and diffracted beams are separated along the points E, L and K and E', K' and L' respectively. The transmitted beam E' (or the diffracted beams K' or L') can be isolated by means of the

selected-area aperture to produce a Bright-Field (or a Dark-Field) LACBED pattern (Figure 6). This LACBED method has a main advantage: in addition to the very large convergence which allows to observe many lines called Bragg lines, the shadow image of the diffracted area is superimposed on the diffraction pattern giving simultaneously information about the direct (the shadow image) and the reciprocal (the Bragg lines) spaces. On figure 6, the shadow image of a dust particle is clearly visible. The patterns also display an excellent quality due to an efficient filtering effect of the inelastic electrons by the SAED aperture. Since the incident beam is defocused with respect to the specimen, the electron dose is low and the LACBED method can be applied to beam sensitive materials.

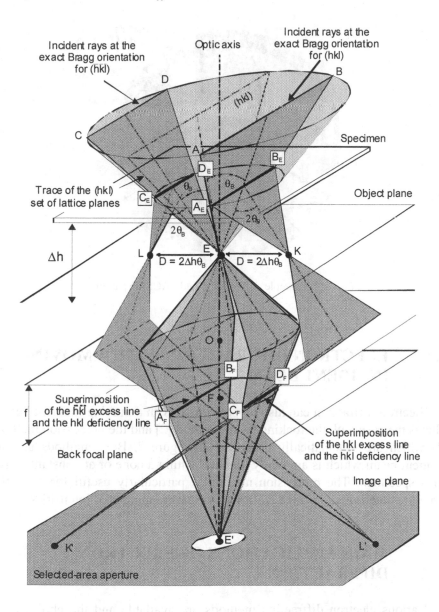

Figure 5. Large Angle Convergent Beam Electron Diffraction (LACBED).

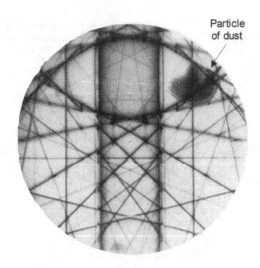

Figure 6. Example of a Bright-Field LACBED pattern.

3.   ELECTRON DIFFRACTION WITH A MOVING INCIDENT BEAM

Electron diffraction can also be performed with a moving incident beam. This is the case with the rocking beam method [2] and the precession method [3] which are schematically described on figure 7. Both methods use an incident beam which is angularly scanned within a cone or at constant angle on a specimen. The precession method is particularly useful for ab-initio structure determination since it allows to measure integrated intensities.

4.   MAIN APPLICATIONS OF ELECTRON DIFFRACTION

Various electron diffraction methods are available and the choice of a pertinent method depends on the type of information which is required and also on the type of specimen.

Electron diffraction performed with a parallel incident beam, i.e. Selected-Area Electron Diffraction is used to obtain good electron micrographs. The "two-beam" condition allows the observation of defects. SAED is also used in High-Resolution Electron Microscopy (HREM) to set a crystal to a zone axis so that the atomic columns are vertical in the microscope. SAED is very useful for the identification of phases and the determination of orientation relationships between two crystals. Due to its very good angular resolution (the patterns are made of sharp spots), it is well adapted to the observations of the fine pattern structure like streaks or satellite reflections.

Nevertheless, this technique has a main disadvantage: the minimum size of the diffracted area, which is selected by means of the selected-area aperture, is about 500 nm. It becomes difficult to prevent some thickness variations and/or some orientation variations in the diffracted area. The SAED patterns are, in fact, average patterns and the diffracted intensities can be strongly affected. For that reason, it is recommended to use Microdiffraction or CBED because the diffracted area is directly defined by the incident beam and can reach a few nanometers with recent microscopes.

Microdiffraction is the pertinent method to identify the crystal system, the Bravais lattices and the presence of glide planes [4] (see the chapter on symmetry determination). For the point and space group identifications, CBED and LACBED are the best methods [5].

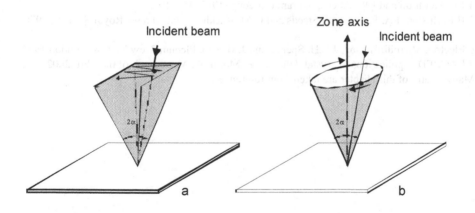

Figure 7. Movement of the incident beam in the beam-rocking (a) and in the electron precession (b) methods.

Quantitative measurement of the intensity for structure factor determination or for charge density measurements can be performed on CBED patterns [6] or from electron precession patterns.

Crystal defects, crystal orientations, accurate lattice parameter measurements, local strain, thin foil thickness can be identified from CBED and LACBED [7].

The pertinent techniques also depend on the specimen. Microdiffraction is well adapted to very small crystals. CBED and LACBED require larger crystals

5.   CONCLUSION

Electron diffraction has a main advantage with respect to the other diffraction techniques: it can be performed at a microscopic and nanoscopic scales in correlation with the image of the diffracted area. The various types of electron diffraction pattern have many applications both in the fields of structure and microstructure characterizations.

References

1 M. Tanaka, R. Saito, K.Ueno and Y. Arada, J. Electr. Micr., 1980, 29, 408

2 J.A. Eades, Ultramicroscopy, 1980, 5, 71-74

3 R. Vincent and P. A. Midgley, Ultramicroscopy, 1994, 53, 271

4 J.P. Morniroli and J.W. Steeds, Ultramicroscopy, 1992, 45, 219

5 B.F. Buxton, J.A. Eades, J.W. Steeds and G.M. Rackham, Phil. Trans. Royal Society, 1976, A281, 181

6 Electron Microdiffraction, J.C.H. Spence and J.M. Zuo, Plenum, New York & London 1992

7 LACBED: Applications to crystal defects, J.P. Morniroli, Monograph of the SFμ, 2002

Many figures of this chapter are taken from reference 7.

SYMMETRY DETERMINATIONS BY ELECTRON DIFFRACTION

Jean-Paul Morniroli

Laboratoire de Métallurgie Physique et Génie des Matériaux, UMR CNRS 8517, USTL & ENSCL, Bâtiment C6, Cité Scientifique, 59650 Villeneuve d'Ascq cedex, France

Abstract: Symmetry determinations performed from Microdiffraction, Convergent-Beam Electron Diffraction (CBED) and Large-Angle Convergent-Beam Electron Diffraction (LACBED) allow the identification of the crystal system, the Bravais lattice and the point and space groups. These crystallographic features are obtained at microscopic and nanoscopic scales from the observation of symmetry elements present on electron diffraction patterns.

Key words: Electron diffraction, Point groups, Space groups

1.   INTRODUCTION

Symmetry determinations allow the identification of very important crystallographic features like the crystal system, the Bravais lattice and the point and space groups. Although it is usually performed from X-ray and neutron diffraction, symmetry determinations can also be obtained from electron diffraction.

Various electron diffraction techniques are available on modern transmission electron microscopes. Selected-Area Electron Diffraction (SAED) and Microdiffraction are performed with a parallel or nearly parallel incident beam and give spot patterns. Convergent-Beam Electron Diffraction (CBED) and Large-Angle Convergent-Beam Electron Diffraction (LACBED) are performed with a focused and defocused convergent beam

T.E. Weirich et al. (eds.), Electron Crystallography, 73–84.

respectively and produce patterns made of disks which contain Bragg lines. They are described in the chapter "Introduction to electron diffraction".

For symmetry determinations, the choice of the pertinent technique among the available techniques greatly depends on the inferred crystallographic feature. A diffraction pattern is a 2D finite figure. Therefore, the symmetry elements displayed on such a pattern are the mirrors m, the 2, 3, 4 and 6 fold rotation axes and the combinations of these symmetry elements. The notations given here are those of the International Tables for Crystallography [1].

2.   IDENTIFICATION OF THE CRYSTAL SYSTEM AND THE BRAVAIS LATTICE

Microdiffraction is the best technique to identify the crystal system. The method consists in finding the Zone-Axis Pattern (ZAP) displaying the highest "net" symmetry. The net symmetry only takes into account the position (and not the intensity) of the diffracted spots on a pattern. It can be easily identified even if the crystal is not perfectly aligned along the zone axis. In order to observe a 3D symmetry, several layers of the reciprocal lattice should be involved meaning that the pattern should, at least, display the First-Order Laue Zone (FOLZ). The Zero-Order Laue Zone (ZOLZ) only gives 2D information. In the example given on figure 1, a 3D-4mm "net" symmetry (characterized by four mirrors at 45° from each other and a four-fold axis) is clearly visible. This "net" symmetry is connected with the crystal system by means of a table given in reference [2].

The Bravais lattice can be identified, on some specific Zone-Axis Patterns, from the observation of the shift between the reflection net located in the ZOLZ and the one located in the FOLZ. This shift is easily observed by considering the presence or the absence of reflections on the mirrors. Thus, in the example given on figure 1, some reflections from the ZOLZ are present on the four m_1, m_2, m_3 and m_4 mirrors. This is not the case in the FOLZ where reflections are present on the m_3 and m_4 mirrors but not on the m_1 and m_2 mirrors. Simulations given in reference [2] allow to infer the Bravais lattice from such a pattern. It is pointed out that Microdiffraction is very well adapted to this determination due to its good angular resolution (the disks look like spots).

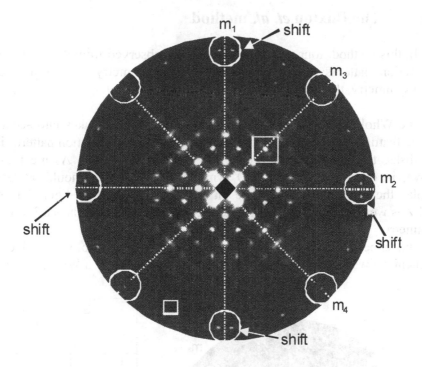

Figure 1. Example of a Zone-Axis Microdiffraction Pattern.

3. IDENTIFICATION OF THE POINT GROUP

Consider a single-crystal; its crystal structure belongs to one of the 32 point groups and prepare a thin foil for electron microscopy with the shape of a perfect parallel slab. Depending on the way this slab is cut from the single-crystal, 31 different types of specimen can be obtained if symmetry elements are taken into account. These 31 types of specimen will give 31 different types of diffraction pattern named diffraction groups.

In an opposite way, if we are able to identify the diffraction group from experimental diffraction patterns, then, we can obtain the point group. This is the basis of the point group determination. To reach this aim, two experimental methods are available: a method proposed by Buxton *et. al.* [3] and a "multi-beam" method proposed by Tanaka *et. al.* [4].

3.1   The Buxton *et. al.* method

In this method, four symmetries should be observed from three different diffraction patterns: the Whole-Pattern (WP) symmetry, the Bright-Field (BF) symmetry, the Dark-Field (DF) symmetry and the +/-g symmetries.

The Whole-Pattern symmetry is the symmetry which takes into account all the features present on a high symmetry zone axis diffraction pattern (i.e. the disks, the lines inside the disk and the Kikuchi lines). As mentioned above, in order to identify a 3D symmetry, the pattern should, at least, display the First-Order Laue Zone. In the example given on figure 2a, this FOLZ is weak, but clearly visible and the Whole Pattern displays a 3D-4mm symmetry.

This experimental symmetry is compared with Figure 2b, which gives in a schematic way, the ten possible symmetries displayed by a Whole Pattern.

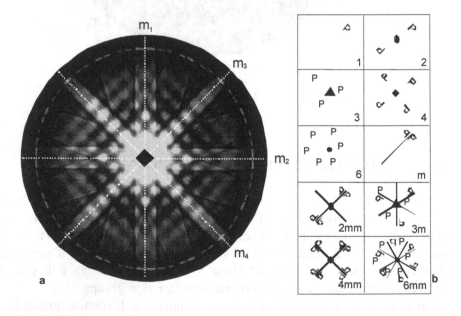

Figure 2. Whole-Pattern symmetries. Experimental pattern (a) and schematic description of the ten possible Whole Pattern symmetries (b).

The Bright-Field symmetry is the symmetry of the transmitted disk (the central disk) of a Zone-Axis Pattern. It is observed on the same ZAP than the one used for the identification of the Whole-Pattern symmetry, but at a higher camera length and a shorter exposure time. Some examples are given on figure 3. The first one (Figure 3a) is the central disk of the Whole Pattern on figure 2a. It only displays a complex set of broad black and white fringes which results from strong interactions of the reflections present in the Zero Order Laue Zone. It gives a 2D-4mm symmetry.

For the other examples (Figures 3b, c , d), in addition to broad black and white fringes, the patterns also display sharp lines which are produced by interactions of the reflections located in the High-Order Laue Zones (HOLZ). The corresponding 3D symmetry is 6mm.

Figures 3b, c, d show the effect of a variation of the accelerating voltage of the microscope on the aspect of a Bright-Field disk. A drastic change occurs but the pattern symmetry remains the same. Of course, it is important to choose a voltage which allows a clear identification of the symmetry elements. In this respect, 200 kV or 250 kV are excellent choices. The ten possible theoretical Bright-Field symmetries are given on figure 3e.

The Dark-Field symmetry is observed inside a hkl diffracted disk (often characterized by its diffraction vector **g**) which is exactly in Bragg orientation. This situation occurs when the hkl Bragg line goes through the center of its hkl diffracted disk.

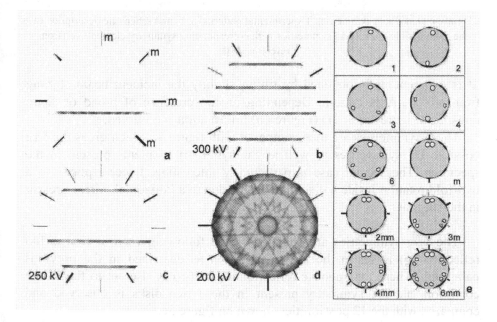

Figure 3. Bright-Field symmetries. Experimental patterns (a-d) and schematic description of the ten possible Bright-Field symmetries (e) (from reference [4]).

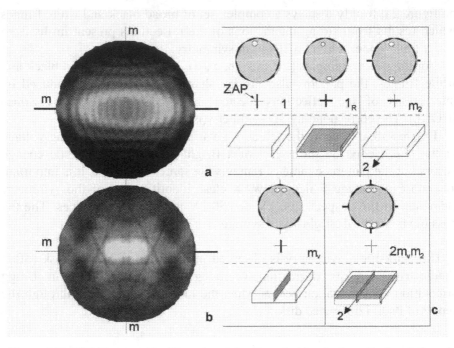

Figure 4. Dark-Field symmetries. Experimental patterns (a, b) and schematic description of the five possible Dark-Field symmetries with corresponding symmetry elements (c)(from reference [4]).

It is experimentally obtained by tilting slightly the incident beam starting from a Zone-Axis Pattern. Depending on the presence of broad or sharp lines, 2D (Figure 4a) or 3D (Figure 4b) information is observed.

These symmetries are then compared with figure 4c which gives the five possible DF symmetries as well as the symmetry elements present in the specimen. The second case is particularly interesting. It corresponds to a two-fold rotation inside a disk (named 1_R) due to a horizontal mirror present in the specimen.

The +/-g symmetries are obtained in the following way: at first, a hkl reflection (**g**) is set to the Bragg condition as described in the previous paragraph. Then, the opposite reflection -h-k-l (**-g**) is also set to the Bragg condition and the symmetry present in these two disks is observed and compared with the 12 possibilities shown on figure 5c.

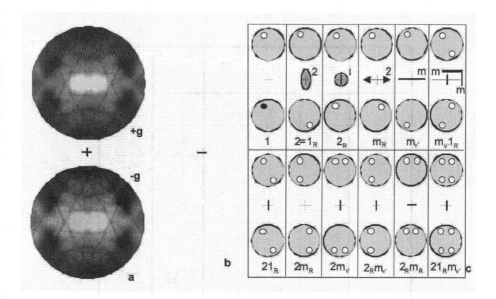

Figure 5. +/-g symmetries. Experimental patterns (a, b) and schematic description of the twelve possible +/-g symmetries (c)(from reference [4]).

It is pointed out, from figure 5c, that the presence of an inversion center (i) in a crystal would give two +**g** and -**g** disks which look exactly the same (symmetry 2_R). On the other hand, the disks are different in the absence of an inversion center (symmetry 1). This important particularity is used to make the distinction, in a very elegant way, between centrosymmetrical and non-centrosymmetrical crystals. One example, obtained from a GaAs specimen, is given on figure 5b. The intensity is different in the two disks.

Once the Whole-Pattern, the Bright-Field, the Dark-Field and the +/-g symmetries are experimentally obtained, the diffraction group can be identified without ambiguity from Table I. This table is valid if 3D information is available (some sharp lines must be present on the patterns). If there is only 2D information available (presence of broad fringes and absence of sharp lines), then, it is equivalent for the specimen to contain a horizontal mirror. In this case, all the diffracted disks display an internal two-fold axis rotation (1_R) and the 31 diffraction groups are reduced to 10 projected diffraction groups listed in the right column on Table I.

Diffraction group	WP	BF	DF	+/-g	Projection Diffraction group
1 1_R	1 1	1 2	1 $2=1_R$	1 1	1_R 1_R
2 2_R 21_R	2 1 2	2 1 2	1 1 2	2 2_R 21_R	21_R 21_R 21_R
m_R	1	m	$\begin{bmatrix}1\\m_2\end{bmatrix}$	$\begin{bmatrix}\overline{1}\,m_R\\1\end{bmatrix}$	$m1_R$
m	m_v	m_v	$\begin{bmatrix}1\\m_v\end{bmatrix}$	$\begin{bmatrix}\overline{1}\,m_v\\1\end{bmatrix}$	$m1_R$
$m1_R$	m_v	2mm	$\begin{bmatrix}2\\2m_vm_2\end{bmatrix}$	$\begin{bmatrix}\overline{1}\,m_v1_R\\1\end{bmatrix}$	$m1_R$
$2m_Rm_R$	2	2mm	$\begin{bmatrix}1\\m_2\end{bmatrix}$	$2\quad 2m_R(m_2)$	$2mm1_R$
2mm	$2m_vm_v$	$2m_vm_v$	$\begin{bmatrix}1\\m_v\end{bmatrix}$	$2\quad 2m_v(m_v)$	$2mm1_R$
2_Rmm_R	m_v	m_v	$\begin{bmatrix}1\\m_2\\m_v\end{bmatrix}$	$2_Rm_v(m_2)\quad 2_Rm_R(m_v)$	$2mm1_R$
$2mm1_R$	$2m_vm_v$	$2m_vm_v$	$\begin{bmatrix}2\\2m_vm_2\end{bmatrix}$	$21_R\quad 21_Rm_v(m_v)$	$2mm1_R$
4 4_R 41_R	4 2 4	4 4 4	1 1 2	2 2 21_R	41_R 41_R 41_R
$4m_Rm_R$	4	4mm	$\begin{bmatrix}1\\m_2\end{bmatrix}$	$2\quad 2m_R(m_2)$	$4mm1_R$
4mm	$4m_vm_v$	$4m_vm_v$	$\begin{bmatrix}1\\m_v\end{bmatrix}$	$2\quad 2m_R(m_2)$	$4mm1_R$
4_Rmm_R	$2m_vm_v$	4mm	$\begin{bmatrix}1\\m_2\\m_v\end{bmatrix}$	$2\quad 2m_R(m_2)\quad 2_Rm_v(m_v)$	$4mm1_R$
$4mm1_R$	$4m_vm_v$	$4m_vm_v$	$\begin{bmatrix}2\\2m_vm_2\end{bmatrix}$	$21_R\quad 2m_R(m_2)$	$4mm1_R$
3 31_R	3 3	3 6	1 2	1 1	31_R 31_R
$3m_R$	3	3m	$\begin{bmatrix}1\\m_2\end{bmatrix}$	$\begin{bmatrix}\overline{1}\\m_R\\1\end{bmatrix}$	$3m1_R$
3m	$3m_v$	$3m_v$	$\begin{bmatrix}1\\m_v\end{bmatrix}$	$\begin{bmatrix}\overline{1}\\m_v\\1\end{bmatrix}$	$3m1_R$
$3m1_R$	$3m_v$	6mm	$\begin{bmatrix}2\\2m_vm_2\end{bmatrix}$	$\begin{bmatrix}\overline{1}\\m_v1_R\\1\end{bmatrix}$	$3m1_R$
6 6_R 61_R	6 3 6	6 3 6	1 1 2	2 2_R 21_R	61_R 61_R 61_R
$6m_Rm_R$	6	6mm	$\begin{bmatrix}1\\m_2\end{bmatrix}$	$2\quad 2m_R(m_2)$	$6mm1_R$
6mm	$6m_vm_v$	$6m_vm_v$	$\begin{bmatrix}1\\m_v\end{bmatrix}$	$2\quad 2m_v(m_v)$	$6mm1_R$
6_Rmm_R	$3m_v$	$3m_v$	$\begin{bmatrix}1\\m_2\\m_v\end{bmatrix}$	$2_R\quad 2_Rm_v(m_2)\quad 2_Rm_R(m_v)$	$6mm1_R$
$6mm1_R$	$6m_vm_v$	$6m_vm_v$	$\begin{bmatrix}2\\2m_vm_2\end{bmatrix}$	$21_R\quad 21_Rm_v(m_v)$	$6mm1_R$

Table I. Identification of the diffraction group from the WP, BF, DF and +/g symmetries. Table from references [3] and [4].

3.2 The Multi-beam method

The previous method has a main disadvantage. It requires the identification of four different symmetries observed on three different patterns (the Zone-Axis, the hkl Dark-Field and the -h-k-l Dark-Field patterns). It is very important that these three patterns come from the same specimen area. This condition is not always easy to fulfill and to solve this problem, Tanaka proposed a method which only requires the observation of a unique diffraction pattern [4].

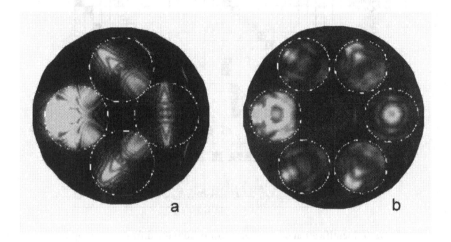

Figure 6. Experimental multi-beam patterns.

This method consists in setting simultaneously four or six disks at the Bragg condition by tilting the incident beam and then, observing the symmetries present inside these Dark-Field disks. The method works with ZAP patterns where the disk in the ZOLZ display a 2mm, a 4mm (Figure 6a) or a 6mm symmetry (Figure 6b). Comparisons with tables given by Tanaka [4] allow the direct identification of the diffraction group.

3.3   Point group determination

The connection between the diffraction group and the point group is obtained from Table II. High symmetry diffraction groups are very useful. One or a few Zone-Axis Patterns are required to identify the point group.

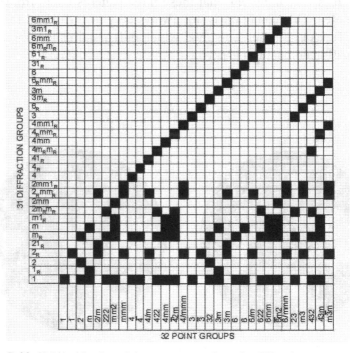

Table II. Identification of the space group from the diffraction groups.
Table from reference [3].

3.4   Space group determination

The space group is usually identified from observations of forbidden reflections due to screw axes and glide planes. This observation is not easy to perform from conventional Selected-Area Electron Diffraction due to the dynamical behavior of electron diffraction. Most of the time, the forbidden reflections appear from multiple diffraction and therefore they cannot be distinguished from the allowed reflections. On a CBED or a LACBED pattern, when the electron beam is exactly parallel to a glide plane or perpendicular to a screw axis (Figure 7a), then the corresponding forbidden disks display a dark line called Gjønnes and Moodie line or line of dynamical absence. Along this zero intensity line, multiple diffractions have exactly the same path but the corresponding waves have opposite amplitude so that they cancel two by two. Two examples of Gjønnes and Moodie lines are given on figures 7b and c. The first one corresponds to a CBED pattern. The second one is observed on a Dark-Field LACBED pattern.

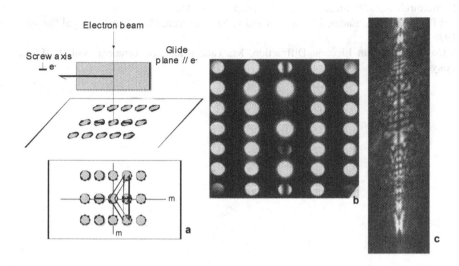

Figure 7. Gjønnes and Moodie (GM) lines. Schematic description of the origin of GM lines (a). CBED (b) and LACBED (c) experimental patterns displaying GM lines.

Glides plane can also be identified from Microdiffraction patterns. Their presence is characterized by a periodicity difference between the reflections present in the ZOLZ with respect to the ones in the FOLZ. This effect is visible on figure 1 where the ZOLZ reflections are placed at the nodes of a centered square net while those of the FOLZ are at the nodes of a smaller non-center square net. The corresponding glide plane is identified from simulations given in reference [2].

4. CONCLUSION

All these crystallographic features are obtained from simple observations of symmetry elements present on the patterns and comparisons with theoretical patterns. Some difficulties may be encountered when the FOLZ is not visible or is weak. In this case, 3D information is not available. A solution to solve this problem consists in tilting the specimen from the ZAP until a part of the FOLZ appears. Other difficulties are encountered when a crystal displays a small departure from symmetry.

References

1 International Tables for Crystallography, Vol. A, Reidel, Dordrecht 1983

2 J.P. Morniroli and J.W. Steeds, Ultramicroscopy, 1992, 45, 219

  3 B.F. Buxton, J.A. Eades, J.W. Steeds and G.M. Rackham, Phil. Trans. Royal Society, 1976, A281, 181

  4 Convergent Beam Electron Diffraction, M. Tanaka and M. Terauchi, Vol. I, JEOL, Tokyo, 1985

ELECTRON DIFFRACTION STRUCTURE ANALYSIS
Specimens and their electron diffraction patterns

VERA KLECHKOVSKAYA
Shubnikov Institute of Crystallography, Russian Academy of Sciences, Leninsky pr. 59, Moskow 119333, Russia

Abstract: Electron diffraction Structure Analysis is generally used to study thin films and finely dispersed crystalline materials and allows the complete structure determinations up to the establishment of the atomic coordinates in the crystal lattice and refinement of atomic thermal vibrations. At present various unknown crystal structures have been determined by the method of electron diffraction structure analysis. In this lecture the first stage of structure analysis - obtaining of appropriate diffraction patterns and their geometrical analysis is exmined.

Key words: electron diffraction, structure analysis, geometrical analysis of electron diffraction patterns

1.   INTRODUCTION

The basic modern data describing the atomic structure of matter have been obtained by the using of diffraction methods – X-ray, neutron and electron diffraction. All three radiations are used not only for the structure analysis of various natural and synthetic crystals – inorganic, metallic, organic, biological crystals but also for the analysis of other condensed states of matter - quasicrystals, incommensurate phases, and partly disordered system, namely, for high-molecular polymers, liquid crystals, amorphous substances and liquids, and isolated molecules in vapours or gases. This tremendous

T.E. Weirich et al. (eds.), Electron Crystallography, 85–96.

amount of material is used for both the solution of problems concerning the relation between the structure of a given substance in the crystalline state and properties, crystal-physical and crystal-chemical generalizations, and the development of solid-state physics of the condensed state in general. Investigation of biological substances has become of exceptional importance for molecular biology and medicine (Fig.1.). The "heart" of this scheme is an atomic structure of matter.

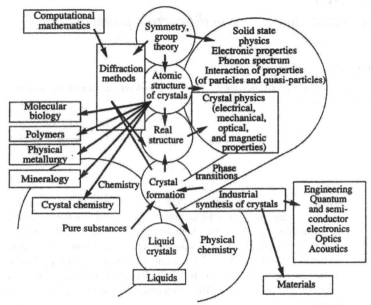

Figure 1. Schematic illustrating branches of modern crystallography, their applications, and the relation of crystallography to the natural sciences.

Analysis of atomic structure of crystals by electron diffraction began its development as a method independent from X-ray analysis by the end of the 1940`s-beginning of the 1950`s (Pinsker, 1949, 1953, Vainstein 1956, 1964). Zinoviyi G. Pinsker, an eminent scientist, made a large contribution to the development of diffraction of short electron and X-ray waves, was born on February 13, 1904, in this years we have his 100th anniversary. He started research in the field of electron diffraction in 1932. In 1944, A.V.Schubnikov invited Pinsker to the newly organized Institute of Crystallography of the USSR Academy of Sciences, where Pinsker organized the electron diffraction laboratory. In the following years, he continued developing his well-known Soviet school of electron diffraction research. Pinsker was the to pay attention to the necessity of developing an independent method - electron diffraction analysis for structure determination. During his many years of working at the Institute of Crystallography, Pinsker, together with his colleagues and numerous students and post-graduates, performed interesting experiments in electron diffraction physics and made valuable contributions to the development of the theory of structure analysis and the geometrical theory of electron diffraction. Especially important studies were performed by Pinsker's student Boris K. Vainshtein (later the head of the Institute of Crystallography), who made an outstanding contribution to the development of Fourier analysis in electron diffraction.

2.   SPECIMENS

Similar to X-Ray and neutron diffraction analysis, electron diffraction structure analysis consists of such main stages as: the obtaining of appropriate diffraction patterns and their geometrical analysis, the precision evaluation of diffraction-reflection intensities, the use of the appropriate formulas for recalculation of the reflection intensities into the structure factors, finally the solution of the phase problem, Fourier-constructions.

Electron Diffraction Structure Analysis (EDSA) is generally used to study thin films and finely dispersed crystalline materials and allows the complete structure determinations up to the establishment of the atomic coordinates in the crystal lattice and refinement of atomic thermal vibrations. However, this possibility often remains only theoretical. The point is that EDSA requires the use of specimens in which the constituent crystallites have some preferred orientation. EDSA differs from X-ray structure analysis in that it deal with the strong interaction of electrons with the substance. This permits investigation of crystals and substances that cannot be obtained as single crystals in a high-dispersion state. Special electron diffraction cameras and electron microscopes are used for such investigations.

The specimens most appropriate for studying in electron diffraction cameras at accelerating voltages ranging within 50-100 kV are crystals of about 10^{-6} cm in length. For larger crystals, the kinematic law of scattering become invalid and the probability of inelastic scattering in these crystals greatly increases, which essentially hinders the transition from the experimental reflection intensities to structure factors. If crystals are too small, two-dimensional diffraction can arise, which also distorts the "kinematical" reflection intensities. The optimum specimen thickness indicated above ranges within 200-600 Å (depending on the atomic number of chemical components in the compound studies). The required crystal dimensions can be obtained by varying the conditions of specimen preparation such as the substrate temperature and the condensation rate. Yet, the situation is rather complicated, because these parameters also influence the orientation of crystallites in the film. As a result, one the main goal of the study, either the crystallite dimensions in the film or their orientation. Crystallite orientation in the film and the film quality provide the formation of different types of electron diffraction patterns. But only three types electron diffraction patterns may be used for EDSA (for an atomic structure determination) - those from polycrystals, textures, mosaic single-crystal films.

3.   ELECTRON DIFFRACTION PATTERNS FOR EDSA

One of the basic features of high-energy ED is the short wavelength of electrons used ~ 0.05 Å (accelerating voltage ~ 100kV). Therefore Ewald`s sphere practically degenerates into a plane, and the electron diffraction pattern (ED) is the planar cross-section of the reciprocal lattice (Fig.2.).

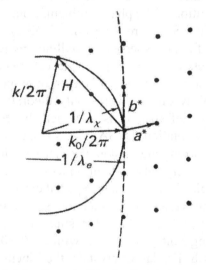

Figure 2. Ewald construction for X-ray (solid sphere) and electron (dotted sphere). (k0, k - wave-vectors, λ - wave-length, a*, b * - parameters of reciprocal unit cell).

Spot-type electron diffraction patterns (fig.3) are formed as a result of diffraction from thin films of mosaic single crystal consisting of a large number of small, almost perfect single-crystal blocks slightly disoriented with respect to one another. These patterns are widely used for determining the symmetry and the unit-cell parameters of crystals and for studying orientation relationships. These patterns are obtained from the specimens prepared by condensation of molecular beams of materials onto heated NaCl cleavages, mica, and other single-crystal substrates or by crystallization of drops of various solutions.

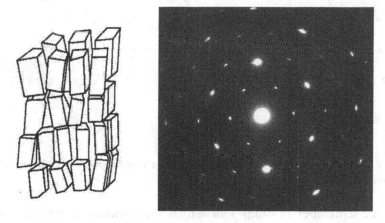

Figure 3. Spot-type ED pattern from the mosaic CdTe film on the mica substrate.

The reciprocal lattice of a mosaic crystal is a three-dimensional periodic system of points, each of which characterized by a vector $H_{hkl} = ha^* + kb^* + lc^*$, where a^*, b^*, c^* are axial vectors and h,k,l, are point indices.

A spot pattern represents a particular plane of the reciprocal lattice of necessity passing through the origin of the coordinates, i.e. through point 000. A spot pattern is most conveniently characterized by the general symbol of the reflection located on it, e.g. (hko), (hkl). If the plane is a coordinate one of the indices must be equal to zero since the point 000 always lies in it (Fig.4). However, if the plane is non-coordinate, then none of its three indices (hkl) is equal to zero, but must a combination of two independent indices h and k (or k and l, or h and l). The reciprocal lattice plane hkl contain the reflections from a definite zone of planes of crystal since each vector of the reciprocal lattice represents a corresponding crystallographic plane. If the zone axis has the indices [u,v,w], then the relationship between (hkl) and [uvw] is expressed by : hu + kv + lw = 0

From a comparison of various spot electron diffraction patterns of a given crystal, a three-dimensional system of axis in the reciprocal lattice may be established. The reciprocal unit cell may be completely determined, if all the photographs indexed. For this it is sufficient to have two electron diffraction patterns and to know the angle between the sections of the reciprocal lattice represented by them, or to have three patterns which do not all have a particular row of points in common (Fig.5). Crystals of any compound usually grow with a particular face parallel to the surface of the specimen support. Various sections of the reciprocal lattice may, in this case, be obtained by the rotation method (Fig.5).

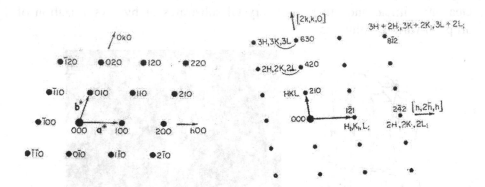

Figure 4. Indexing of an electron diffraction pattern representing a coordinate and non-coordinate planes of reciprocal lattice.

There are a number of studies in which the structure determination was performed solely by the spot-type patterns. The use of spot-type patterns proved to be very successful in the studies of organic compounds consisting of light atoms. This direction is actively being developed by D.L.Dorset.

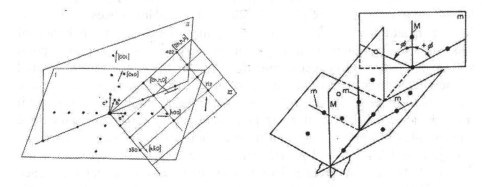

Figure 5. Interrelationship between three reciprocal lattice sections, i.e. between three ED patterns of different zones (a). Schematic representation of the rotation method. Period c* is common for all ED patterns obtained by the rotation method. The other period of these patterns is determined by the structure of the net perpendicular to the axis of rotation (b).

Polycrystal-type (rings) electron diffraction patterns (Fig.6) are especially valuable for precision studies – checking on the scattering law, identification of the nature of chemical bonding, and refinement of the chemical composition of the specimen – because these patterns allow the precision measurements of reflection intensities. The reciprocal lattice of a polycrystal is obtained by "spherical rotation" of the reciprocal lattice of a single crystal around a fixed 000 point; it forms a system of spheres placed one inside the other and has the symmetry ∞:∞.m. It is also important for structure investigation by electron diffraction that polycrystalline patterns do not exhibit a "dead zone", all the reflections being uniformly revealed.

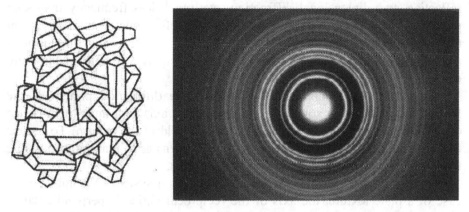

Figure 6. Electron diffraction pattern from polycrystalline Sb_2I_3 film

The character of the pattern does not alter when the angle of inclination of the specimen to the beam is changed. Thus the geometry of a polycrystalline pattern is a set of lengths, $\mathbf{H}_{hkl}$, i.e. set of inter-planar distances d_{hkl} characteristic of the crystal lattice. Polycrystal-type electron diffraction patterns provided a complete three dimensional set of diffraction reflections, however, two or more reflections, with different *hkl* can overlap in one ring of the pattern, especially in the cases where the material studied has large lattice parameters.

As in X-ray diffraction, cubic, tetragonal, orthorhombic and hexagonal unit cells can be determined by establishing simple ratios between the *d* values. The quadratic form for orthorhombic (cubic, tetragonal, hexagonal) crystals is:

$$1/d^2_{hkl} = h^2/a^2 + k^2/b^2 + l^2/c^2$$

Every substance is characterised by a definite set of *d* values. Therefore, apart from their use in investigations of the structure of new, unknown, substances, polycrystalline patterns may be used also in phase analysis, when from the set of *d* values a given specimen can be identified as one or other (previously known) phase or mixture of phases.

Oblique texture electron diffraction patterns (Fig.7) are main experimental material for the electron diffraction structure analysis since they have a large number of reflections, so that only one pattern can provide an almost three dimensional set of diffraction reflections. The formation of textures in specimens is achieved by the use of orienting agents. As in the preparation of single crystal films, use can be made of orienting supports, mechanical action or even the application of an electrical field. Electron diffraction patterns from textures, i.e. from aggregates of a large number of regularly oriented crystals, show the undesirable effects of secondary reflection, two-dimensional diffraction, etc. much less frequency than spot patterns. The simultaneous presence of all reflections in patterns from texture facilitates the evaluation of intensities. The above-mentioned advantages make this electron diffraction extremely valuable in structure determinations.

One of the shortcomings of oblique-texture diffraction pattern is the formation of the "dead zone" around the texture axis and, thus, the absence of reflections located outside the interference field of the pattern. To reduce of dead zone, one has to obtain diffraction patterns as to maximum possible tilt angles of the texture axis to the beam (60-80°).

The reciprocal lattice of single crystal is a system of points. In the case of a plate texture, the axis of the reciprocal lattice is perpendicular to the specimen support. When a plate texture specimen is perpendicular to the electron beam, the diffraction pattern becomes a system of concentric rings (equivalent to the rotation of single crystal about the texture axis).

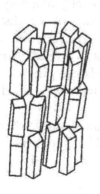 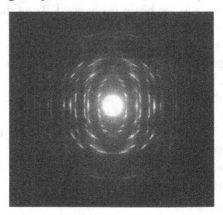

Figure 7. Oblique texture electron diffraction pattern from the thin film of the hexagonal modification of In_2Se_3.

If there are layer lines on the pattern (for orthogonal lattice), for a zero layer line, $R_{hk} = H_{hk0}$. Thus having a set of values

$$R^2_{hk} = h^2A^2 + k^2B^2 + 2hkAB \cos\gamma'; \quad R = r (L\lambda)^{-1},$$

Where r represents the distances on the pattern, we must determine constants A,B and γ' of the two-dimensional lattice (Fig.8) In real situations, reflections on texture patterns are in the sharp of arcs. This. may be explained as follows. The intersection of the rings of the reciprocal lattice with a plane gives, in the ideal case, a point. However with real textures, because of a certain disorder in the crystal orientation relative to the texture axis, reciprocal lattice rings become spherical belts with a centre at point 000, and intersection of them produces arcs (Fig.9 a,b).

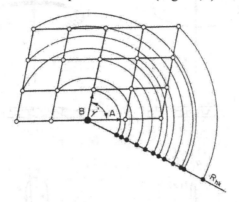

Figure 8. Projection net and the corresponding set of R_{hk} values

Figure 8 c and d compare an ideal texture with a real one possessing a certain disorder. The character of the disorder may be described by distribution function $f(\alpha)$ of the angular orientation of crystal axes in the texture (Fig.9 e,f). Its type is reflected in the character of the arc, i.e. in the angular width: moreover, the angular width of the arc depends also on the angle of inclination of the specimen to the beam. In the ideal case $f(\alpha)$ equals zero everywhere except when angle $\alpha = 0$; this corresponds to the axis of the texture being parallel to the beam. In the real texture, $f(\alpha)$ has a maximum at $\alpha = 0$, although sometimes a very narrow one. (In the case of a polycrystal the distribution function $f(\alpha)$ of crystals over solid angles is spherical.).

The best formed plate textures are found in crystals with a layer lattice, and generally in all crystals having the form of thin plates. Diffraction pattern (Fig.7) indicates a texture of this type, and was obtained from crystals in the shape of thin hexagonal plates. The specific role of the oblique-texture type electron diffraction patterns have in the study of clay minerals having layer structures (B.B.Zviagin, 1964, 1967).

The experimental data obtained from these patterns allows the determination of the atomic structure of both high- and low-symmetric crystals. Since the crystals of a plate texture are all oriented with a particular plane parallel to the support, the properties of the reciprocal lattice require that its points will be distributed exclusively along straight lines perpendicular to the support (Fig.10), independent of the symmetry of the crystals forming the texture. As a result the rings of the reciprocal lattice lie on coaxial cylinders whose axis is texture axis. This distribution of the rings is the most important characteristic of the reciprocal lattice of plate textures.

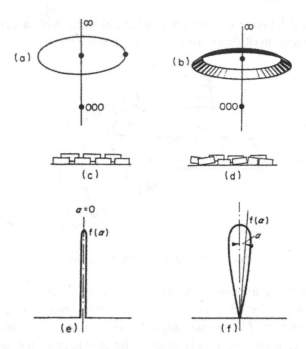

Figure 9. Formation of circular scattering regions (rings) in the reciprocal lattice of a texture, and relationship between their shape and the structure of the specimen. Transition from a point to a ring (a) for an ideal texture without disorder (c), having a distribution function (e); (d), (f) – corresponding diagrams for a real texture with some disorder.

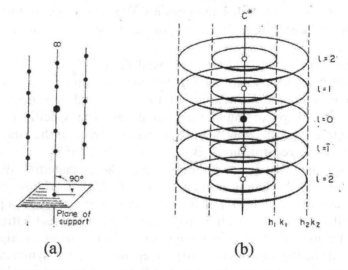

Figure 10. Distribution of reciprocal lattice points of a plate texture along straight lines parallel to the texture axis and perpendicular to the face lying on the support (a) and distribution of circular scattering regions of the reciprocal lattice of a texture on coaxial cylinders.

For the reciprocal lattice of plate texture, the distribution of points along vertical straight lines, parallel to axis z, is characteristic. An important part is played by distance of the hkl point from the origin 000 (spherical coordinates), i.e. by the modulus of vector H_{hkl}:

$$H_{hkl} = x^2 + y^2 + z^2 = R^2 + z^2$$

Let the reciprocal lattice of the texture be intersected by a plane (Ewald`s sphere) at an angle φ so that $(90° - \varphi)$ is the angle formed by this plane and the axis z. Thus the normal position of the film perpendicular to the electron beam corresponds to $\varphi = 0°$, the rotation of the specimen away from this position - to an increase of φ. When the texture axis is perpendicular to the beam (and the film parallel to it), $\varphi = 90°$. This case is realized when a reflection pattern is recorded from a plane texture. In practice, the unit cell is found from measurements, of the magnitude of H, which is independent of the angle of rotation. If the angle $\varphi \neq 90^0$ we have ellipses with the minor semi-axis R and the major semi-axis R/cos φ. Thus, the distribution of reflections along ellipses is characteristic of oblique texture patterns from crystals with any given symmetry (Fig.11).

If unit cell is orthogonal there are layers lines on oblique texture electron diffraction pattern. These lines occur when certain reciprocal lattice planes lie perpendicular to the texture axis. In this case period c* may be more accurately determined by measuring the minor semi-axis R of any ellipse (in the presence of layer lines it is measured directly, since there is a zero line with l=0) and H of any reflection on that ellipse (preferably with a large l):

$$c* = \{\sqrt{(H^2 - R^2)}\}/l ; \quad c = L \, \lambda l/\sqrt{(r^2_{hkl} - r^2_{hk0})}$$

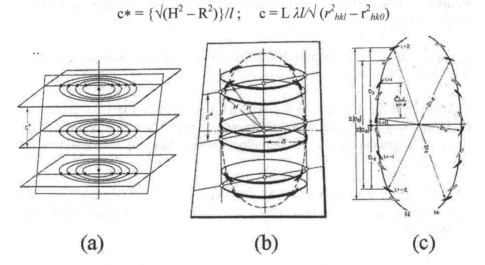

(a)   (b)   (c)

Figure 11. (a) formation of layer lines in the reciprocal lattice of a texture for a orthogonal unit cell. (b) the doubling of number of circular scattering regions in the reciprocal lattice of a texture and therefore the number of reflections on an ellipse of a pattern for a non-orthogonal unit cell. (c) measurement of a values of 2r and 2 | D | on a texture pattern.

The distances 2r between symmetrical *hkl* and *hkl* with minus spots can be more accurately measurement than r. When they are substituted for r, $L\lambda$ must also be doubled. On the basis of various measurements of r_{hkl} and r_{hk0} on one or on several ellipsis, the value c may be determined from equestion:

$$c = l/\eta \sin\varphi = lL\lambda/ D \sin\varphi ,$$

if we make direct measurements of the vertical coordinates (D) of reflections above the zero layer line, related to heights η in the reciprocal lattice by the general relationship:

$$D = \eta L\lambda .$$

4.   CONCLUSION

Thus, mosaic single crystal, polycristal, texture electron diffraction patterns provide valuable material for complete structural investigations. A complete determination of the unit cell of any crystal can be made from electron diffraction patterns, particularly of plate textures.

References

Pinsker Z.G. (1953) Electron diffraction. London: Butterwords.

Vainshtein B.K. (1964) Structure analysis by electron diffraction. Oxford: Pergmon Press.

Zvyagin B.B. (1967) Electron diffraction analysis of clay mineral structures. New-York, Plenum.

Vainshtein B.K., Zvyagin B.B., Avilov A.S. (1992) Electron diffraction structure analysis. In: Electron diffraction techniques Vol. 1 (Ed. J. Cowley), IUCr Monographs on Crystallography, Oxford Science Publications, pp. 216-312.

Klechkovskaya V.V., Imamov R.M. Electron diffraction structure analysis – from Vainshtein to our days. (2001) Crystallography Reports, v.46, N4. pp.598-613.

QUANTITATIVE ELECTRON DIFFRACTION STRUCTURE ANALYSIS (EDSA)

Theory and practise of determining the electrostatic potential and chemical bonding in crystals

A. S. Avilov

Institute of Crystallography Russian Academy of Sciences, Moscow

Key words: Electron diffraction structure analysis (EDSA), quantitative electron diffraction, electrostatic potential, chemical bonding

1.   INTRODUCTION

Investigation of structure and properties of crystal is one of the most important problems in solid state physics and chemistry. Thus study of the features of electron diffraction (ED) and their relation to the inner crystalline field and establishment of their link to physical properties is one of the major requests of modern structure analysis (SA).

Electron crystallography has many attractive properties which make it an important method for solving the indicated problem.

1. Due to the strong interaction with substances electron diffraction gives the strongest continuous signal from the smallest volume of matter. This is very important for nanosciences and nanotechnologies which deal with fine-grained crystallites or thin films which cannot be solved by X-ray diffraction (XRD).

2. Interaction of incident electrons with the electrostatic potential (ESP) gives the possibility to reconstruct the potential from transmission electron diffraction (ED) experiments. ESP and the electron density determine all physical properties of crystals (e.g. energy of electrostatic interaction, characteristics of the electrostatic field in crystals, dipole, quadruple and other momentum of nuclear, diamagnetic susceptibility,

97

static electron polarizability and many others). Thus determination of the ESP distribution from ED experiments is very important task.

3. Transmission ED at small angles is very sensitive to ionicity, which allows to study the electron density and chemical bonding in crystals.
4. Localization of light atoms in the presence of heavy ones is easier with electrons than with XRD. This solves the problems of hydrogen localization.
5. Great intensity of the signal.

The first investigations of the atomic structure of crystals were made by Russian scientists around 1940. They used an electron diffraction chamber with a simple optical system that contained only a condenser lens to form a wide plane parallel primary beam. Thin polycrystalline films were the first objects which have been studied with this setup. These investigations became the basis for the following direction of structure analysis of polycrystalline thin films, named "electron diffraction structure analysis (EDSA)". Later in 1950ies structure analysis of single crystals had begun to develop, which was carried out using very small electron probes formed in an electron microscope. Special techniques have been developed for this: CBED (convergent beam electron diffraction), SAED (selected area electron diffraction), CV (critical voltage), ND (nanodiffraction) and same others.

EDSA of thin polycrystalline films has several advantages: *First* of all the availability of a wide beam (100-400 µm in diameter) which irradiates a large area with a large amount of micro-crystals of different orientations [1, 2]. This results into a special type of diffraction patterns (DP) (see Fig.1). Thus it is possible to extract from a single DP a full 3D data set of structure amplitudes. That allows one to perform a detailed structure analysis with good resolution for determining structure parameters, reconstruction of ESP and electron density.

The second advantage is the small size of the micro-crystallites which form the thin film. Accordingly kinematical or quasi-kinematical scattering are more likely and effects of diffuse scattering are small (important for background substraction).

Due to the large probe, the current density is sufficiently low to minimize radiation damage. This is in particular very important for investigations of organic and metal-organic substances.

Several methods of quantitative structure analysis are known in electron crystallography. These are:

-   QCBED (determination of space-group, cell constants, strain mapping, charge-density measurement)
-   quantitative EDSA (precise structure analysis of organic and inorganic crystals, quantitative reconstruction of ESP and electron density and their analysis)

- ultra-high vacuum transmission ED from reconstructed surfaces of thin films
- cryomicroscopy in biology. Most successful area is a combination of electron diffraction and electron microscopy imaging. (Solving structures of proteins - organic monolayer crystals at 0.3 nm resolution)
- *Semi-quantitative* method is selected area electron diffraction (SAED) (Symmetry, lattice constants, organics and inorganics).

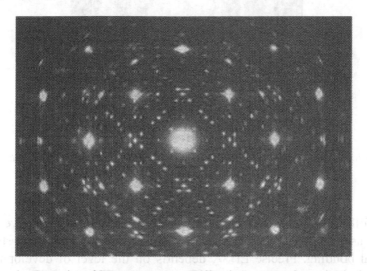

Figure 1a. Examples of ED patterns: a - Diffraction pattern of mosaic single crystals, microcrystallites in the film are a little disoriented around definite directions.

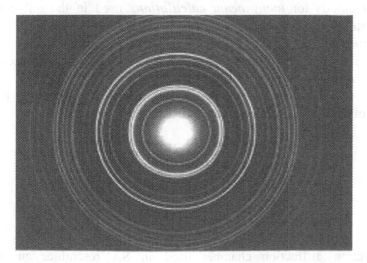

Figure 1b. DP of polycrystalline film with full disorientation's of microcrystallites.

Figure 1c. DP of textured type films. Micro-crystallites are oriented along one direction named "texture axis"

In the following we will consider the present status of quantitative EDSA and its use for the reconstruction of electrostatic potential and study of chemical bonding. Precise EDSA depends on the level of development in several areas:

- *precise quantification* of electron DPs
- methods for *many beam calculations* used in the refinement of structure parameters
- means to account for *inelastic scattering*
- methods for *modeling* ESP on the base of experimental information and the estimation of its real accuracy;
- methods for *treatment of uninterrupted ESP distribution* in terms of conception of physics and chemistry of solids;

2.   EXPERIMENTAL TECHNIQUE

2.1   Apparatus

The electron diffraction chamber used in SA resembles an electron microscope. However, it has simplified electron optics which consists of two condenser lenses for collimating the electron beam (upper) and for focusing on the screen (lower) (Fig. 2). The specimen holder and the sizes of the chamber at the sample allow displacements in two mutually perpendicular directions by ± 10 mm, sample tilt by ± 90⁰ and azimuth rotation by 360⁰ (for the transmission and reflection modes). Furthermore special devices for heating (up to 1200 K) and cooling (down to 4 K) the sample can easily attached. The diffraction chamber has an electron energy filter of the integral type (for the removal of inelastic scattered electrons) and a special system which enables precise measurements of the DP intensities (see below).

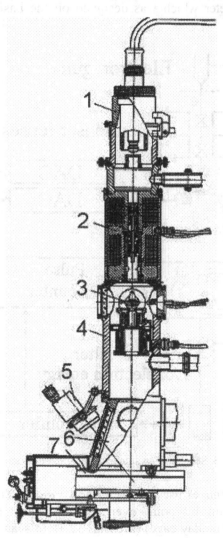

Figure 2. The scheme of the electron diffraction camera EMR-102 (produced by SELMI Ltd. (Sumy, Ukraine). 1 - Electron gun, 2-Two condenser lenses, 3 - Specimen holder, 4 – Chamber, 5 - Optical microscope, 6 – Tubule, 7 - Photo-chamber.

2.2   Measurement of intensities

The precision structure analysis of crystals and quantitative reconstruction of ESP requires the maximum possible set of structure amplitudes, which would provide a precise scaling factor, information on the thermal atomic motion and the high resolution of the ESP with good statistical accuracy being at least 1-2%. Such an experimental data set can be collected by an electron diffractometer which was designed on the basis of EMR-102 [3] (Fig. 3).

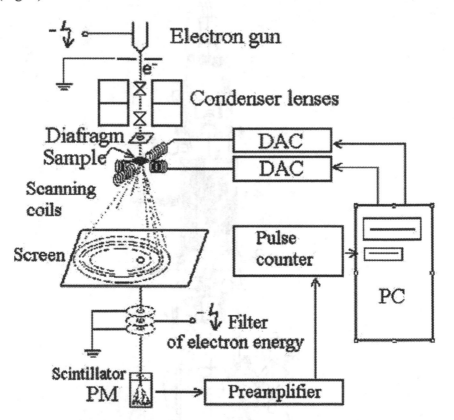

Figure 3. Scheme of an automated electron diffractometer.

The diffractometer makes use of a well known measurement principle - the DP is scanned point by point over a static detector (scintillator with photomultiplier or Faraday cage) (see Fig. 3). Both scanning and measuring the intensities is controlled by a computer (operated in accumulation mode). In contrast to CCD cameras and imaging plates, which are used in the constant time mode measuring (low intensity values have a larger error) this system collects the data with a constant statistical error over the whole range of angles. For the improvement of the accuracy a special statistical treatment of signal and regime, i.e. quasimonitoring are used (for details see [3]). A set of program modules have been developed which handle the measurement and treatment of the experimental intensities, refinement of structural models (κ-model) and Fourier reconstruction of ESP- maps.

2.3 Calculation of the structure amplitudes from the intensities

Depending on the type of the DP there exist different formulas for the transmission case which relate the intensity with the structure amplitude in the kinematical approximation [1,2]. The key-formulas for *integral* intensity are:

$$I_{hkl} / J_0 S = \lambda \left| \Phi_{hkl} / \Omega \right|^2 L$$

$$L_m = t \, d_{hkl} / \alpha \quad \text{and} \quad L_t = t \lambda \, p / (2\pi R' \sin \varphi)$$

Where J_0 and S are the current density of the primary beam and the area of the irradiated sample, λ is the wave length, Φ_{hkl} the structure factor amplitude, Ω the volume cell, L a factor that takes the microstructure of sample into account (L_m – for a mosaic single crystalline film, L_t – for a texture film), t is the sample thickness, d_{hkl} the interplanar spacing, α represents the mean angular distribution of the microcrystallites in the film, p is a multiplicity factor (accounts for the number of reflections of coincidence), R' is a horizontal coordinate of a particular reflection in DP from textures and φ is the tilt angle of the sample. In the case of polycrystalline films, a *local* intensity is usually measured and the corresponding relation is:

$$I'_{hkl} = J_0 \lambda^2 |\Phi_{hkl} / \Omega|^2 V' \, d^2_{hkl} \Delta p / 4\pi L\lambda$$

Where V' denotes the irradiated sample volume and Δ is the entrance slit of the measuring unit.

The most important question for the calculation of the structure amplitudes from the intensities is that for the validity of the kinematical approximation. Due to the strong interaction of fast electrons with matter the effects of dynamical scattering become more pronounced with increasing size of the microcrystallites in the film. In order to justify application of the kinematical equations it is necessary that the diffracted intensity is much less than the intensity of the primary beam: $I_{hkl} \ll J_0S$. According to two-beam theory [2] the following condition should be satisfied in the limiting case:

$$A = \lambda \, |\Phi_{hkl}| \, t_{el} / \Omega \leq 1 \qquad (1)$$

where t_{el} represents the *exctinction length* (which determines the transition to dynamical scattering), and Ω is the unit cell volume. In the kinematical case we find $|\Phi_{hkl}|^2 \sim I_{hkl}$, in the two-beam case $|\Phi_{hkl}| \sim I_{hkl}$, and for many-beam diffraction there exist complex relations between $|\Phi_{hkl}|$ and I_{hkl}.

According to the criterion in (1), the critical thickness t_{el} increases with higher accelerating voltages and/or larger cell volumes and smaller structure factor amplitude. Usually reflections from the zero order laue zones rather belong to the regime of dynamical scattering. Approximate thickness limits for using the kinematical approximation are given in [1]. Generally one can say that the range of suitable sample thicknesses is much smaller for inorganic substances than for organic materials with light elements and large cells. For medium complex structures composed of atoms with middle atomic numbers Z the critical thickness is about 300 to 500 Å, for the most simple structures with hard atoms this value is about 100 Å or less.

Thus strong dynamical scattering can only be avoided if samples of small thicknesses are used in the investigation. The actual thickness values can be estimated using relation (1) for every structure. Another way to reduce strong dynamical diffraction is the "hollow cone" precession technique. As shown by Vincent and Midgley (Ultramicroscopy, 1994, **53**, 271) the precession data are "more kinematical" due to the effect of angular-integrating.

2.4   Dynamical corrections

Aside a few specimens with predominantly kinematical scattering, many specimens investigated by EDSA show pronounced dynamical scattering. In these cases suitable corrections must be applied to link $|\Phi_{hkl}|$ and observed I_{hkl}. For the latter case one has to use successive approximations, i.e. evaluation of parameters from weak "kinematical" reflections which are then used to apply dynamical corrections to strong "dynamical" reflections. Without such corrections the residual R-factor of the structure amplitudes is usually about 20 % or more, while suitable corrections lead to R-factors in the range of 5 - 2 %.

In order to single out the "dynamic" reflections, the observed $|\Phi_{hkl}|$ must be normalized to the theoretical values which have been calculated for a preliminary structural model. The normalization is carried out on the basis of the non-strong reflections at a medium and great diffraction angles which are assumed as "kinematical". Prior to normalization, one may evaluate the degree of the "dynamic nature" of the reflections by comparing the curves of the mean experimental intensity I_{hkl} $(\sin\theta/\lambda)_{average}$ with Σ $|\Phi_{hkl}|$ and Σ $|\Phi_{hkl}|^2$ averaged over the same angular interval [2].

Corrections for primary extinction in the two-beam approximation (the Blackman correction) uses the following relation between the observed and calculated (kinematical) reflection intensity [2]:

$$D = I_{hkl,obs}/I_{hkl,calc} = A^{-1} \int_0^A Jo(2x)dx .$$ (2)

Where $J_0(2x)$ is the Bessel function of zero order. This relation depends only on A and is valid for all types of electron DP. For practical purposes the graph of the D(A) function (the "Blackman curve") is plotted (Fig. 4), and the values of A and thickness values t are determined for each D value according to (2).

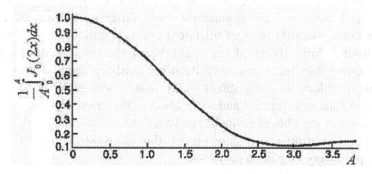

Figure 4. "Blackman curve" used for two-beam corrections

The t-values obtained for different reflections are averaged and corrected $D_{cor.}$ values are obtained corresponding to t_{av}. The corrected intensities are than $I_{hkl,obs.,cor.} = I_{hkl,obs.}/ D_{hkl,cor.}$ Finally the values of $|\Phi_{hkl,cor.}|$ are obtained from $I_{hkl,obs.,cor.}$ using the kinematical formula for different specimen types.

Many EDSA studies have shown that, even for small crystals, the observed intensities sometimes can considerably deviate from the kinematical values. This is the case for reflections corresponding to higher orders of strong reflections. These deviations cannot be described by two-beam scattering which implies existence of systematic many-beam interactions. This is a situation when higher and lower orders of a given reflection are excited simultaneously.

This effect might be interpreted by the Bethe "dynamic potential" approximation, which does not take into account the crystal orientation (as in the Blackman correction case) nor crystal thickness parameters. In terms of this approach, the effect of weak beams can be included in two-beam theory by replacing the potential coefficients, v_h, by :

$$U_{0,h} = v_h - \Sigma_g \, v_g \, v_{h-g}/(\kappa^2 - k_g^2);$$

for which the conditions of the weak excitations should be satisfied:

$$v_g \, /(\kappa^2 - k_g^2) << 1, \, v_{h-g}/(\kappa^2 - k_g^2) << 1.$$

Here indexes h and g denote the three indexes hkl. The interaction between weak beams is ignored. The approximation enables to estimate the effect of a weak wave on the field of the strong waves and is sufficient for taking into account the systematic interactions within a series of reflections containing one strong reflection h_o.

In the case of thicknesses larger than mentioned above the intensities must be calculated according to the more general many-beam theory. The calculation should include summation over different groups of crystals having a certain distributions of thickness and orientation. A method based on the matrix formulation of the many-beam theory was developed for partly-oriented thin films and have been successfully applied samples [2]. The main problem in using direct many-beam calculation is to find the distribution functions for size and orientation of the microcrystals. However, it is not always possible to refine these functions in the process of intensity adjustment. Additional investigation of the micro-structure by electron microscopy is very helpful in such case.

Inelastic scattering arises due to energy losses of diffracted electrons mainly consumed in phonon and plasmon excitation and interband transitions. For rather thick crystals, e.g. as used for convergent-beam electron diffraction, one should also take into account absorption. This is done by introducing a complex crystal potential. The major contribution to the imaginary part of the structure factor comes from thermal diffuse scattering (TDS), which can be quantitatively taken into account. Using a system equipped with an energy filter (filter resolution about 2-3 eV), one can measure diffraction patterns which are formed by quasi-elastically scattered electrons only. This, in turn, allows a more accurate evaluation of the thermal motion of atoms in crystals. Without going into details of this problem, we should emphasize that, in the general case, TDS leads to underestimation of the heights of Bragg maxima and is pronounced in the vicinity of these peaks and between them. This causes problems since a smooth background line can only be calculated when thermal diffuse scattering is low. Moreover, it will be shown by some examples that neglecting absorption for very thin polycrystalline films of light-atom materials has no real effect on determining structure amplitudes.

3. APPLICATIONS

3.1 Refinement of position of hydrogen atom in the structure of brucite [4]

A few important and widely discussed crystal-chemical problems concern the structure of brucite $Mg(OH)_2$, portlandite $Ca(OH)_2$ and related compounds. In particular studying the specific location of hydrogen atoms, the role of hydrogen bonds, and possible phase transition in such structures is interesting. The brucite structure (SG $P3m1$) is built by $Mg(OH)_6$ octahedra, sharing their edges and forming layers parallel to the (001) plane. Hydroxyl groups (OH)⁻ are located on threefold axes, with each hydrogen atom being equidistant from three oxygen atoms in the adjacent layer. This is a rather disadvantageous situation for forming hydrogen bonds between the layers. Neutron diffraction studies (including substitution of hydrogen by deuterium) indicate a statistical distribution of the hydrogen atoms over three positions with the same probability. It should be mentioned that single crystal X-ray diffraction is insensitive to detect the above effects. Thus the brucite structure was studied by the precise EDSA method. Intensities of reflections from a textured film were measured by the electron diffractometer with statistical accuracy of about 1-2%. For several first order reflections the primary extinction Blackman corrections were made at the average thickness of crystallites t_{av} = 440 Å. The final reliability factor on the structure amplitudes is R= 2.7%.

The calculated difference Fourier map of the electrostatic potential passing through the hydrogen atom (without oxygen) shows a striking anisotropy in the distribution of the H potential (Fig. 5a, b). It can be seen, that the isolines are elongated on three symmetrical connected directions (along the mirror planes m) to the nearest atoms O in the adjacent layer. This can be a result of shifts of the hydrogen atoms from the equilibrium position on threefold axes towards positions coinciding with directions of the preferred thermal motion of this atom. From the crystal-chemical point of view this structure model with "splitted" position of H is the most reasonable.

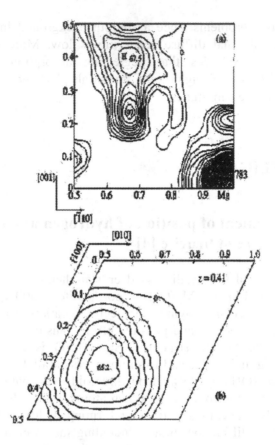

Figure 5. Difference ESP maps for brucite without oxygen; (a) 110 plane; (b) 001 plane through the hydrogen atom. Interval between isolines 10 Volts.

3.2　Quantitative investigations of the electrostatic potential and electron density by EDSA.

Quantitative analysis of ESP is important for several reasons. The precise knowledge of the ESP is necessary to allow comparison of atomic potentials in different structures, analysis of composition (partial occupancy) and chemical bonding (crystal formation, structure property relations).

3.2.1　About the method of the reconstruction of the ESP

Fourier methods are used in usual structure analysis [1]. An application of this method is based on the validity of the decomposition of potential in Fourier series:

$$\varphi(xyz) = (1/\Omega) \sum_{hkl=-\infty}^{+\infty} \Phi_{hkl} \exp[-2\pi i(hx + ky + lz)]$$

where $\Phi_{hkl} = |\Phi_{hkl}| \exp(i\alpha)$.

Here is α the phase value of the amplitude. Application of Fourier method in EDSA had until recently mainly a qualitative or at best semiquantitative character. This was mainly caused by insufficient determination of the intensities of the reflections (5 – 10 %) [1]. Another important reason is connected with the reconstruction of the potential by Fourier-maps. Since Fourier series suffers from angle limitations, the accuracy in the determined atomic coordinates and a peak heights from the potential map is limited [1]. Evidently the effect of termination of Fourier series is minimal for the difference syntheses when the most part of false maximums is subtracted, but distortions of the peak maxima and other peculiarities of ESP distributions in the internuclear area remain. Hence a method free from the above mentioned drawbacks is required for quantitative reconstruction of the potential and for investigations of the inner-crystalline electrostatic field. This is performed by a method of *analytical reconstruction* of the ESP on the model. The parameters of the model are refined by comparison of the trial structure amplitudes with the experimental amplitudes. The calculation of the ESP is realized in direct space by using Hartree-Fock wave functions. Analytical methods are free from many faults, e.g. errors caused by truncation of Fourier series, inaccuracies in calculating structure amplitudes from intensities and noise caused by experimental errors. For the following analysis the static ESP has been calculated which leads to atomic peaks in the potential map that are free from thermal smearing. This approach allows quantitatively to establish features of the ESP in the inter-nuclear area, to calculate the intensity of the electric field and its gradient and to make a topological analysis of the ESP. Moreover reconstruction of the ESP on the analytical model allows comparison of the experimental results with theoretical calculations (e.g. by *ab initio* methods). This is important for the estimation of reliability and accuracy.

Amongst known structural models which could be used for the reconstruction of ESP and electron density the multipole model proposed by Hansen and Coppens is the most suitable for EDSA [5]:

$$\rho(r) = P_{core}\, \rho_{core}(r) + P_{val}\, \kappa^3\, \rho_{val}(\kappa r) + \sum \kappa'^3 R_l(\kappa'^3 r) \sum P_{lm}\, y_{lm}(r/r) \qquad (3)$$

Where ρ (r) is the electron density of each pseudo atom, $\rho_{core}(r)$ and ρ_{val} (r) are the core and spherical densities of the valence electron shells, P_{val} and P_{lm} (multipoles) describe the electron shell occupations, κ and κ' denote the spherical deformation and y (r/r) is a geometrical function. The parameters κ, κ', P_{val} and P_{lm} are refined during adjustment of the experimental and models structure amplitudes.

The first two terms in equation.(3) describe a spherical deformation being characteristic for the ionic bonding and is known in the precise X-ray structure analysis as κ-model:

$$\rho\,(\,r\,) = P_{core}\,\rho_{core}\,(\,r\,) + P_{val}\;\kappa^3\,\rho_{val}\;(\kappa\,r).$$

This equation can be transformed by Fourier conversion and Mott formula to the relation determining an electron structure amplitude:

$$\Phi\,(g) = (\Omega\pi\;\mid g\mid^2)^{-1}\;\sum\,\{Z - [\,f_{core}(g) + P_{val}\;f_{val}\,(g/\kappa\;)]\}.$$

This relation has been used for the precise EDSA studies of ionic compounds with NaCl type structure.

The third term in equation (3) $\sum \kappa'^3\,R_l\,(\kappa'^3\,r)\,\sum P_{lm}\;y_{lm}\,(r/r)$ is the non-spherical part, describing space anisotropy of the electron density being characteristic for the covalent bonding. It was used for the ED study of Ge (see below), at that radial functions $R_l\,(\,\kappa''r) = r\;\exp\,(-\kappa''\,\xi r)$, and $\xi =2.1$ a.u. were calculated theoretically [6].

3.2.2   Binary ionic crystals LiF, NaF, and MgO

For the study of these crystals high quality polycrystalline films were prepared by the vacuum evaporation method or by blazing of Mg in air (for MgO). The intensities were measured by help of the electron diffractometer with an accuracy of about 1-2%. Structure factor amplitudes were calulated from intensities using the kinematical formulas (see above). Extinction Blackman corrections were applied to several first order reflections. After evaluation of suitable thermal parameters, the parameters of the κ-model were refined using equation (3). The results of the refinement, which have been calculated using the MOLDOS programs, and the theoretical calculations from the Hartree-Fock method are compiled in Table 1. The obtained low residual factors *R* testify the high quality of the obtained results.

Theoretical calculations estimate the accuracy of the experimental results were obtained by the following. The program CRYSTAL95, which is based on the non-empirical Hartree-Fock method, was used to calculate the 3D periodical crystal. Broadened atomic basis 6-11G+, 8-511G, 7-311G*, 8-511G* и 8-411G* for Li+, Na+, F-, Mg2+, and O2- corr. were used initially and then optimized to determine the minimum crystal energy. The accuracy of such calculations for an infinite three-dimensional crystal is about 1%. Theoretical electron density X-ray structure factor amplitudes were then calculated, converted into electron amplitudes and used for the refinement of the models parameters [5].

Table 1. Quantitative data of structural κ – model for the ionic crystals LiF, NaF, and MgO[a]

compound	atom	Experimental electron diffraction structure factor amplitudes				Hartree-Fock structure factor amplitudes		
		P_v	κ	R%	R_w %	P_v	κ	R%
LiF	Li	0.06(4)	1[b]	0.99	1.36	0.06(2)	1[b]	0.52
	F	7.94(4)	1[b]			7.94(2)	1.01(1)	
NaF	Na	0.08(4)	1[b]	1.65	2.92	0.10(2)	1[b]	0.20
	F	7.92(4)	1.02(4)			7.90(2)	1.01(1)	
MgO	Mg	0.41(7)	1[b]	1.40	1.66	0.16(6)	1[b]	0.31
	O	7.59(7)	0.960(5)			7.84(6)	0.969(3)	

[a] Structural κ - models were as follows: LiF: $\rho_{cation} = \rho_{1s}(r)$
$+ P_{val} \kappa^3 \rho_{2s}(\kappa r)$, $\rho_{anion} = \rho_{1s}(r) + P_{val} \kappa^3 \rho_{2s,2p}(\kappa r)$; NaF and MgO: $\rho_{cation} = \rho_{1s,2s,2p}(r) +$
$P_{val} \kappa^3 \rho_{3s}(\kappa r)$, $\rho_{anion} = \rho_{1s}(r) + P_{val} \kappa^3 \rho^{2s,2p}(\kappa r)$
[b] Parameters were not refined

It can be seen from Table 1 that the compounds LiF and NaF have almost pure ionic character. Magnesium oxide MgO, however, has a more complicated bonding character with considerable contribution of the covalent component.

ESP and electron density maps were obtained by analytical calculations using the program MOLPROP with parameters for the κ-model Characteristic peculiarities in the ESP along bonding lines are seen in the (110) map for MgO (Fig. 6). All peculiarities in the ESP distribution can be seen, i.e. in the form of maximums on the nuclei and at one-, two-, and three-dimensional minimums in the inter-nuclear areas.

Moreover it is possible to analyze the features of the model distributions of the ESP in terms of a gradient field $\nabla\varphi(r)$, i.e. determining a vector of the intensity of the *classical electrostatic field* $E(r) = -\nabla\varphi(r)$. The characteristics of $\nabla\varphi(r)$ and curvature $\nabla^2\varphi(r)$ do not depend on the mean inner potential φ_0, which is connected with the form and sizes of crystals. The points in which $\nabla\varphi(r) = 0$ are named the "critical points" (CP). CPs corresponding to one- and two-dimensional minimums are denoted as (3,-1) and (3,+1), and a three-dimensional maximum and minimum are (3,-3) and (3,+3) corr..

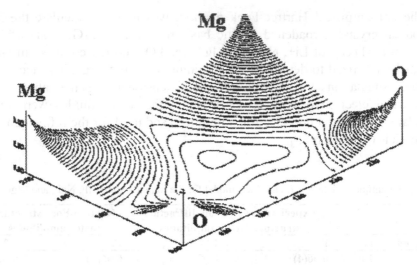

Figure 6. ESP – map along (110)- plane reconstructed on the kappa-model's parameters
(ESP values are given in logarithmic scale).

Here 3 is the number of a non-singular, non-zero eigenvalue λ_i of the *Hessian matrix* for the ESP, and the second number in brackets is the sum of algebraic signs of λ_1, λ_2 and λ_3.

The pairs of the gradient lines in the field E(r), beginning in the CP (3,-1) and ending at two neighboring nuclei are determined eigenvectors, corresponding only positive value λ_3 of Hessian in this point. They form lines which connect adjacent nuclei, and along the ESP is minimal with respect to either lateral shift. The electrostatic field E(r) that acts on the positive unit charge along the internuclear line is directed to the critical point (3,-1) and changes its direction in it.

The nuclei of neighboring atoms and molecules in crystals are separated by the E(r) by "zero-flux" surfaces S (r):

$$E (r) \cdot n (r) = - \nabla \varphi (r) \cdot n (r) = 0, \quad \forall r \in S (r),$$

where n(r) is the unit normal to this surface. These surfaces define the electrically neutral *bonded* pseudoatoms in a statistic equilibrium at the account for only classic Coulomb interactions. Inside these surfaces the nuclear charge is fully screened by the electronic charge.

Such analysis sufficiently supplements with topological analysis of the electron density proposed earlier by Bader, where the condition $\nabla \rho (r) \cdot n (r) = 0$ allows to pick out bonded pseudo-atoms in crystal, formed as a result of action of all these factors. So joint analysis of the gradient fields $\nabla \rho (r)$ and $\nabla \varphi (r)$ is a direct way for the study of influence of the exchange and correlation of electrons on the structure forming interactions in crystals.

Let consider now the ESP-distribution along main crystallographic directions, reconstructed by the κ-model. The ESP and electron density for LiF as characteristic ones are shown in Fig.7.

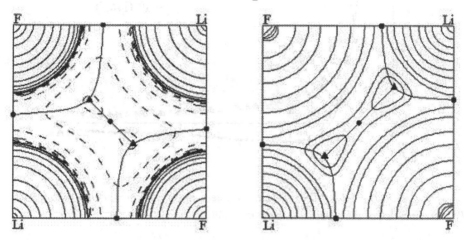

Figure 7. . Distributions of ESP (left) and ED for (100) plane of LiF. CPs (3,-1) are denoted by dotes, (3,+1) – by triangles. The lines of the intersection of the "zero-flux" surfaces with the plane of the figure are shown.

The ESP along the cation-anion line for all studied crystals has smooth character with axial one-dimensional minimum at the distance 0.928 Å (for LiF) to the anion position. This minimum reflects evidently an acquisition of the excess an electron charge by anions in the crystal. Really an existence of these areas is a result of a specific dependence of potential of negative (non-point) anion on the nucleus distance. Compare for example LiF and NaF (Fig.8). The negative minimums of ESP of remote from crystal ions F, calculated with parameters P_{val} and κ, obtained from ED experiment for LiF and NaF (see Table 1), are displaced on the distances 1.06 and 1.03 Å from nucleus corr. This is determined by the mutual sizes of the atomic pair. Positive ESP of relatively large ion Na falls more slowly than potential of Li. But the size of the elemental cell of NaF is more. As a result a distances from the positions anions to axial minimums in crystals LiF and NaF are near each other.

One-dimensional maximums are observed in the center of line anion-anion in the plane (100) (symmetry of position *24d*): they fluently join pairs of two-dimensional minimums lying at the same line nearer to positions of anions being situated at the same line at the position *48h* (Fig.9).

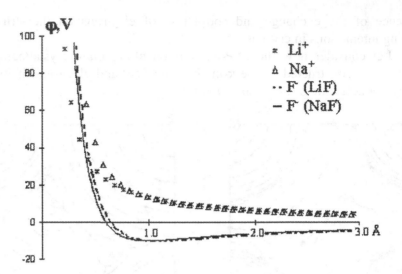

Figure 8. Distributions of ESP of single (remoted from crystal) non-point ions Li+, Na+, F+, reconstructed on the parameters of the κ- model

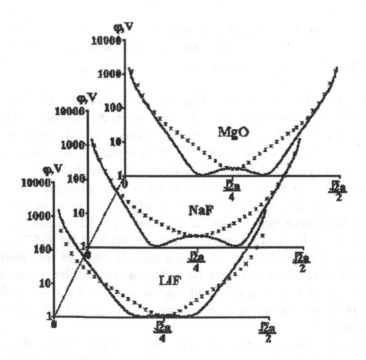

Figure 9. Distribution of ESP in binary compounds is along cation-cation (dotted) and anion-anion (solid).

These minimums reflect the fact of an acquisition of excess electron charge by the fluorine atoms in the crystal and their existence is a consequence of a specific dependence ESP of negative charged atom on the distance (Fig.8). The reliability of existence of two-dimensional minimums in crystals is confirmed by direct calculations of the ESP by the Hartree-Fock method. An analysis shows the more negative charge of isolated ions (deeper negative minimum of potential). At the same time the minimum position shifts to the nucleus. The κ-parameter influences the characteristics of the minima only marginally.

Comparison of forms of atomic fragments limited by the "zero flux" surfaces in ESP and electron density (Fig. 7) displays the role of different factors in the formation of the crystal structure. So in crystals with NaCl-type structure the exchange and correlation of electrons decrease the size of the cation and enlarge the size of the anion which leads to the structure-forming interactions "anion-anion" in the (001) plane of the electron density maps. In ESP-maps the big cations and small anions are seen.

It can be concluded from the ESP and electron density maps for LiF (as characteristic) that both set of critical points do not coincide (Fig.7). A nuclear potential gives also the picture of CPs differing from one in electron density. These observations reflect a well known theoretical statement that the electron density (and energy) of a many-electron system is not determined fully by the inner-crystalline electrostatic field.

The values of the ESP at the nuclear positions, as obtained from the electron and Hartree-Fock structure amplitudes for the mentioned crystals (using a κ-model and corrected on self-potential) are given in table 2. An analysis shows that the experimental values of the ESP are near to the *ab initio* calculated values. However, both set of values in crystals differ from their analogs for the free atoms [5]. It was shown earlier (Schwarz M.E. Chem. Phys. Lett. 1970, 6, 631) that this difference in the electrostatic potentials in the nuclear positions correlates well with the binding energy of 1s-electrons. So an ED-data in principle contains an information on the bonding in crystals, which is usually obtaining by photoelectron spectroscopy.

Table 2. Values of the Electrostatic potentials (V) at the nuclear positions in crystals and free atoms and corresponding mean inner potentials (φ_0)

| Compound | Atom | EDSA κ - model | Hartree-Fock (crystal) | | free atoms | φ_0 |
			direct space	reciprocal space		
LiF	Li	-158(2)	-159.6	-158.1	-155.6	7.07
	F	-725(2)	-726.1	-727.2	-721.6	
NaF	Na	-968(3)	-967.5	-967.4	-964.3	8.01
	F	-731(2)	-726.8	-727.0	-721.6	
MgO	Mg	-1089(3)	-1090.5	-1088.7	-1086.7	11.47
	O	-609(2)	-612.2	-615.9	-605.7	

3.2.3   Direct calculation of physical properties from ED data

The parameters of the model were used for the calculation of the diamagnetic susceptibility χ_d and the electron static (low frequency) polarizability $\alpha(0)$ (Ivanov-Smolensky G.A. *et al*, Acta Cryst. 1983, A39, 411). For ionic bonding and spherical symmetry of atoms the classical Langeuven equation is valid. So with account to the symmetry we will obtain:

$$\chi_d = - (\mu_0\, e^2\, N_A\, a^2\, /\, 4m)\, [\, N/96 + 1/(2\pi^2)\, \sum (-1)^{h/2}\, F_{(h00)}\, /\, h^2\,]$$

N − number of electrons in cubic cell
N_A − Avogadro's number
a − lattice constant of the unit cell
μ_0 − magnetic constant value
$F_{(h00)}$ − X-ray structure amplitude (converted to electron amplitude using the Mott- formula) of *h00* reflection.

For $\alpha(0)$ it is possible to use the Kirkwood relation between the number of electrons in molecules and the mean square radius-vector of electrons in atoms:

$$\alpha(0) = 16\, \pi\, a^4\, /(a_0\, N_e)\, [\, N_e\, /96 + 1/(2\, \pi^2)\, \sum (-1)^{h/2}\, F_{(h00)}\, /(2\, \pi^2\, h^2)]^2$$

N_e − number of electrons in molecular unit
a_0 − Bohr-radius

The received values in Table 3 agree well with the data of independent measurements. This proves that the obtained parameters from the electron diffraction experiment are very reliable and give a *physical significant characteristics* of properties of crystals.

Table 3. Values of diamagnetic susceptibility χ_d and static electron polarizability $\alpha(0)$

Compound	χ_d $(x10^{-10}$ $M^3/mole)$		$\alpha(0)$ $(x10^{-30}$ $M^3)$	
	EDSA	Magnetic measurements	EDSA	Optical measurements
LiF	1.37(4)	1.31	12.3(8)	11.66
NaF	2.02(5)	1.93	15.6(9)	15.10
MgO	2.10(4)	2.31	23.2(8)	18.61

3.2.4 Germanium crystal — covalent bonding [6]

A traditional object for testing new methods is germanium. The chemical bonding in Ge was studied by X-ray diffraction, Hartree-Fock methods and density functional theory for which good agreement of electron densities was reached. Hence the ESP calculated from electron diffraction were also compared with the above results obtained from other methods.

Diffraction patterns from thin polycrystalline Ge films were measured by the electron diffractometer. After refinement of scale and thermal factors and corrections for the primary extinction within the two-beam approximation, the parameters κ' (spherical decompression of valence electron shell) and multipoles P_{32-} and P_{40} (anisotropy of electron density) were calculated (Table 4). The residual factor R calculated from the experimental and theoretical amplitudes (the latter were calculated by the LAPW method, Lu Z.W., *et al.* Phys.Rev. 1993, B47, 9385) is 2.07% and proofs the high quality of the experimental.

Table 4. Results of the multipole model refinement of germanium (single crystal) with electron diffraction data

	Electron diffraction	Refinement with LAPW structure factors
κ'	0.922(47)	0.957
P_{32-}	0.353(221)	0.307
P_{40}	-0.333(302)	-0.161
R(%)	1.60	0.28
R_w(%)	1.35	0.29
GOF	1.98	-

Analysis of the multipole parameters show that the probability of their statistical significance is at least 70%. The obtained value $\kappa' = 0.922$ indicates about 8% decompression of the spherical part of the valence shell in a crystal. This value is in accordance with earlier obtained results (accuracy about 4.5 – 10 %).

the obtained multipole parameters were used for the calculation of the statistical electron density and ESP (Fig.10) and Laplacian of the electron density - $\nabla^2\rho(r)$ (Fig.11). Generally the location of the CPs in the electron density and ESP does not coincide, but for the Ge crystal (due to symmetry) the electron density and potential are homeomorphous, i.e. in this particular case the location of critical points is equivalent (see Fig. 10).

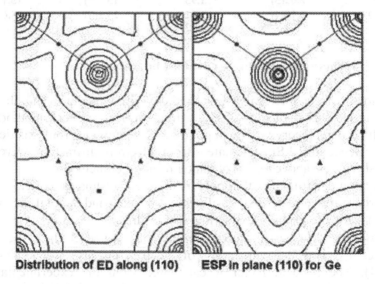

Distribution of ED along (110)   ESP in plane (110) for Ge

Figure 10. Fragment of structure of Ge along plane (110). Critical points (3,-1) are denoted by dotes, (3,+1) – by triangles. Three-dimensional minimum (3,+3) – by squares.

It should be noted that the experimental set of structure amplitudes for Ge was obtained up to $\sin\theta/\lambda = 1.72$ Å. This allowed not only to refine more exactly the scale and temperature factors but also provided high resolution in the electron density, the ESP maps and the Laplacian of the electron density. For example the inner electron shells of the core in Ge can be seen in figure 11.

The topological characteristics of CPs on bonds gives a quantitative explanation of the known effect that the formation of a Ge crystal is accompanied by shifting the electron density towards the Ge-Ge bonding line. This is can be seen by comparing the parameters of the curvature of the electron density λ_i at the critical point (3,-1) with analogous parameters for a "procrystal" (a set of noninteracting spherical atoms placed at the same positions as the interacting atoms in the crystal) (see Table 5). However, the curvature of the electron density along the Ge-Ge bond shows only a small shift of electrons towards the atomic positions.

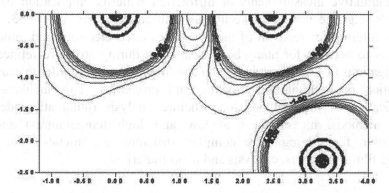

Figure 11. Laplacian of electron density for Ge - $\nabla 2 \rho (r)$ for small fragment in Ge curvature of electron density along Ge-Ge bond shows only small shift of electrons to the atomic positions.

Table 5. Topological characteristics of the electron density in Ge at the bond (3,-1), cage (3,+3) and ring (3,+1) critical points. First row presents the ED results, second row presents the calculations based on model parameters, obtained by LAPW data. Characteristics of the CP (3,-1) for the procrystal: ρ =0.357, $\lambda_1=\lambda_2$=-0.65, λ_3= 1.85, $\nabla^2\rho$= 0.55

Critical point type and Wyckoff position	ρ (eÅ^{-3})	λ_1(eÅ^{-5})	λ_2 (eÅ^{-5})	λ_3 (eÅ^{-5})	$\nabla^2\rho$
Bond critical point, 16c	0.575(8)	-1.87	-1.87	2.04	-1.70(5)
	0.504	-1.43	-1.43	1.68	-1.18
Ring critical point, 16d	0.027(5)	-0.02	0.013	0.013	0.25(5)
	0.030	-0.02	0.014	0.014	0.26
Cage critical point, 8 b	0.024(5)	0.05	0.05	0.05	0.15(5)
	0.022	0.05	0.05	0.05	0.15

4. CONCLUSIONS

The modern state of EDSA in combination with topological analysis of the ESP and electron density allows to obtain reliable and quantitative information about chemical bonding and properties.

The electrostatic field in a crystal is good characterized, the determining factor is however, the introduction of cations in the ESP of NaCl type structures. Thus precise EDSA data for calculating the distribution of ESP adds to the physical picture about interaction of atoms and molecules.

Further development of EDSA involve development of improved techniques for quantitative measurements of diffraction patterns, application of new methods, e.g. the precession technique (Ultramicroscopy, 1994, **53**, 271), energy-filtering for removal of inelastical scattered electrons and improved methods to account for many beam scattering during structure refinement. Investigations of ESP distribution and chemical bonding in order to establish a relation between atomic structure and properties. This includes also modification of the methods for structure analysis (automatic indexing, direct methods, measurement at low and high temperatures) and its application for solving more complex structure, e.g. metal-organic and organic films, polymers, catalysts and nano-materials.

References

1. Vaishtein, B.K. (1964). Structure Analysis by Electron Diffraction. Oxford: Pergamon Press.

2. Vaishtein, B.K., Zvyagin, B.B., Avilov, A.S. (1992). Electron Diffraction Techniques, Vol.1, pp.216-312. Oxford University Press.

3. Avilov A.S., Kuligin, A.K., Pietsch, U., *et al*, Scanning System for High Energy Electron Diffraction. J.Appl.Cryst. (1999), **32**, 1033.

4. Zhukhlistov, A.P., AvilovA.S., Ferraris, P. *et al*, Statistical Distribution of Hydrogen over Three Positions in the Brucite $Mg(OH)_2$ Structure from Electron Diffractometry Data. Crystallogr.Rep. (1997) **42**, 774

5. Tsirelson, V.G., Avilov, A.S., Lepeshov, G.G., et al, Quantitative Analysis of the Electrostatic Potential in Rock-Salt Crystals Using Accurate Electron Diffraction Data. J.Phys.Chem, (2001), **B105**, 5068.

6. Avilov, A.S., Lepeshov, G.G., Pietsch, U., Tsirelson, V.G. Multipole Analysis of the Electron Density and Electrostatic Potential in Germanium by High-resolution Electron Diffraction. J.Phys.Chem.Solids. (2001), **62**, 2135.

Acknowledgements: This work was supported by the Russian Foundation for Basic Research (grants 98-03-32654, 01-03-33000, 04-02-16241), U.S. Civilian Research and Development Foundation (grant RP1-208), Deutsche Forschungsgemeinschaft DFG (grant Pi217/13-2), INTAS (grant 97-32045).

QUANTIFICATION OF TEXTURE PATTERNS

Peter Oleynikov, Sven Hovmöller, Xiaodong Zou
1Stockholm University, Structural Chemistry, SE – 106 91 Stockholm, Sweden

Key words: texture patterns, electron diffraction, intensities extraction.

Abstract: The program TexPat was developed for quantification of texture patterns in order to facilitate, speed up and improve the accuracy of texture pattern analysis. The program introduces new approaches for automated detection of centre and symmetry axes and simplifies the process of indexing and calculating the unit cell parameters. The main algorithm of the program uses the symmetry properties of the texture pattern images. The successive steps help to process the reflections of the pattern using the peak shape extracted from well-separated peaks. The program generates a list of unit cell parameters, all processed reflections with Miller indices and their integrated intensities. The quality of the data obtained by TexPat is evaluated numerical and found to be sufficient for crystal structure determination.

1.   INTRODUCTION

Structure analysis of crystals by electron diffraction was first performed in 1937 by a group of Soviet crystallographers headed by Z.G. Pinsker and B.K. Vainshtein. Soon after, Boris Zvyagin joined the group at Institute of Crystallography in Moscow. Much of their work has been reviewed in 1967 by Zvyagin [1]. Zvyagin made many of the important advances in the field, including quantitative structure analyses of clay minerals such as kaolinite, celadonite, muscovite and phlogopite-biotite, based on oblique texture electron diffraction intensities. These determinations had been made in three dimensions using low voltage electrons (40 to 50 kV). Nevertheless, oxygen positions were also visualized in the final potential maps of these layer silicate structures.

T.E. Weirich et al. (eds.), Electron Crystallography, 121–142.
© 2006 *Springer. Printed in the Netherlands.*

Many compounds, including clay minerals, form needle- or plate-shaped crystals. With finely dispersed minerals, the electron diffraction method can give a special kind of diffraction pattern, the texture pattern, which contains a two dimensional distribution of a regularly arranged set of 3D reflections [2]. Specimens of fine-grained lamellar or fiber minerals, prepared by sedimentation from suspensions onto supporting surfaces or films, form textures in which the component microcrystals have a preferred orientation. Texture patterns of lamellar crystals tilted with respect to the electron beam are called oblique texture electron diffraction patterns [1].

Obviously, it is possible to use X-ray diffraction for structure refinement of tabular shaped microcrystals. However, oblique texture electron diffraction patterns have the following advantages: firstly, they can provide data from the full 3D diffraction pattern in a single exposure, secondly, there is a possibility of obtaining almost perfectly oriented samples, which, owing to the minimization of their overlapping, produce diffraction patterns with good resolution of reflections. Extremely small crystals can and should be used – thinner than what is needed for X-ray powder diffraction.

One big problem, which arises mainly in crystals with rather large unit cells, is the overlapping of reflections. This prevents an accurate measurement of the local integrated intensities of a large number of reflections. A solution to this problem can be the 2D pattern decomposition method, which is based on the same principles as in X-ray powder diffraction. This method takes into account the dependence of intensities on the particle orientation function and the size of microcrystals. It is therefore necessary to establish the mathematical formalism that describes the diffraction pattern taking into account these parameters.

The scattering intensities for electrons are about 10^6 times higher than for X-rays. This clearly favors electrons over X-ray for studying extremely small sample volumes. On the other hand, this very strong interaction of electrons with matter produces dynamical effects (multiple diffraction). Nevertheless, Vainshtein [3] has shown that for samples of small disoriented crystals (with thickness 100 to 300 Å) containing moderately heavy atoms, the averaging of intensities (owing to the thickness distribution of microcrystals and their disorientation) decreases the role of dynamic effects so that the kinematic approximation essentially is valid. This conclusion is supported by a large number of investigations, which have been carried out using oblique texture electron diffraction ([1], [3], [4]).

However, quite complicated algorithms are needed for extracting the data from texture patterns. The analysis of texture patterns must be performed in a different way compared to regular electron diffraction patterns, due to different geometrical settings. Firstly, the centre of the texture pattern must be obtained and a background correction performed. Secondly, the main symmetry axes, which pass through the centre of the texture pattern, must be detected since the plate can be arbitrarily rotated during the digitization procedure. The last steps have the same aim as in electron diffraction structure analysis: determination of unit cell parameters and integration of intensities of indexed reflections. None of the steps in the analysis are trivial. The present work is aimed at facilitating and speeding up the analysis, using a computer program TexPat which was developed for this purpose. The set of algorithms presented in this work is general and works on all types of patterns with 2*mm* symmetry.

2. ELECTRON DIFFRACTION PATTERNS: GENERAL DESCRIPTION

The purpose of this chapter is to describe the basic geometrical properties of texture patterns, which lead to an understanding of possible methods needed for extracting the quantitative information from diffraction patterns of this kind.

2.1 Different types of electron diffraction patterns

The type of diffraction pattern depends on the degree of orientation and the distribution of crystals (if not a single) in the specimen. The following types of electron diffraction patterns may be distinguished:
- spot patterns with reflections in the shape of spots, obtained from a single crystal Figure 1a);
- texture patterns with ring or arc reflections, obtained from crystals uniformly oriented on the specimen support and having one degree of freedom, i.e. with so one face parallel to the specimen support but with random azimuthal distribution of the crystals (Figure 1b);
- polycrystalline patterns with ring reflections, obtained from randomly distributed crystals (Figure 1c), a so-called powder.

2.2 Texture patterns

Geometrically, electron diffraction patterns of crystals can be approximated as sections of the reciprocal lattice, since the Ewald sphere can be regarded as a plane (i.e. the radius of the Ewald sphere, $1/\lambda$, is much larger than the lengths of low-index reciprocal lattice vectors).

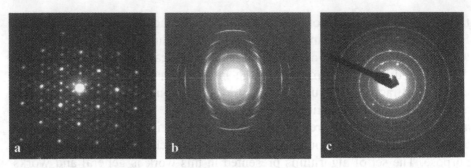

Figure 1. Electron diffraction pattern examples: a – spot pattern from a single crystal, b – oblique texture pattern tilted by 60°, c – powder ring pattern.

2.2.1   Powder electron diffraction patterns

If a specimen contains a number of crystals, then their reciprocal lattices are combined into a single reciprocal lattice with a common origin. If the number of crystals in the sample is high enough, then the reciprocal lattice will have a system of spheres with common origin. An electron diffraction pattern is a representation of the cross section of the reciprocal lattice by the Ewald sphere, which passes through the common origin and has the centre on the vector parallel to the electron beam. For polycrystalline samples, the diffraction pattern becomes a system of concentric rings (Fig. 1c). Such a pattern can be observed from any direction. The radial distribution of the rings is the most important characteristic of polycrystalline samples.

2.2.2   Zero tilt textures

If a specimen consists of numerous crystals rotated at random around a certain axis, it is called a texture and the axis mentioned is the texture axis. When a plate texture specimen is perpendicular to the electron beam (i.e. zero degree tilt), the diffraction pattern becomes a system of concentric rings (Fig. 2a, b) similar to that from a polycrystalline sample. Actually there is a difference: a zero tilted texture pattern has only rings coming from *hk0*-reflections (if c^* is the texture axis), while a powder pattern has rings from all *hkl*-reflections. A rotation of the reciprocal lattice mentioned above will generate a ring from each point (excluding those lying on the axis of rotation) (2c). As the result of rotation, the rings of the reciprocal lattice lie on coaxial cylinders whose axis is the texture axis.

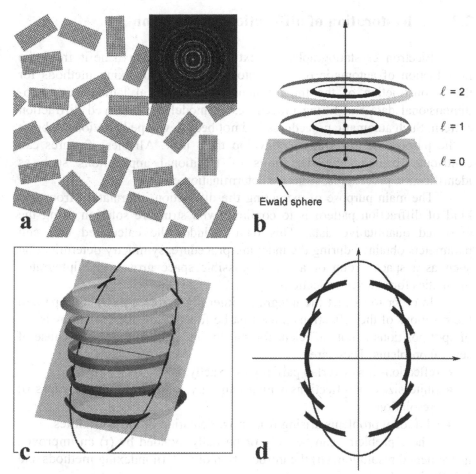

Figure 2. Formation of ring and oblique texture patterns. a – several randomly rotated artificial crystallites and its Fourier transform (inset); b – reciprocal space with rings and zero tilt Ewald sphere; c – 60° tilt of the Ewald sphere (reflection centers lie on the ellipse); d – the diffraction pattern as it is seen on the image plane.

2.2.3 Tilted textures

A tilting of the texture support plane with respect to the electron beam by an angle φ corresponds to the Ewald sphere cutting out a section through cylinders (2c, d). This intersection gives ellipses, along which the reflections are grouped. The distribution of reflections along ellipses is the characteristic of a texture pattern with a given symmetry.

2.3   Restoration of diffraction information

Electron crystallography of textured samples can benefit from the introduction of automatic or semi-automatic pattern indexing methods for the reconstruction of the three-dimensional reciprocal lattice from two-dimensional data and fitting procedures to model the observed diffraction pattern. Such automatic procedures had not been developed previously, but it is the purpose of this study to develop them now. All these features can contribute to extending the limits of traditional applications such as identification procedures, structure determination etc.

The main purpose of extracting the diffraction information from any kind of diffraction pattern is to continue with structure solution using the extracted quantitative data. This data includes the calculated unit cell parameters obtained during the indexing procedure, symmetry determination such as a space group or a set of possible space groups and integrated intensities for indexed reflections.

In order to extract the integrated intensities from a diffraction pattern the positions of the reflections must first be found. However, the knowledge of spot positions is not sufficient for starting intensity extraction because of several problems. These are:

- reflections may overlap partially or exactly;
- finite sizes of reflections which can vary with d and cause a loss of resolution;
- additional problems arising from misorientation of the crystallites.

These problems can be fully or partially avoided by (i) an improved instrumental resolution, (ii) the introduction of pattern indexing methods and (iii) the introduction of fitting methods.

The modeling of electron diffraction by the pattern decomposition method, for which no structural information is required, can be successfully applied for extraction of the diffraction information from the pattern. Several parameters can be refined during the procedure of decomposition, including the tilt angle of the specimen; the unit cell parameters; peak-shape parameters; intensities. The procedure consists of fitting, usually with a least-squares refinement, a calculated model to the whole observed diffraction pattern.

The calculated intensity $z(x_i, y_j)$ at any point x_i, y_j of a diffraction pattern is expressed as a function of the integrated intensity I_k of the reflections contained in the pattern and a normalized analytical peak shape function $PS(x, y)$ is used to model the individual profiles. It is given by

$$z(x_i, y_j) = \sum_k I_k \cdot PS(x_i - x_k, y_j - y_k) + b(x_i, y_j),   \tag{1}$$

where $b(x_i, y_j)$ is the intensity of the background and the sum is over all k reflections contributing to the intensity at the point (x_i, y_j).

2.3.1   Peak shape

The normalized peak-shape function PS introduced by equation (1) must be determined in order to figure out the dependence of PS on several crystallite parameters, such as average size of crystallites, misorientation of crystallites in the sample etc. These parameters lead to a broadening of reflections, which must be taken into account.

2.3.1.1   Azimuthal asymmetry of reflections

The intersection of the Ewald sphere with a ring of the reciprocal lattice would give a point in the ideal case. However with real textures, the crystals of the sample are small and not perfectly parallel to the support film, so there is a certain variation in the crystal orientation relative to the texture axis. This can be described by a misorientation parameter α, which is usually in the range 2 to 4°. In this case, the region of appreciable scattering is not concentrated at points of the reciprocal lattice but extends around the reciprocal nodes. The rings, formed by the superposition of such scattering areas of individual microcrystals, are then smeared out into bands with a width of α. All bands have the same common centre in the reciprocal lattice origin. When the Ewald sphere intersects these bands, azimuthally elongated arcs are produced (as seen in Fig. 3). The smaller the α, the more perfect is the texture.

The type of distribution function is reflected in the character of the arc, i.e. its angular width. Normally a simple form, for example a Gaussian distribution, can approximate it. The elongation of an arc on a texture pattern within the same ellipse depends also on the distance R' from the vertical axis (Fig. 3). This distance depends on the tilt angle of the textured specimen. There is an asymmetry in the azimuthal peak-shape, which is especially large for reflections close to the major axes of the ellipses. An example of azimuthal asymmetry is shown in Fig. 3. It can be described by two different halfwidths: shorter (α_S) and longer (α_L) (as shown in Fig. 3). The Ewald sphere intersects a spherical band differently on the two sides of the band's maximum in reciprocal space and this causes an asymmetric elongation in the azimuthal direction.

We need to calculate coordinates of three points, in order to calculate both the half widths α_S and α_L of a reflection: the peak maximum and both ends – defined for example as the two points inside which we have 95% ($\pm 2\sigma$ if Gaussian) of the full reflection area.

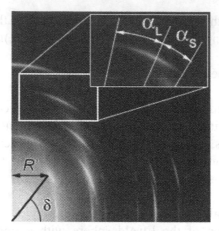

Figure 3. Azimuthal distortion of an arc-shaped reflection (lizardite 1T).

2.3.1.2   Radial broadening and asymmetry of reflections

The radial asymmetry of reflections increases with longer reciprocal vector H. The radial peak broadening contains contributions from different factors, including domain size, crystallite imperfections (so-called 'size' broadening) and also lattice distortions ('strain' broadening). The finite column length of coherently scattering domains, where this length is parallel to the diffraction vector, causes 'size' broadening [7]. Since 'size' broadening is strongly influenced by the form of domains, there can be significant differences in line breadth throughout a pattern. The broadening is then said to be 'anisotropic'. 'Strain' broadening due to lattice distortions, dislocations, stacking faults, grain boundaries *etc*. can generally be described as Gaussian, but it need not necessarily be symmetric. In practice either or both categories of broadening may be significant.

The radial broadening of reflections is practically independent of the shape of the distribution function $f(\alpha)$ [8]. Thus this broadening can be treated independently.

It is important to take into account the radial broadening and asymmetry of reflections for several reasons. Firstly, the peak-shape of reflections depends on the radial broadening. The broadening must be taken into account when extracting the intensities, because of the dependency of integrated intensities on the peak-shape. Secondly, the average crystallite thickness can be estimated from the radial broadening effect. Knowledge of the thickness is essential for correcting for dynamical effects after the intensities have been extracted.

2.3.2 Reciprocal coordinates of a reflection

The indexing procedure of any diffraction pattern requires the knowledge of positions of some reflections on the pattern. The relations between the position of each peak on a two-dimensional diffraction pattern and the corresponding point in reciprocal space can be established only after a successful indexing. The combined use of the reciprocal coordinates and the determined peak-shape are essential for extracting integrated intensities from diffraction data.

A reciprocal lattice-vector H for any point in reciprocal space can be represented by equation

$$H = ha* + kb* + lc*,$$ (2)

where $a*$, $b*$ and $c*$ are main vectors of the reciprocal lattice, not necessarily orthogonal, h, k and l are some integer numbers.

If the tilt of the texture is in the $y*z*$-plane (due to cylindrical symmetry it does not matter in which direction we tilt the sample, so the x_C* coordinate of the centre of the Ewald sphere can be assigned to zero) then the Ewald sphere equation can be written as:

$$x*^2 + (y* + R_E \cdot \sin\varphi)^2 + (z* - R_E \cdot \cos\varphi)^2 = R_E^2,$$ (3)

where R_E is the radius of the Ewald sphere, φ is the tilt angle of the textured sample and $(x*, y*, z*)$ are the coordinates of a point on the Ewald sphere.

The vertical position $z*$ of the hkl-ring is given by equation:

$$z* = ha* \cos\beta* + kb* \cos\alpha* + lc*$$ (4)

$z*$ determines the height of the reflection above the minor axis on the diffraction pattern.

Then the coordinates of the intersection of the Ewald sphere and the hkl-reflection sphere of radius H are:

$$y* = \frac{z* \cdot \cos\varphi}{\sin\varphi} - \frac{H^2}{2 \cdot R_E \cdot \sin\varphi},$$ (5)

$$x* = \pm\sqrt{H^2 - y*^2 - z*^2}.$$ (6)

and z^* as described by (4) (z^* can be negative or positive). The sign of the y^*-coordinate is the same as the sign of the z^*-coordinate, but $y^* < 0$ when $z^* = 0$ due to the curvature of the Ewald sphere:

$$y^*_{z^*=0} = -\frac{H^2}{2 \cdot R_E \cdot \sin\varphi}. \tag{7}$$

Oblique texture patterns have almost perfect $2mm$ symmetry and thus the whole set of diffraction spots is represented by the reflections in one quadrant. The arcs are exactly symmetrically placed relative to the major axis, being sections of the same spherical band in reciprocal space. The reflections on the lower half of the pattern are sections of reciprocal lattice rings, which are Friedel partners and thus equivalent to those giving the reflections of the upper half assuming a flat surface of the Ewald sphere. Actually, if the curvature of the Ewald sphere is taken into account, the upper and lower parts of a texture pattern will differ slightly.

2.3.3   Ewald sphere curvature

What is the order of the extra component, which appears in (5) due to the curvature of the Ewald sphere? For an electron microscope operating at 250 kV the wavelength of the electron wave is 0.022 Å. Then the radius of the Ewald sphere is 45.46 Å^{-1}. At 1 Å resolution the value of H is 1 Å^{-1} and the extra component in (5) is

$$\frac{H^2}{2 \cdot R_E \cdot \sin\varphi} \approx 0.011 \text{ Å}^{-1}$$

On an image with a scale factor of 500 to 800 this additional component leads to significant shift of the reflection position (6 to 9 pixels). The shift of the reflection centre caused by the Ewald sphere curvature is seen easily while analyzing the positions of reflections lying on the minor axis.

2.3.4   Indexing texture patterns

The purpose of indexing texture patterns is the geometrical reconstruction of the three-dimensional reciprocal lattice from the two-dimensional distribution of H spacings. One advantage of texture patterns is the possibility to determine all unit cell parameters of a crystal unambiguously and index all the diffraction peaks from only a single texture pattern. In some cases, it is even possible to distinguish different crystal systems and determine the space group of the crystal from one texture pattern. A full and detailed description of how to index texture patterns can be found in [1].

The indexing is based on the direct relationship between the texture pattern and the reciprocal lattice. For a reflection *hkl* with the reciprocal lattice vector $\boldsymbol{H}_{hkl}$:

$$d_{hkl} = 1 \,/\, |\boldsymbol{H}_{hkl}| = 1 \,/\, H_{hkl} \tag{8}$$

Indexing a texture pattern is often rather complicated, but several characteristic features facilitate the indexation (see [142]).

The projection of $\boldsymbol{H}$ on the plane formed by two reciprocal vectors $a*$ and $b*$ can be expressed in terms of real unit cell parameters a, b and γ [1]:

$$H_{a*b*} = \frac{1}{\sin\gamma} \sqrt{\frac{h^2}{a^2} + \frac{k^2}{b^2} - 2 \cdot \frac{h}{a} \cdot \frac{k}{b} \cdot \cos\gamma} \tag{9}$$

Since diffraction pattern has a scale factor $L\lambda$ all distances measured on the pattern should be recalculated into reciprocal. Following Zvyagin's notation one measures B_{hk} values on the texture pattern, which correspond to reciprocal values H_{a*b*} by

$$B_{hk} = H_{a*b*} \cdot L\lambda = \frac{L\lambda}{\sin\gamma} \sqrt{\frac{h^2}{a^2} + \frac{k^2}{b^2} - 2 \cdot \frac{h}{a} \cdot \frac{k}{b} \cdot \cos\gamma} \tag{10}$$

where $L\lambda$ is the scale factor for the given diffraction pattern.

Equations (9) and (10) depend only on three unit cell parameters a, b and γ and two Miller indices h and k. Therefore it is possible to calculate the values of a, b and γ unit cell parameters automatically from a set of B_{hk} values. This also means that all spherical bands with the same h and k but different l indices will lie on the same cylinder with radius H_{a*b*} in reciprocal space (Fig. 2b).

Since the cross-section of a cylinder by the nearly flat Ewald sphere is an ellipse, equation (10) evaluates the length of the minor axis of this ellipse.

The other three unit cell parameters c, α and β can be found by analyzing the distribution of reflections on the ellipses.

The distance between the line where the reflection *hkl* is located and the minor ellipse axis is equal to the projection of the reciprocal lattice vector $\boldsymbol{H}_{hkl}$ onto the $c*$ axis (which is the major ellipse axis) and can be expressed from (4) as

$$D_{hkl} = \frac{L\lambda}{\sin\varphi}\left[h \cdot a^* \cdot \cos\beta^* + k \cdot b^* \cdot \cos\alpha^* + l \cdot c^*\right] \qquad (11)$$

where $L\lambda$ is the scale factor of the pattern and φ the tilt angle of the texture specimen. Note that the scale factor $L\lambda/\sin\varphi$ appeared here in comparison with equation (4). As described above only a quarter of the image is unique, so we can consider only those reflections with $D_{hkl} \geq 0$. According to (11) this will yield a cell with $\alpha, \beta \geq \pi/2$.

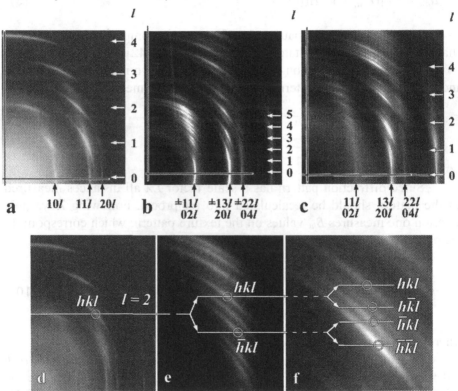

Figure 4. Indexing texture patterns from crystals of different symmetries (see the text for explanations). Assigning the indices to the reflections is shown in (a) – (c). The splitting of the reflections *hkl* for different symmetries is shown in (d) – (f). (a), (d) trigonal (lizardite 1T). (b), (e) monoclinic (muscovite 2M$_1$). (c), (f) triclinic (kaolinite).

All six unit cell parameters of a lattice can be obtained from these two equations (10)–(11) by measuring the D_{hkl} and B_{hk} values of at least three independent reflections with known Miller indices, all with different *h* and *k* and, at least one of which with $l \neq 0$ in order to calculate c^*.

Equations (10) and (11) are simplified for several kinds of symmetries (Fig. 4 a-f) as described in [142].

The position of any reflection with given Miller indices *hkl* can be calculated once all unit cell parameters have been determined. This allows starting the last but very important procedure in the analysis of a texture diffraction pattern – estimating the integrated intensities of the diffraction maxima.

2.4   Integration of the diffraction maxima

Extraction of the intensities from diffraction data and the subsequent derivation of structure factors is one of the most important stages in structure analysis.

There are two methods for texture pattern intensity measurements described in [3]. The first approach is to perform integration of a small region at the centre of the arc-shaped reflection. It can be done by hand and has been used by Vainshtein, Zvyagin and others.

Some theoretical approaches developed in [8] introduce the basis for using two-dimensional peak-fitting procedures in order to perform the integration of whole diffraction maxima.

The X-ray and neutron powder diffraction methods successfully use two approaches connected to extraction of intensities (see [14]):

(i) Pattern decomposition method for extracting the profile parameters for individual Bragg reflections without reference to a structural model;

(ii) The Rietveld method, where the integrated intensities of the reflections are calculated from the atomic parameters in the current model.

The decomposition method can be applied directly to process any texture pattern since it (i) does not require the knowledge about the structure and (ii) does not take into account the dynamical effects. The last is quite important because it is quite difficult to describe analytically and implement practically, the correction for dynamical effects before a structure model is obtained. However, the decomposition method has some disadvantages, although they can be avoided in principle. The main disadvantage is the treatment of nearly or exactly overlapping reflections; overlapping reflections must be treated as a single peak, otherwise the minimizing procedure can sometimes give negative intensities on its output. This is a problem for all fitting and minimizing methods.

The lack of software for the decomposition method in two dimensions in application to electron diffraction texture patterns was avoided by Zvyagin and Zhukhlistov by using one-dimensional methods of intensity extraction. Later the decomposition method for one-dimensional radial profiles (starting at the centre of the diffraction pattern and passing through the specified reflection) of simple structures was applied. The big disadvantage of this method is the lack of algorithms, which take into account overlapping reflections also in the azimuthal direction.

2.4.1   Measuring local intensity

The square of the structure factor amplitude can be calculated from the "local intensity" I_{hkl}' [1], [4] as

$$|F_{hkl}|^2 \propto \frac{I_{hkl}'}{m \cdot d_{hk0} \cdot d_{hkl}} \qquad (12)$$

where d_{hkl} and d_{hk0} are the d values of reflection hkl and $hk0$ respectively and m is the multiplicity factor for the reflection hkl. Equation (12) uses the values of d_{hkl} and d_{hk0} in order to take into account the shape of the reflection hkl.

This approach with "local" intensities does not depend on the shape of the reflection but does depend on the region of integration. However, it is the simplest in implementation and can be done by hand, because it does not require the knowledge of the peak shape.

The method can also be used for very elongated arcs. In this case it is easier to measure the local intensity, for example the I_{hkl}' value in the centre of the arc.

The overlapping of peaks must be taken into account while using this approach and this is not a trivial task.

2.4.2   Measuring integrated intensity

The second method is to integrate the intensities under the whole arc-shaped reflections. The separation of overlapping peaks can be done automatically, once the peak shape has been parameterized from a few well-separated peaks. Occasionally there is a possibility of full overlapping when two or more reflections have the same or almost (within given precision) the same geometrical properties, like d_{hkl} and D_{hkl} values (see equations (8) and (11)). Only then the overlapping reflections cannot be separated.

For oblique texture patterns, when the tilting angle φ is taken into account we have (in relative values):

$$|F_{hkl}|^2 \propto \frac{I_{hkl} \cdot R_{hkl}'(\varphi)}{m} \qquad (13)$$

where F_{hkl} is the structure factor for the reflection hkl, $R'_{hkl}(\varphi)$ is the distance between the reflection and the major axis of the texture pattern (shown on Fig. 3), I_{hkl} the integrated intensity of the whole arc and m is multiplicity factor of the reflection. This relation holds for kinematical scattering.

This method cannot be applied manually, and so it was not possible to it previously for estimating peak-shape parameters for all reflections simultaneously, due to the lack of software.

In conclusion we can make a general note: the design of the intensity extraction procedure should take into account the differences in equations (12) – (13) as discussed above. In order to compare the results obtained by different methods one should calculate the modulus of structure factors $|F_{hkl}|$ or $|F_{hkl}|^2$ first and then compare these calculated values.

3.   APPLICATION: PROCESSING TEXTURE PATTERN OF BRUCITE

Every software package aimed at extracting the information from data obtained from any kind of diffraction experiment gives certain information as output, namely

- The set of Miller indices hkl for reflections present on the diffraction image or profile;
- Unit cell parameters a, b, c, α, β and γ;
- Extracted magnitudes of structure factors $|F_{hkl}|$ (or $|F_{hkl}|^2$) for indexed reflections;
- Space group or symmetry class.

In the case of texture patterns none of the steps is trivial.

The space group (or the set of possible groups) can be obtained during the analysis of all hkl data, gained from the step of pattern indexing procedure. The extraction of structure factor magnitudes can be done after the indexing procedure. Thus, the indexing must be performed first. This procedure requires knowledge of the following geometrical parameters:

- Centre of the texture pattern;
- Main axes slope on the image;
- Tilt angle of the textured specimen in respect to the primary electron beam;
- Peak shape extracted from good non-overlapping reflections;
- Distribution of ellipses on the image.

All these steps are implemented in the program TexPat.

Here we will describe an example of data extraction during processing of texture pattern of brucite and show the results of indexing, determination of unit cell parameters and extraction of intensities.

3.1   Experimental techniques

Texture patterns were taken at the Institute for Geology of Ore Deposits, Petrography, Mineralogy and Geochemistry in Moscow using a 400 kV electron diffraction camera and recorded on glass plates or imaging plates (DITABIS, Germany). Glass plates were digitized off-line using a Kite CCD camera (Calidris, Sollentuna, Sweden) with a 12-bit grey scale. Each scan produced a digital image with a size of 1280x1024 pixels. The images were saved in tiff format since it allows storing the data in up to 16-bit grey scale. Imaging plates were scanned by DITABIS imaging plate scanner producing files with a size of 3600x3200 pixels and resolution 17.5 μm with 16-bit grayscale.

3.2   Centre detection

The first critical step in the analysis of any texture pattern is to determine the position of the centre (*000* reflection), since all the other steps are dependent on a correctly placed centre. The centre can be positioned automatically or manually.

Alternatively, the user can choose four symmetry related reflections, with the same d-value, to determine the centre from those points (four-clicks method). Then the centre of the texture pattern is at the centre of the circle containing those four reflections (see Fig. 5a).

The difference between automatically calculated centre position and the centre calculated from the user-defined four reflections was found to be ~½ pixel in each *x*- and *y*-directions on texture patterns with the size of 1024x1280 pixels.

The centre can be calculated during calculation of the tilt angle also, where the centre can be estimated from a single ellipse.

The *y* coordinate of the centre calculated from points located on the same ellipse differs from the *y* coordinate calculated from the circle (containing 4 opposite reflections) by 0.5 to ~2 pixels.

This difference in *y*-coordinate can be explained by the shift of the centre in *y*-direction caused by the curvature of the Ewald sphere. The curvature of the Ewald sphere does not affect the position of the centre in the four-clicks method, but leads to a shift of the ellipse centre when calculating this centre from a single ellipse. Thus the four-clicks method is better for the centre calculation.

It was found that the difference between the centre coordinates calculated from different sets of symmetry related reflections in the four-clicks method is less than 0.1 pixel which shows the quality of centre calculations by this simple method.

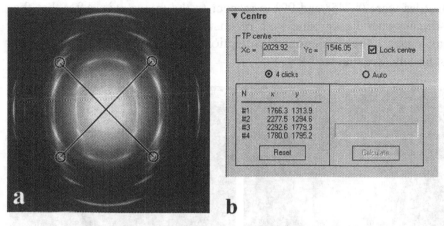

Figure 5. Finding the centre (a) by choosing 4 symmetry related reflections; (b) the dialog box for centre calculations.

Summarizing the results we can make a conclusion that all the methods of the centre calculation are almost equal in the precision. The maximum difference between the results obtained using described methods was 2 pixels, which is less than 0.003 Å (the scale factor was about 750 to 900 pixels/Å) for the given texture patterns. Such small difference has no significant influence on further calculations, which was proved by calculations performed on these texture patterns.

3.3 Background estimation

The presence of the central spot (the primary beam) and diffuse rings I_{diff} from the film support brings significant errors into estimated intensities. The shape of the primary beam I_{beam} can be approximated by one of several peak-shape functions such as pseudo-Voigt, Gaussian or Lorentzian [16]. The diffuse background can be described by a polynomial function of order 12. Then equation (1) becomes

$$I(x,y) = I_{beam}(x,y) + I_{diff}(x,y) + \sum_k I_{peak}(x,y), \qquad (14)$$

where I is the intensity value in a point (x, y) and the sum is done over k reflections.

In the texture pattern described in this chapter, the background was approximated by a Gaussian curve from three different radial profiles, since there can be some deviations from the average radial profile in different directions which can lead to over- or underestimations of the background. Each profile was averaged over 3° in the azimuthal direction; the directions of profiles were 0°, 45° and 90° in respect to the minor axis. After that the two-dimensional background was reconstructed and subtracted from the initial image. The background correction improves the texture pattern substantially (see Fig. 6a, b).

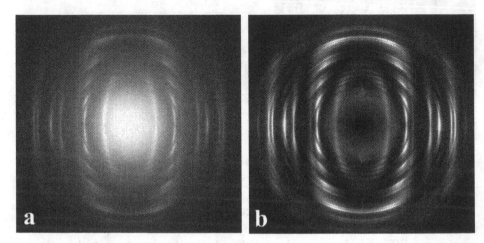

Figure 6. Background estimation of a texture pattern (a) before and (b) after background subtraction.

3.4 The axis slope and the tilt angle

The diffraction spots in oblique texture patterns fall on sets of ellipses due to the geometry of the texture patterns (Fig. 6a). The user can define any single ellipse by marking the reflections falling on that ellipse. Using these points the ellipse parameters such as the length of the minor and major axes, centre and tilt can be calculated. This method gives good results in the tilt angle calculations.

The ratio between the major (e_1) and minor (e_2) ellipse axes is determined by the tilt angle, i.e. the angle between the texture support plane and the plane normal to the electron beam (Fig. 6b). The tilting angle φ can be calculated from:

$$\varphi = \arccos(e_2 / e_1), \tag{15}$$

Another way to define the direction of the minor axis is to manually locate it by choosing a single reflection lying on the axis. The disadvantages of this method are (i) the difficulties with a reflection location on complex images due to overlapping problems and (ii) the curvature of the Ewald sphere can give significant error (up to 0.5°) when picking a reflection which is far from the centre of the diffraction pattern.

3.5 Extraction of intensities

The results given in the output of the pattern decomposition method contain the integrated intensities, which can be used directly as input data for structure determination software.

Reliability factors for estimated structure factors were calculated using the equation:

$$R = \frac{\sum \left| \left| F_{exp} \right| - \left| F_{calc} \right| \right|}{\sum \left| F_{exp} \right|} \tag{16}$$

An R-value of ≤ 10% was obtained in case of brucite. This is quite accurate considering that the data comes from electron diffraction.

3.6 Example of the analysis of texture patterns: brucite

This section presents results of processing of electron diffraction texture pattern of brucite and structure calculations from gained data.

The structure of the mineral brucite $Mg(OH)_2$ (space group $P\overline{3}m1$) is formed by $Mg(OH)_6$ octahedra, which are connected by edges, forming layers parallel to the (001) plane (see Fig. 7). Hydroxyl groups OH⁻ lie on the 3-fold axes and each H-atom is equidistant from three oxygen atoms of the next layer, which is not a perfect position for forming hydrogen bonds between layers.

The electron diffraction texture pattern was obtained using a tilt angle of ~60° and recorded on image plates scanned with a resolution of 17.5 μm/pixel (at 250 kV).

The centre was calculated using the 4-click method. The minor axis was defined by picking the reflection 300 lying on the minor axis. The reflection was chosen for two reasons (i) it has high intensity and (ii) the distance from the centre is quite large. This allows defining the minor axis direction with high precision.

The background was approximated by a Gaussian curve as described in section 3.3.

Two-dimensional *hk* indices were assigned to the set of ellipses and the unit cell parameters a, b and γ were calculated from this two-dimensional indexing. Three different *hkl* indices were assigned manually to three reflections 010, 111 and 030 using the knowledge about the *hk* indices of ellipses. The unit cell parameters obtained after refinement are given in Table . The ratio between a and c from TexPat agrees with previous reports ([16], [17]) to within 0.2%.

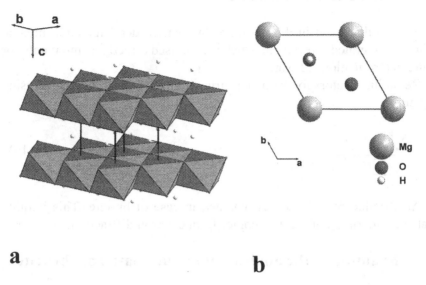

a

b

Figure 7. The structure of brucite. Gray octahedra correspond to $Mg(OH)_6$ octahedra (a). The projection along c-axis (b)

Table 1. Unit cell parameters for brucite (space group $P\bar{3}m1$).

Source	$a/\text{Å}$	$c/\text{Å}$	Ratio c/a
[16] [1]	3.14979(4)	4.7702(7)	1.5144
Zhukhlistov, [17]	3.149(2)	4.769(2)	1.5144
TexPat [2]	3.149(3)	4.775(4)	1.5162

[1] – from neutron diffraction
[2] – scaled to the a parameter given by [17]

The intensities were extracted from the diffraction pattern by the decomposition method. The intensities of reflections, which coincide exactly on the diffraction pattern were treated as single during the decomposition and were afterwards split according to the corresponding values of $|F_{theor}|^2$. The list of extracted $|F_{obs}|^2$ is given in Appendix D. The structure was refined in the program SHELX96 [18] using atomic scattering factors for electrons for neutral atoms. The total number of independent intensities used in the calculations was 79 with maximum $l = 7$ (8 ellipses, the resolution limit is $\sin\theta/\lambda = 0.85$ Å^{-1}, $d = 0.6$ Å). More independent reflections were extracted using TexPat, than those extracted by Zhukhlistov [19]. Sometimes the Ewald sphere intersects the spherical bands away from their maximum – this can be seen as tails of reflections. In order to extract the intensity from the tails of such reflections the peak-shape must be estimated with high precision. This became possible due to introduction of the two-dimensional decomposition method into the intensity extraction procedure.

Table 2. Refined atomic coordinates for the brucite structure.

Atom	x/a	y/b	z/c
Mg	0	0	0
O	1/3	2/3	0.2181(6) / 0.2205(3) [1]
H	1/3	2/3	0.425(7) / 0.418(4) [1]

[1] The first coordinate corresponds to the refined data, the second coordinate is taken from [17].

The structure was refined with anisotropic non-hydrogen atoms. The final R-factor was 0.142. The refined atomic coordinates are listed in Table . The atomic coordinates agreed with those published by [17] to within ~0.03 Å for the oxygen atom and ~0.08 Å for the hydrogen atom. The Mg atom is in a special position (0,0,0) so its position is not refined.

4. CONCLUSION

The results obtained in the present work show that the developed methods can be applied successfully to oblique texture electron diffraction patterns. The program TexPat has been designed to produce accurate lattice parameters and intensities. Indexing oblique texture electron diffraction patterns from structures with symmetry lower than orthorhombic remains quite difficult task due to the geometrical properties of texture patterns. This difficulty is overcome by the introduction of semi-automatic algorithms for the indexing.

A two-dimensional pattern decomposition method has been developed and implemented for this purpose. Since the pattern decomposition method does not require an initial structure model, the method can be successfully applied to texture patterns from samples with unknown structure.

Once the unit cell is known, pattern decomposition methods are used for extracting integrated intensities, which can then be used to solve the structure (if it is unknown) or refine it, if a model is known.

Thus the analysis of texture patterns expands the present possibilities of electron crystallography for accurate structure characterization of crystalline materials, which can form microcrystalline textured samples.

References

[1]   Zvyagin, B.B.: Electron diffraction analysis of clay mineral structures. Plenum Press, New York, 1967.

[2]   Zvyagin, B.B., Zhukhlistov, A.P., Nickolsky, M.S. Z. Kristallogr. **218** (2003) 316-319.

[3]   Vainshtein, B.K.: Structure analysis by electron diffraction. Pergamon Press, 1964.

[4]   Vainshtein, B.K., Zvyagin B.B., Avilov A.S.: Electron Diffraction Structure Analysis. In: *Electron Diffraction Techniques* (Ed. D. Dorset), vol. 1, p. 117-312. Published by IUCr 1992.

[5]   Zou, X.D., Sukharev, Y., Hovmöller, S., Ultramicroscopy **49** (1993) 147-158.

[6]   Zou, X.D., Sukharev, Y., Hovmöller, S., Ultramicroscopy **52** (1994) 436-444.

[7]   Guinier, A. *X-ray Diffraction*. (1963) San Francisco: Freeman.

[8]   Plançon, A., Tsipurski, S.I., Drits, V.A., J. Appl. Cryst. **18** (1985) 191-196.

[9]   Oleynikov, P., Hovmöller, S., Zou, X.D., Zhukhlistov, A.P., Nickolsky, M.S., Zvyagin, B.B. Z. Kristallogr. **219** (2004) 12-19.

[10]   Tang, L., Laughlin, D.E., J. Appl. Cryst. **29** (1996) 411-418.

[11]   Giacovazzo, C., Monaco, H.L., Viterbo, D., Scordari, F., Gilli, G., Zanotti, G., Catti, M., *Fundamentals of Crystallography*. (1992) IUCr: Oxford University Press.

[12]   Langford, J.I., Louër, D., Scardi, P. J. Appl. Cryst. **33** (2000) 964-974.

[13]   Langford, J.I., Wilson, A.J.C., J. Appl. Cryst. **11** (1978) 102-113.

[14]   Louër, D. Acta Cryst. A**54** (1998) 922-933.

[15]   Rietveld, H.M., J. Appl. Cryst. **32** (1969) 65-71.

[16]   Fraser, R.D.B., Macrae, T.P., Suzuki, E., Tulloch, P.A., J. Appl. Cryst. **10** (1977) 64-66. Catti, M., Ferraris, G., Hull, S., Pavese, A. Phys. Chem. Minerals. **22** (1995) 200-206.

[17]   Zhukhlistov, A.P., Avilov, A.S., Ferraris, D., Zvyagin, B.B., Plotnikov, V.P. Crystallography Reports **42** (1997) 774-777.

[18]   Sheldrick, G.M., SHELX96. Program for the Refinement of Crystal Structures, University of Göttingen, Göttingen, Germany, 1996.

[19]   Zhukhlistov, A.P., Zvyagin, B.B., Kristallografiya (Rus) **43**, 6 (1998) 1009–1014.

[20]   Collins, D.R., Catlow, C.R.A., **77** (1992) 1172-1181.

[21]   Güven, N., Z. Kristallogr. **134** (1971) 196-212.

[22]   Press, W.H., Vetterling, W.T., Teukolsky, S.A., Flannery, B.P.: Numerical Recipes in C++, 2nd edition. Cambridge University Press 2002.

[23]   Marquardt, D.W., SIAM J. Appl. Math. **11** (1963) 431-441.

[24]   Levenberg, K., Q. Appl. Math. **2** (1944) 164-168.

[25]   Izumi, F., Rigaku J., **6**, No.1 (1989) 10-19.

QUANTITATIVE CONVERGENT BEAM ELECTRON DIFFRACTION
Measurements of Lattice Parameters and Crystal Charge Density

J.M. Zuo

Department of Material Science and Engineering and F. Seitz Materials Research Laboratory, University of Illinois at Urbana-Champaign, 1304 West Green Street, Urbana, Illinois 61801, U.S.A

Abstract: This chapter introduces quantitative convergent-beam electron diffraction from quantitative electron diffraction intensity recording to quantitative structural retrieval by electron diffraction intensity refinement and pattern matching. It is shown that structure information, such as unit cell parameters and electron structure factors, can be obtained from experimental diffraction intensities by optimizing the fit between the experimental and theoretical intensities through the adjustment of structural parameters in a theoretical model. While the principle of refinement is similar to the Rietveld method in X-ray powder diffraction, its implementation in electron diffraction is powerful since it includes the full dynamic effect. The accuracy of this technique will be demonstrated through the accurate determination of lattice parameters for strain mapping and charge density for the study of crystal bonding.

Key words: Electron Diffraction, CBED, Charge Density, Strain Mapping

1.   INTRODUCTION

      Recent progress in electron diffraction has significantly broadened its applications from a primary a microstructure characterization tool to an accurate structure analysis technique that traditionally belongs almost exclusively to the domain of X-ray and neutron diffraction. This development is timely since the focus of modern materials feature size is increasingly on nanoscale structures, where the electron high spatial

T.E. Weirich et al. (eds.), Electron Crystallography, 143–168.

resolution and strong interaction with matter are of advantage. It is the purpose of this chapter to introduce quantitative convergent-beam electron diffraction (CBED) and highlight its application in materials science.

The development of quantitative electron diffraction (QED) is relatively new at the convergence of several microscopy technologies. The development of field emission gun in 70's and its adoption in conventional transmission electron microscopes (TEM) brought high source brightness, small probe size and coherence to electron diffraction. The significant impact is the ability to record diffraction patterns to obtain crystallography information from very small (nano) structures. Electron energy-filter, such as the in-column Ω-filter, allows inelastic background from plasmons and higher electron energy losses to be removed with an energy resolution of a few electron volts. The development of array detectors of CCD camera or imaging plates enables parallel recording of diffraction pattern and quantification of diffraction intensities over a large dynamic range that was not available to electron microscopy before. The post-specimen lenses of the TEM give the flexibility of recording electron diffraction patterns at different magnification. Last but just as important, the development of efficient and accurate algorithms to simulate electron diffraction and modeling structures on a first-principle basis using fast modern computers has significantly improved our ability to interpret experimental diffraction patterns.

For readers learning electron diffraction, there are a number of books on electron diffraction for materials characterization [1,2,3,4]. Most of these books focus on crystals since many polycrystalline materials are perfect single crystals in an electron microscope because of the small electron probe. Full treatment of the dynamic theory of electron diffraction is given in several special topic books and reviews [5,6,7,8].

This chapter is organized in 6 sections. Section 2 describes the geometry of (CBED). Section 3 covers the theory of electron diffraction and the principles for simulation using the Bloch wave method. Section 4 introduces the experimental aspect of quantitative CBED including diffraction intensity recording and quantification and the refinement technique for extracting crystal structural information. Application examples and conclusions are given in section 5 and 6.

2. CONVERGENT BEAM ELECTRON DIFFRACTION

Electron optics in a microscope can be configured for different diffraction modes from a parallel beam illumination to convergent beams. There are three basic modes of electron diffraction: selected area electron diffraction (SAED), nano-area electron diffraction (NED) and CBED. Variations from these three techniques include large-angle CBED [9], convergent-beam imaging [10], electron nanodiffraction [11] and their modifications [12]. For nanostructure characterization, the recently developed NED [13,14] and electron nanodiffraction techniques are particularly relevant. NED uses a small parallel beam to record diffraction pattern at high angular resolution. The beam size is tens of nanometers. In electron nanodiffraction, a small electron probe of sizes from a few Å to a few nanometers is placed directly onto the sample. Diffraction pattern thus can be obtained from localized area as small as a single atomic column, which is very sensitive to local structure and probe positions [15]. The technique was developed by Cowley [11] and others in the late 1970s that uses a scanning transmission electron microscope (STEM). Readers interested in these techniques can find their description and applications in above references.

CBED is formed by focusing electrons to form a small probe at the specimen (fig. 1). Compared to selected area electron diffraction, CBED has two main advantages for studying *perfect crystals* and the local structure:

1. The pattern is taken from a much smaller area with a focused probe; The smallest electron probe currently available in a high-resolution FEG-STEM is close to 1 Å. Thus, in principle and in practice, CBED can be recorded from individual atomic column. For crystallographic applications, CBED patterns are typically recorded with a probe of a few to tens of nanometers;

2. CBED patterns record diffraction intensities as a function of incident-beam directions. Such information is very useful for symmetry determination and quantitative analysis of electron diffraction patterns.

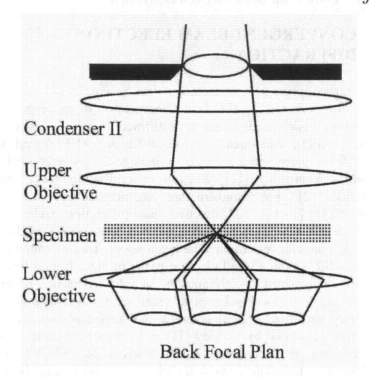

Condenser II

Upper
Objective

Specimen

Lower
Objective

Back Focal Plan

Figure 1. Schematic diagram of convergent-beam electron diffraction.

An example of CBED pattern is given in fig. 2. The pattern consists of disks. Each disk can be divided into many pixels, each pixel approximately represents one incident beam direction. For an example, let us take the beam P in fig. 2. This particular beam gives one set of diffraction pattern shown as the full lines. The diffraction pattern by the incident beam P is the same as the selected area diffraction pattern with a single parallel incident beam. For a second beam P', which comes at different angle compared to P, the diffraction pattern in this case is displaced from that of P by α/λ with α as the angle between the two incident beams.

Experimentally, the size of the CBED disk is determined by the condenser aperture and the focal length of the probe-forming lenses. In modern microscopes with an additional mini-lens placed in the objective prefield, it is also possible to vary the convergence angle by changing the strength of the mini-lens. Underfocus the electron beam also gives a smaller convergence angle, however, it leads to a bigger probe, which can be an issue for specimens with a large wedge angle.

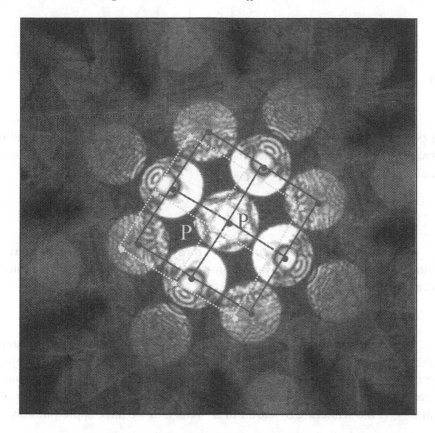

Figure 2. An example of zone-axis CBED pattern, from spinel along [001]. The pattern was recorded using LEO 912 energy-filtering electron microscope at 120 kV.

The advantage of being able to record diffraction intensities over a range of incident beam directions makes CBED readily accessible for comparison with simulations. Thus, CBED is a quantitative diffraction technique. In past 15 years, CBED has evolved from a tool primarily for crystal symmetry determination to the most accurate technique for strain and structure factor measurement [16]. For defects, large angle CBED technique can characterize individual dislocations, stacking faults and interfaces. For applications to defect structures and structure without three-dimensional periodicity, parallel-beam illumination with a very small beam convergence is required.

3.   THEORY OF ELECTRON DIFFRACTION

3.1   Electron Atomic Scattering

Electron interacts with an atom by the Coulomb potential of the positive nucleus and electrons surrounding the nucleus. The relationship between the potential and the atomic charge is given by the Poisson equation:

$$\nabla^2 V(\vec{r}) = -\frac{e[Z\delta(\vec{r}) - \rho(\vec{r})]}{\varepsilon_o} \qquad (1)$$

If we take a small volume, $d\vec{r} = dxdydz$, of the atomic potential at position $\vec{r}$, the exit electron wave from this small volume is approximately given by:

$$\phi_e \approx (1 + i\pi\lambda U dxdydz)\phi_o \qquad (2)$$

Here we take $U = 2meV(\vec{r})/h^2$, which is treated as constant within the small volume. This is known as the weak-phase-object approximation. For a parallel beam of incident electrons, the incident wave is described by the plane wave $\exp(2\pi i \vec{k}_o \cdot \vec{r})$. For high-energy electrons with E>>V, the scattering by the atom is weak and we have approximately:

$$f(s) = \frac{2\pi m n e}{h^2} \quad V(\vec{r}') e^{4\pi i \vec{s} \cdot \vec{r}'} d\vec{r}' \qquad (3)$$

The potential is related to the charge density, Fourier transform of electron charge density is commonly known as the X-ray scattering factor. To relate electron scattering factor to the X-ray, which gives

$$f(s) = \frac{me^2}{8\pi\varepsilon_o h^2} \frac{(Z - f^x)}{s^2} = 0.023934 \frac{(Z - f^x)}{s^2} (\overset{\circ}{A}) \qquad (4)$$

The X-ray scattering factor in the same unit is given by

$$f^x(s) = \frac{e^2}{mc^2} \quad f^x = 2.82 \times 10^{-5} f^x (\overset{\circ}{A}) \qquad (5)$$

For typical value of s~0.2 1/Å, the ratio $f/(e^2/mc^2)f^x \sim 10^4$. Thus electrons are scattered by an atom much more strongly than X-ray.

Electron distribution in an atom depends the atomic electronic structure and bonding with neighboring atoms. At sufficiently large scattering angles, we can approximate atoms in a crystal by spherical free atoms or ions. Atomic charge density and the Fourier transform of charge density can be calculated with high accuracy. Results of these calculations are published in literature and tabulated in the international table for crystallography. Tables optimized for electron diffraction applications are also available [17].

3.2 The Geometry of Electron Diffraction From Perfect Crystals

Electrons diffract from a crystal under the *Laue* condition $\vec{k} - \vec{k}_o = \vec{G}$, with $\vec{G} = h\vec{a}^* + k\vec{b}^* + l\vec{c}^*$. Each diffracted beam is defined by a reciprocal lattice vector. Diffracted beams seen in an electron diffraction pattern are these close to the intersection of the Ewald sphere and the reciprocal lattice. A quantitative understanding of electron diffraction geometry can be obtained based on these two principles.

3.2.1 The Cone of Bragg condition

The Bragg condition defines a cone of angles, θ_{hkl}, normal to the (hkl) planes. Alternatively, we use the Laue condition to specify the Bragg diffraction:

$$k_o^2 - \left|\vec{k} + \vec{G}\right|^2 = 0 \tag{6}$$

$$2\vec{k} \cdot \vec{G} + G^2 = 0 \tag{7}$$

and

$$k_G = -G/2 \tag{8}$$

Thus irrespective to the wavelength of electrons, the wave vector along the plane normal direction is half the reciprocal lattice vector length. The Bragg angle becomes increasingly smaller as the electron energy increases, the wavelength decreases and the wave vector along the plane increases.

3.2.2   Zone axis

A typical zone axis diffraction pattern is shown in fig. 2. The diffraction pattern can be indexed based on the two shortest G- vectors $\vec{g}$ and $\vec{h}$. The zone axis is along the direction determined by

$$\vec{g} \times \vec{h} = \left(h_1\vec{a}^* + k_1\vec{b}^* + l_1\vec{c}^*\right) \times \left(h_2\vec{a}^* + k_2\vec{b}^* + l_2\vec{c}^*\right)$$

$$= (k_1l_2 - k_2l_1)\vec{b}^* \times \vec{c}^* + (l_1h_2 - l_2h_1)\vec{c}^* \times \vec{a}^* + (h_1k_2 - h_2k_1)\vec{a}^* \times \vec{b}^*$$

$$(9)$$

The zone axis indices are taken such as

$$\vec{z} = u\vec{a} + v\vec{b} + w\vec{c} \tag{10}$$

with [u,v,w] the smallest integer satisfying

$$u/v/w = (k_1l_2 - k_2l_1)/(l_1h_2 - l_2h_1)/(h_1k_2 - h_2k_1) \tag{11}$$

By definition, a zone axis is normal to both g and h and other reciprocal lattice vectors in the plane defined by these two vectors. The reciprocal lattice plane passing through the reciprocal lattice origin is called the zero-order zone axis. A G-vector with $\vec{z} \cdot \vec{G} = n$ with $n \neq 0$ is said to belong to a high order Laue zones, which separate to upper Laue zones (n>0) and lower Laue zones (n<0).

A diffraction pattern with diffraction spots belonging to both zero-order Laue zone and higher order Laue zones can be used to determine the three-dimensional unit-cell of the crystal.

3.2.3   The zone axis coordinate and the line equation for Kikuchi and HOLZ lines

Given a zone axis, we can define an orthogonal zone axis coordinate, $(\hat{x}, \hat{y}, \hat{z})$, with the z parallel to the zone axis direction. Let us take x along the g direction and y normal to the x-direction. Expressing the condition for Bragg diffraction (equation 6) in this coordinate, we have

$$2\vec{k} \cdot \vec{g} + g^2 = 2\left(k_x g_x + k_y g_y + k_z g_z\right) + g^2 = 0 \tag{12}$$

and

$$k_y = -\frac{g_x}{g_y}k_x + \frac{2g_z - g^2}{2g_y}|k_z| \tag{13}$$

Here

$$|k_z| = \sqrt{k_o^2 - k_x^2 - k_y^2} \approx k_o = 1/\lambda \tag{14}$$

The approximation holds for high-energy electrons of small wavelength and the typical diffraction viewing angles. Within this approximation, beams, satisfying the Bragg condition, form a straight line.

3.2.4 The excitation error

The deviation of the electron beam from the Bragg condition is measured by the distance from the reciprocal lattice vector to the Ewald sphere along the zone axis direction, which approximately is defined by

$$S_g = \left(k_o^2 - |\vec{k} + \vec{g}|^2 \right)/2|k_o| \tag{15}$$

This parameter is important for the interpretation of diffraction contrast images and electron diffraction intensities. From (E29), let us take

$$k_G = -g/2 + \Delta \tag{16}$$

The positive Δ corresponds to a tilt towards the center of Kikuchi band. Then we have

$$S_g = \lambda\left(\left(-\frac{g}{2} + \Delta\right)^2 - \left(\frac{g}{2} + \Delta\right)^2\right)/2 = -\lambda g\Delta \approx -g\delta\theta \tag{17}$$

Here $\delta\theta$ is the deviation angle from the Bragg condition. S is negative for positive Δ and positive for negative Δ.

The starting point for understanding CBED is the Ewald sphere construction. Fig.3 shows one example. By the requirement of elastic scattering, all transmitted and diffracted beam are on the Ewald sphere. Let's take the incident beam P, which satisfies the Bragg condition for g. For an incident beam P', to the left of P, the diffracted beam also moves to the left, which gives a positive excitation error. Correspondingly, a beam, moving to the right of P, gives a negative excitation error. The excitation error varies according to equation 17. Generally, the excitation error varies across the disk and along the g direction for each diffracted beams, The range of excitation errors within each disk is proportional to the length of g and the convergence angle. Consequently, excitation error changes much faster for a HOLZ reflection than a reflection in ZOLZ close to the direct beam.

For high order reflections with a large g, the rapid increase in the excitation error away from the Bragg condition results a rapid decrease in diffraction intensity. Under the kinematical condition, the maximum intensity occurs at the Bragg condition, which appears as a straight line within a small convergence angle.

3.2.5   The Geometry of a CBED pattern

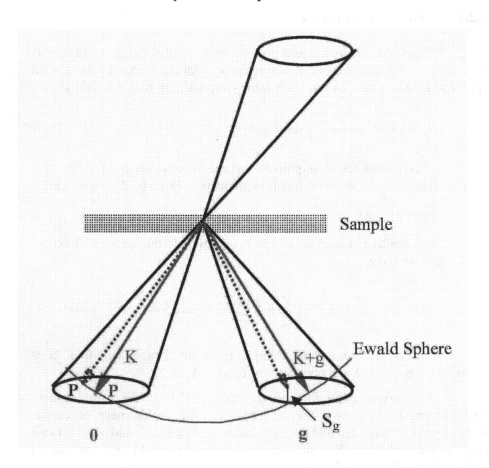

Figure 3. The excitation and Ewald sphere construction of CBED.

These lines, named as high order Laue zone (HOLZ) lines, are very useful for measuring lattice parameters and local strain. The sensitivity of the lines to the lattice parameters comes from the large angle scattering. For example, let's start with the Bragg condition for a cubic crystal:

$$\sin\theta = g\lambda/2 = \sqrt{h^2 + k^2 + l^2}\,\lambda/2a \tag{18}$$

A small change in a gives

$$\delta\theta \approx \sqrt{h^2 + k^2 + l^2}\,\lambda\delta a/2a^2 = 0.5g\lambda\delta a/a \tag{19}$$

Thus the amount change in the angle is proportional to the length of g. The position of the lines, which moves relative to each other with a small change in the lattice parameter [18].

3.3   Electron Dynamic Theory – The Bloch wave method

Electron dynamic scattering must be considered for the interpretation of experimental diffraction intensities because of the strong electron interaction with matter for a crystal of more than 10 nm thick. For a perfect crystal with a relatively small unit cell, the Bloch wave method is the preferred way to calculate dynamic electron diffraction intensities and exit-wave functions because of its flexibility and accuracy. The multi-slice method or other similar methods are best in case of diffraction from crystals containing defects. A recent description of the multislice method can be found in [8].

For high-energy electrons, the exchange and correlation between the beam electron and crystal electrons can be neglected, and the problem of electron diffraction is reduced to solve the Schrödinger equation for an independent electron in a potential field:

$$\left[K^2 - \left(\vec{k} + \mathbf{g}\right)^2\right]C_g + \sum_h U_{gh}C_g = 0 \tag{20}$$

With

$$K^2 = k_o^2 + U_o \tag{21}$$

And

$$U_g = U_g^C + U_g^{''} + iU_g^{'} \tag{22}$$

Here $U_g = 2m|e|V_g/h^2$ is the optical potential of the crystal for beam electrons, which consists of the crystal potential U^C, absorption U' and a correction to the crystal potential due to virtual inelastic scattering U". The most important contribution to U' and U" for |g| >0 comes from inelastic phonon scatterings. Details on the evaluation of absorption potential can be found in the references of Bird and King [19], Weickenmeier and Kohl [20] for atoms with isotropic Debye-Waller factors and Peng for anisotropic thermal vibrations[21]. The U" term is significantly smaller than U', which is neglected. For high-energy electrons, the exchange and correlation between the beam electron and crystal electrons can be neglected.

For convergent electron beams with a sufficiently small probe, the diffraction geometry can be approximated by a parallel crystal slab whose surface normal direction is **n**. To satisfy the boundary condition, letting

$$\mathbf{k} = \mathbf{K} + \gamma\mathbf{n} \tag{23}$$

Here γ is the dispersion of wavevector inside the crystal. Inserting equation 4 into 1 leads to inside the square bracket of equation 1. Neglecting the backscattering term of γ^2, we obtain from equation 1

$$2KS_g C_g + \sum_h U_{gh} C_h = 2K_n\left(1 + \frac{g_n}{K_n}\right)\gamma C_g \tag{24}$$

Here $K_n = \mathbf{K}\cdot\mathbf{n}$ and $g_n = \mathbf{g}\cdot\mathbf{n}$. Equation 6 reduces to an eigen equation by renormalizing the eigenvector:

$$B_g = \left(1 + \frac{g_n}{K_n}\right)^{1/2} C_g \tag{25}$$

The zone axis coordinate system can be used for specifying the diffraction geometry: the incident beam direction and crystal orientation. In this coordinate, an incident beam of wavevector **K** is specified by its tangential component on x-y plan $\vec{K}_t = k_x\hat{x} + k_y\hat{y}$, and its diffracted beam at $\mathbf{K_t}+\mathbf{g_t}$, for small angle scatterings. For each point inside the CBED disk of **g**, the intensity is given by

$$I_g(k_x, k_y) = |\phi_g(k_x, k_y)|^2 =$$
$$\left|\sum_i c_i(k_x, k_y) C_g^i(k_x, k_y)\exp\left[2\pi i\gamma^i(k_x, k_y)t\right]\right|^2 \tag{26}$$

Here the eigenvalue γ and eigenvector C_g are obtained from diagonalizing equation 5. And c^i is the first column of the inverse eigenvector matrix, or excitation coefficient as determined by the incident beam boundary conditions. The solution of equation 5 converges with the increasing number of beams included in the calculations. Solving equation 5 straightforwardly with a large number of beams is often impractical, since matrix diagonalization is very time consuming. The computer time needed to diagonalize an NxN matrix is proportional to N^3. A solution to this is to use Bethe potential:

$$U_g^{eff} = U_g - \sum_h \frac{U_{g-h}U_h}{2KS_h} \tag{27}$$

and

$$2KS_g^{eff} = 2KS_g - \sum_h \frac{U_{g-h}U_{h-g}}{2KS_h} \tag{28}$$

Here the summation is over all weak beams h. The Bethe potential accounts for the perturbational effects of weak beams on the strong beams. The criterion for a weak beam we use is

$$\left| \frac{KS_g}{U_g} \right| \geq \omega_{max} \tag{29}$$

which is selected for the best theoretical convergence [22]. An initial list of beams is selected using the criteria of maximum g length, maximum excitation error and their perturbational strength. Additional criteria are used for selecting strong beams. These criteria should be tested for theoretical convergence.

Fig. 4 shows an example of simulated CBED pattern using the Bloch wave method described here for Si [111] zone axis and electron accelerating voltage of 100 kV. The simulation includes 160 beams in both ZOLZ and HOLZ. Standard numerical routine was used to diagonalize a complex general matrix (for a list of routines freely available for this purpose, see [23]). The whole computation on a modern PC only takes a few minutes.

Figure 4. A simulated CBED pattern for Si[111] zone axis at 100 kV using the Bloch wave method described in the text.

4.   EXPERIMENTAL ANALYSIS

4.1   Experimental Diffraction Pattern Recording

The optimum setup for quantitative electron diffraction is a combination of a flexible illumination system, an imaging filter and a 2-D array detectors with a large dynamic range. The three diffraction modes described in section 2 can be achieved through a three-lenses condenser system. There are two types of energy-filters that are currently employed, one is the in-column Ω-energy filter and the other is the post-column Gatan Imaging Filter (GIF). Each has its own advantages. The in-column Ω-filter takes the full advantage of the post specimen lens of electron microscope and can be used in combination with detectors such as film or IP, in addition to CCD camera. For electron diffraction, geometric distortion, isochromaticity, and angular acceptance are the important characteristics of filters [24]. Geometrical distortion complicates the comparison between experiment and theory and is best corrected by experiment. Isochromaticity defines the range of electron energies for each detector position. Ideally this should be the same across the whole detector area. Angular acceptance defines the maximum range of diffraction angles that can be recorded on the detector without a significant loss of isochromaticity.

Current 2-D electron detectors include slow-scan CCD cameras and imaging plates. The performance of the CCD camera and IP for electron recording has been measured [26]. Both are linear with large dynamic range. At low dose range, the CCD camera is limited by the readout noise and dark current in CCD. IP has a better performance at low dose range due to the low dark current and readout noise by the photo-multiplier. At medium and high dose, the IP is limited by a linear noise due to the granular variation in the phosphor and instability in the readout system. The CCD camera is limited by the linear noise in the gain image, which can be made very small by averaging over several reference images obtained with a blank illumination. The performance of CCD varies from one to other, which makes the individual detector characterization necessary.

Neither of the CCD camera and IP has the ideal resolution of a single pixel, and in both detectors additional noises are introduced in the detection process. The recorded image can be generally expressed as

$$I = Hf + n \qquad (30)$$

Here the application of H on image f denotes the convolution between the point spread function (PSF) (encoded in H) and image, which plus the noise n is the readout image. The effects of PSF can be removed partially by deconvolution. The PSF is experimentally characterized and measured by the amplitude of its Fourier transform the modulated transfer function (MTF). However, the direct deconvolution of recorded image using the measured MTF leads to an excessive amplification of noise. We have found two deconvolution algorithms that are particularly effective in overcoming the difficult with noise amplification. One uses a Wien filter and the other is the Richardson-Lucy algorithm [25].

The noise in the experimental data can be estimated using the measured detector quantum efficiency (DQE) of the detector

$$\mathrm{var}(I) = mgI / DQE(I) \qquad (31)$$

Here I is the estimated experimental intensity, var denotes the variance, m is the mixing factor defined by the point spread function and g is the gain of the detector [26]. This expression allows an estimation of variance in experimental intensity once DQE is known, which is especially useful in the χ^2-fitting, where the variance is used as the weight.

4.2   The Refinement Technique

Crystal structure information, such as unit cell parameters, atomic positions and crystal charge distribution, can be obtained from experimental diffraction intensities using the refinement method [27,28,29]. The refinement method works by comparing the experimental and theoretical intensities and optimizing for the best fit. During optimization, parameters in the theoretical model are adjusted in search of the minimum difference between experiment and theory. Multiple scattering effects can be taken into consideration by simulation during refinement. Previously, strong electron multiple scattering effects have made it difficult to use kinematical approximations for structure determinations in the way of how x-ray and neutron diffraction work except a few special cases [30]. In case of accurate structure factor measurement, electron interference from multiple scattering can enhance the electron diffraction sensitivity to crystal potential and thickness and improve the accuracy of electron diffraction measurement.

The refinement is automated by defining a goodness of the fit (GOF) parameter and using numerical optimization routines to do the search in a computer. One of the most useful GOF's for direct comparison between experimental and theoretical intensities is the χ^2

$$\chi^2 = \frac{1}{n-p-1} \sum_{i,j} \frac{1}{\sigma_{i,j}^2} \left(I_{i,j}^{\ exp} - c I_{i,j}^{\ Theory} \left(a_1, a_2, ..., a_p \right) \right)^2 \quad (32)$$

Here, I^{exp} is the experimental intensity (in unit of counts) measured from an energy-filtered CBED pattern and i and j are the pixel coordinate of the detector and n is the total number of points. I^{theory} is the theoretical intensity calculated with parameters a_1, *to* a_p and c is the normalization coefficient. The other commonly used GOF is the R-factor

$$R = \sum_{i,j} \left| I_{i,j}^{\ exp} - c I_{i,j}^{\ Theory} \left(a_1, a_2, ..., a_p \right) \right| / \sum_{i,j} \left| I_{i,j}^{\ exp} \right| \quad (33)$$

The χ^2 is best when the differences between theory and experiment are normally distributed and when the variance σ is correctly estimated. The optimum χ^2 has a value close to unity. χ^2 smaller than 1 indicates an over-estimation of the variance. On the other hand, σ is not needed for the evaluation of R-factor, which measures the residual difference in percentage.

The disadvantage of R-factor is that the same R factor value may not indicate the same level of fit depending on the noise in the experimental data. The other difference is that the R-factor is based on an exponential distribution of differences. This makes the R factor a more robust GOF against possible large differences between theory and experiment. The exponential distribution has a long tail compared to the normal distribution. R-factor is used extensively in crystallographic methods.

Large differences between experiment and theory are often the indication of systematic errors, such, deficiency in theoretical model (as in the case kinematics approximation for dynamically scattered electrons) an measurement artifacts (such as uncorrected distortions).

Another useful definition of GOF is the correlation function as defined by

$$R = \sum_{i,j} I_{i,j}^{\ exp} I_{i,j}^{\ Theory} \left(a_1, a_2, ..., a_p\right) / \sqrt{\sum_{i,j} I_{i,j}^{\ exp\,2}} \sqrt{\sum_{i,j} I_{i,j}^{\ Theory\,2}} \quad (34)$$

This measures the likeness of two patterns, which is useful for pattern matching as we will see in the application of lattice parameter and strain measurements.

To model diffraction intensities, detector effects and the background intensity from thermal diffuse scattering must be included. A general expression for the theoretical intensity considering all of these factors is

$$I_{i,j}^{Theory} = \iint dx' dy' t(x', y') C(x_i - x', y_j - y') + B\left(x_i, y_j\right) \quad (35)$$

Here, the diffracted intensity t is convoluted with the detector response function C plus the background B. The intensity is integrated over the area of a pixel. For a pixilated detector with fixed pixel size, the electron microscope camera length determines the resolution of recorded diffraction patterns. There are two contributions to the detector response function, one is the finite size of the pixel and the other is the point spread in the detector. Point spread function can be measured and removed, or deconvoluted, numerically. Procedures for doing this have been published [26]. At sufficient large camera length, we can approximate C by a delta function for a deconvoluted diffraction pattern. The background intensity B, in general, is slow varying, which can be subtracted or approximated by a constant.

To calculate theoretical intensities, an approximate model of potential is needed. For structure refinement, we need an estimate of cell sizes, atomic position and Debye-Waller factor. In case of bonding charge distribution measurement, crystal structure is first determined very accurately. The unknowns are the low order structure factors, absorption coefficients and experiment parameters describing the diffraction geometry and specimen thickness. The structure factor calculated from spherical atoms and ions can be used as a sufficiently good starting point. Absorption coefficients are estimated using Einstein model with known Debye-Waller factors [19]. In refinement, unknown parameters are adjusted for the best fit.

The refinement process is divided into two steps. In the first step, the theoretical diffraction pattern is calculated based on a set of parameters. The pattern can be the whole, or part of the, experimental diffraction pattern. In some cases, a few line scans across the experiment diffraction contain enough data points for the refinement purpose. In the second step, the calculated pattern is placed on top of the experimental pattern and try to locate the best match by shifting, scaling and rotation. Both steps are automated by optimization. The first step optimizes structural parameters and the second step is for experimental parameters. The experimental parameters include the zone axis center (in practice the Kt for a specific pixel), length and angle of the x-axis of the zone axis coordinate used for simulation, specimen thickness, intensity normalization and backgrounds. Given a set of calculated theoretical intensities, their correspondents in experimental pattern can be found by adjusting experimental parameters without the need of dynamical calculations. Thus considerable computation time is saved and while the number of parameters in each optimization cycle is reduced. The structural parameters can be individual structure factors, lattice parameters or atomic positions. Each leads to different applications, which is discussed in [16].

For the fit using χ^2 as GOF parameter, the precision of measured parameters is given by

$$\sigma_{a_k}^2 = \chi^2 D_{kk}^{-1} \tag{36}$$

Here

$$D_{kl} = \sum_{i,j} \frac{1}{\sigma_{i,j}^2} \left(\frac{\partial I_{ij}^{Theory}}{\partial a_k} \right) \left(\frac{\partial I_{ij}^{Theory}}{\partial a_l} \right) \tag{37}$$

Derivative of intensity against structure parameters and thickness can be obtained using the first order perturbation method [31]. The finite difference method can also be used to evaluate the derivatives. Estimates of errors in refined parameters can also be obtained by repeating the measurement. In case of CBED, this can also be done by using different regions of the pattern. In general, the number of experimental data points in CBED far exceeds the number required for an accurate estimation.

5.   APPLICATIONS TO MATERIALS SCIENCE

5.1   Crystal charge density

A major application of QED is the accurate determination of crystal charge density. The scientific question here is how atoms bond to form crystals, which can be addressed by accurate measurement of crystal structure factors (Fourier transform of charge density) and from that to map electron distributions in crystals.

Accurate measurements of low order structure factors are based on the refinement technique described in section 4. Using the small electron probe, a region of perfect crystal is selected for study. The measurements are made by comparing experimental intensity profiles across CBED disks (rocking curves) with calculations, as illustrated in fig. 5. The intensity was calculated using the Bloch wave method, with structure factors, absorption coefficients, the beam direction and thickness treated as refinement parameters.

There are two approaches to map crystal charge density from the measured structure factors; by inverse Fourier transform or by the multipole method [32]. Direct Fourier transform of experimental structure factors was not useful due to the missing reflections in the collected data set, so a multipole refinement is a better approach to map charge density from the measured structure factors. In the multipole method, the crystal charge density is expanded as a sum of non-spherical pseudo-atomic densities. These consist of a spherical-atom (or ion) charge density obtained from multi-configuration Dirac-Fock (MCDF) calculations [33] with variable orbital occupation factors to allow for charge transfer, and a small non-spherical part in which local symmetry–adapted spherical harmonic functions were used.

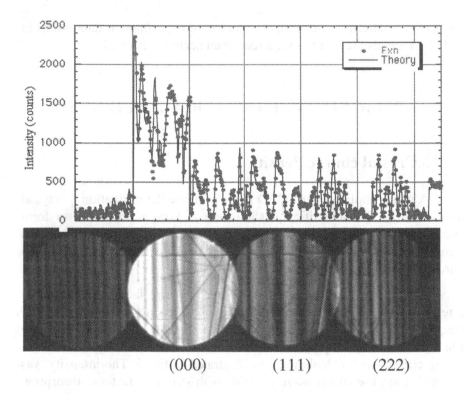

Figure 5. A record Si CBED pattern with (111) and (222) at Bragg condition and the intensity profile (cross) and the theoretical fit (continuous curve). The structure factors of (111) and (222) reflections are obtained from the best fit.

To illustrate the principle of multipole technique, we use Cu_2O as an example. Cu_2O has a cubic structure with no free internal parameters (only Ag_2O is isostructural). The copper atoms are at the points of a f.c.c. lattice with oxygen atoms in tetrahedral sites at (1/4,1/4,1/4) and (3/4,3/4,3/4) of the cubic cell. The resulting arrangement of Cu-O links is made up of two interpenetrating networks. The simplest description of Cu_2O using an ionic model with closed-shell Cu^+ and O^{2-} ions is known to be inadequate. It fails to explain the observed linear 2-coordination of Cu.

The electron density of Cu_2O can be expressed by:

$$\rho^a(\mathbf{r}) = \rho^s(\mathbf{r}) + \sum_{lm} P_{lm\pm} N_{lm\pm} R_l(r) y_{lm\pm}(\theta, \phi) \qquad (38)$$

here $\rho_{cu}^s = \rho_{Cu^+}^s + (1-q)\left(\rho_{Cu}^s - \rho_{Cu^+}^s\right)$ and $\rho_O^s = \rho_O^s + q\left(\rho_{O^{2-}}^s - \rho_O^s\right)$ with q as the charge transfer from Cu to O; $R_l(r) = r^{n_l} \exp(-\alpha r)$ (n_3=3; n_4=4) is the radial function with population coefficient $P_{lm\pm}$; $N_{lm\pm}$ is the density-normalization coefficient.

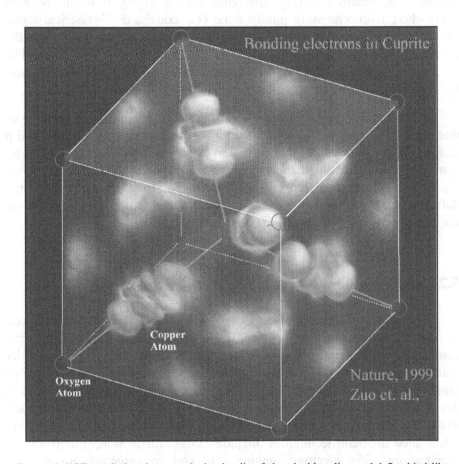

Figure 6. A 3D rendering that reveals the details of chemical bonding and dz2 orbital-like holes in Cu2O. The amount of charge redistribution is very small and its detection requires a high degree of experimental accuracy. In this picture, the small charge differences between the measured crystal charge density derived from convergent-beam electron diffraction (CBED) and that derived from superimposed spherical O2- and Cu+ ions are shown. The red and blue colors represent excess electrons and holes, respectively.

The multipole model reduces the crystal electron density to a number of parameters, which can be fitted to experimental structure factors. For Cu_2O, structure factors for the (531) and higher-order reflections out to (14,4,2) were taken from X-ray measurements. Weak (ooe) (with o for odd and e for even) and very weak (eeo) reflections were also taken from X-ray work. Fig. 6 shows a three-dimensional plot of the difference between the static crystal charge density obtained from the multipole fitting to experiment, and superimposed spherical O^{2-} and Cu^+ ions calculated by the MCDF method. The O^{2-} ion was calculated using a Watson sphere of 1.2 Å radius. The electron density difference shown in fig. 6 would be zero everywhere if cuprite were purely ionic (i.e. consisted of spherical ions). There is little variation around oxygen in both the experimental and the theoretical results, which suggests that an O^{2-} anion description is valid. The most significant difference between experiment and theory is around Cu, and the charge in the interstitial region.

The non-spherical charge density around Cu^+ can be interpreted as due the hybridization of d electrons with higher-energy unoccupied s and p states. Among these states, hybridization is only allowed for d_z^2 and 4s by symmetry, and when this happens part of the d_z^2 state becomes unoccupied ("d hole"). These states are responsible for the spatial distribution of the deficiency in the map shown in fig. 6. The complementary empty states are important for EELS, which probe empty states.

5.2   Strain measurement by Pattern matching

Strain, and the related stresses, in nanometer-sized structures is one of major issues in materials science. In semiconductor technologies, as the complexity of devices continues to increase with the push toward smaller feature sizes, strain that comes from the incorporation of a wide range of dissimilar materials and their geometries becomes increasingly important in device designs [34]. This demands experimental measurement of strain and stress in materials at nanometer-resolution with high-sensitivity. Existing strain measurement techniques include X-ray diffraction, micro-Ramen spectroscopy [35], electron imaging [36] and convergent-beam electron diffraction (CBED) [18,37,38,39] Both X-ray diffraction and Raman spectroscopy have the spatial resolution in the orders of microns. X-ray diffraction measures the lattice parameter by Bragg diffraction with accuracy to 10^{-4}. Dark-field electron imaging maps the variation of a single lattice planes and can, in principle for specimens of constant thickness and composition, be sensitive to strain of 10^{-3} at resolution of several nanometers. CBED measures the local lattice parameter by detecting the shift of high-order Laue zone (HOLZ) lines using a sub-nanometer sized electron probe. HOLZ lines are formed by electron diffraction from lattice planes of large reciprocal lattice vectors in the upper Laue zones, which are very sensitive to a small lattice strain. This, combined with the very small electron probe, makes CBED a versatile technique for strain mapping in nanoscale devices[40].

Previous CBED measurements of strain use the kinematical approximation to simulate HOLZ patterns [37,38,39]. The strain is measured by matching simulated and experimental HOLZ lines based on the χ^2 procedure [37]. In this procedure, the dynamic effect of electrons interacting strongly with matter is treated by a constant shift in the momentum dispersion surface of the incident electrons. However, this approximation is only valid in a sparse off zone axis orientation where the dispersion surface is relatively flat. This limits both the orientation that strain can be analyzed and introduce experimental uncertainties that requires robust verification and calibration. We have developed a pattern matching method for lattice parameter measurement using CBED by optimizing the correlation between experimental and theoretical patterns using equation 34. Parameters used for simulation include the lattice parameters, sample thickness, incident beam electron directions, and diffraction pattern recording related parameters. Both experimental and theoretical diffraction patterns are processed by a line detection algorithm to enhance the geometric information that is sensitive to lattice parameters. Diffraction intensities, in general, are very sensitive to crystal potential. But diffraction geometry is well described by approximating atoms in crystals as spherical neutral atoms. We calculate the electron dynamic intensity of the direct beam using the Bloch wave method (section 3).

The pattern matching between experiment and theory was carried using the general-purpose EXTAL program developed by one of the authors for electron crystallographic refinements [41]. By including the dynamic and thickness effects in the simulation, the method is general, applicable to any orientations and gives the absolute value in lattice parameter measurements.

Fig.7 shows an example of the type of fit we obtain between experiment and theory. The experiment pattern was recorded in a CCD camera, and energy filtering was not used. The experimental and theoretical patterns are processed by a line detection program. To save computation times, only selected areas of the experimental pattern are matched. The areas are selected based on their sensitivity to lattice parameters.

Application of the pattern matching technique for silicon device characterization is described in ref. [42].

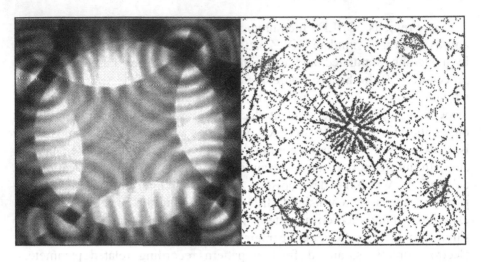

Figure 7. Lattice parameter measurement for NiO at −170°C, which has a rhomboherdal distortion. The best fit (shown at right) gives a=2.9522 Å and a=60.05°.

6.   CONCLUSIONS AND FUTURE PERSPECTIVES

In conclusion, this chapter describes the practice and theory of quantitative CBED. It is demonstrated that the information obtainable from electron diffraction with a small probe and strong interactions complements with other characterization techniques, such as X-ray and neutron diffraction that samples a large volume and real space imaging by HREM with a limited resolution. The recent developments in electron energy-filtering, 2-D digital detectors and computer-based image analysis and simulations have significant improved the quantification of electron diffraction. Examples were given to demonstrate the remarkable accuracy of electron diffraction with application for crystal charge density and strain mapping.

The future for electron diffraction is very bright for two reasons. First, electron diffraction pattern can be recorded selectively from individual nanostructure at sizes as small as a nanometer using the electron probe forming lenses and apertures, while electron imaging provides the selectivity. Second, electrons interact with matter much more strongly than X-ray and Neutron diffraction. These advantages, coupled with quantitative analysis, enable the structure determination of small, nonperiodic, structures that was not possible before.

Acknowledgement: The work described here would not be possible without the support of funding from NSF (at Arizona State University) and

DOE BES (at UIUC) and some of work described here are the results of close collaboration with Profs. J.C.H. Spence and M. O'Keeffe, and Drs. M. Kim, Bin Jiang and R. Holmestadt.

References

1. J.W. Edington, Practical Electron Microscopy in Materials Science, Monograph 2, Electron Diffraction in the Electron Microscope, MacMillan, Philips Technical Library, (1975)
2. P. Hirsch et al., Electron Microscopy of Thin Crystals, p. 19, R.E. Krieger, Florida (1977)
3. D.B. Williams and C. B. Carter, Transmission Electron Microscopy, Plenum, New York (1996)
4. B. Fultz and J. Howe, Transmission Electron Microscopy and Diffractometry of Materials, Springer Verlag, New York (2002)
5. J.M. Cowley, Diffraction Physics, North-Holland, New York (1981)
6. J.C.H. Spence and J.M. Zuo, Electron Microdiffraction, Plenum, New York (1992)
7. Z.L. Wang, Elastic and Inelastic Scattering in Electron Diffraction and Imaging, Plenum, New York (1995)
8. E. J. Kirkland, Advanced Computing in Electron Microscopy, Plenum Press, New York, (1998)
9. Tanaka M, Terauchi M and Kaneyama T Convergent-Beam Electron Diffraction, JEOL, Tokyo, (1988)
10. J.P. Morniroli, Electron Diffraction, Dedicated Software to Interpret LACBEDPatterns USTL, Lille, France (1994)
11. J.M. Cowley, Micros. Res. Tech. 46, 75 (1999)
12. L.J. Wu, Zhu YM, Tafto J, Phys. Rev. Lett. 85, 5126 (2000)
13. M. Gao, J.M. Zuo, R.D. Twesten, I. Petrov, L.A. Nagahara and R. Zhang, Appl. Phys. Lett. 82, 2703 (2003)
14. J.M. Zuo, M.Gao, J.Tao, B.Q. Li, R. Twesten and I. Petrov, Microscopy Research Techniues, Accepted, (2004)
15. J.C.H. Spence and J.M. Cowley, Optik 50, 129 (1978)
16. J.M. Zuo, Materials Transactions JIM 39, 938 (1998)
17. L.M. Peng et al, Acta Cryst. A52, 257 (1996)
18. J.M. Zuo, Ultramicroscopy 41, 211 (1992)
19. D.M. Bird and Q.A. King, Acta Cryst. A46, 202 (1990)
20. A. Weickenmeier and H Kohl, Acta Cryst. A47, 590 (1991)
21. L.M. Peng, Acta Cryst. A53, 663 (1997)
22. J.M. Zuo and A.L. Weickenmeier, Ultramicroscopy 57, 375-383 (1995)
23. http://www.netlib.org/
24. H. Rose in Energy filtering Transmission Electron Microscopy, Edited by L. Reimer, Springer, Berlin (1995)
25. BH. Richardson J Opt Soc Am 62, 55–59 (1972)
26. J.M. Zuo, Micros. Res. Tech., 49, 245 (2000)
27. J.M. Zuo, And J.C. Spence, Ultramicroscopy 35, 185-196 (1991)
28. J.M. Zuo, Ultramicroscopy 41, 211-223 (1992)

29. J.M. Zuo, Acta Cryst. A 49, 429-435 (1993)
30. L.D. Marks, Phys. Rev. B 60, 2771 (1999)
31. J.M. Zuo, Acta Cryst. A 47, 87 (1991)
32. P. Coppens, X-ray Charge Densities and Chemical Bonding, Oxford, New York, (1997)
33. Rez, D., Rez, P. and Grant, I., Acta Cryst. A50, 481 (1994); Acta Cryst. A 53: 522 (1997)
34 . A. Steegen, K. Maex, Materials Science & Engineering R-Reports 38, 1 (2002)
35. I. De. Wolf, H. Maes, S.K. Jones, J. Appl. Phys., 79, 7148 (1998)
36.   J. Demarest, R. Hull, K.T. Schonenberg, K.G.F. Janssens, Applied Physics Letters 77, 412 (2000)
37. J.M. Zuo, Ultramicroscopy 41, 211 (1992)
38. S.J. Rozeveld and J.M. Howe, Ultramicroscopy 50, 41 (1993)
39. R. Balboni, S. Frabboni and A. Armigliato, Phil. Mag. A77, 67 (1998)
40. J.C.H. Spence and J.M. Zuo, Electron Microdiffraction, Plenum Press, New York (1992)
41. J.M. Zuo, Materials Transactions JIM 39, 938-946 (1998)
42. Miyoung Kim, J.M. Zuo, G.S. Park, Appl. Phys. Lett. 84, 2181 (2004)

NEW INSTRUMENTATION FOR TEM ELECTRON DIFFRACTION STRUCTURE ANALYSIS:
ELECTRON DIFFRACTOMETRY COMBINED WITH BEAM PRECESSION

Stavros Nicolopoulos[3], Alexey Kuligin[2], Kiril Kuligin[2], Boulahya Khalid[1], Gregory Lepeshov[2], Jean Luc DelPlancke[1], Anatoly Avilov[2], Maximilian Nickolsky[4] and Arturo Ponce[5]

[1]Lab de Electrochimie, Université Libre de Bruxelles, Dpt. Science des Materiaux, 1050 Bruxelles (Belgium). [2] Insitute of Crystallography, RAS Lenniskii Prospekt 59, Moscow, 117333 (Russian Federation). [3]Universidad Politecnica de Valencia /ITQ Avda de los Naranjos s/n,Valencia (Spain), [4] IGEM –RAS,Leninskii Prospekt, Moscow, [5] University of Cadiz, Depart of Material Sciences, Puerto Real (Spain).

Abstract :   Electron diffractometry has been used in the past in combination with electron diffraction cameras (EDC) and has been prooved very succesfull for electron crystalography work; by measuring electron diffraction intensities with high precision (up to 1 %) structure resolution of many minerals with even refined light atoms like H,O positions has been resolved this way. We are reporting for the first time a new generation of electron diffractometry equipment (combined with beam precession technique to avoid dynamical interactions). The device could be pontentially interfaced to any commercial TEM in order to measure precisely ED data and resolve structures of nm size . The technique has a strong potential for electron crystallography as any voltage TEM may be used (> 100 kv);preliminary results are presented with a case study where light lithium atoms are revealed in a spinel structure.

Key words: electron diffraction, precession method, electron diffractometry, structure determination

1.   INTRODUCTION

In the modern world, the importance of new materials with new and specific properties is getting higher than ever. For example, in chemical engineering designing new catalysts for Industrial applications with

T.E. Weirich et al. (eds.), Electron Crystallography, 169–183.

specified properties is not possible without understanding the atomic structure. Therefore, in modern materials science/engineering is essential to know and understand the atomic structure/physical properties of materials at microscopic level (few tenths of Angstroms). In order to reveal the atomic structure of new advanced industrial materials we alternatively use

a) X-Ray techniques with quite limited success, as often industrial powders contain a lot of different phases and crystals are normally too small to give precise structure determination or

b) transmission electron microscopes (TEM) through high resolution imaging (HREM) and electron diffraction, usually working from 100-300 KV and magnifying images up to x1.000.000.

Although TEM-HREM is a more useful technique as can reveal a picture of the atoms in the materials, however this picture is difficult to interpret. On the other hand, radiation damage from the electron beam is an issue (in HREM observations), as long periods of TEM observation (>1 min) usually can alter/damage seriously the structure of many industrial applications materials like organic chemical compounds, zeolites, polymers, etc.

Advanced materials structure analysis by electron diffraction in a TEM presents a lot of advantages over conventional X-Ray diffraction: the size of studied crystallites in TEM can be very small (even tens of Angstroms), therefore individual phases in Industrial powders (nm size) can be examined.

On the contrary, using X-Ray diffraction only an average structure can be obtained, usually over a few thousands of particles, where many phases may be present. New syntetised materials are often in powder form and usually present poor crystallinity for precise X-Ray structure determination. Even modern high intense X-Ray synchrotron radiation sources can account only for crystals bigger from several cubic micrometers.

On the other hand, electron diffraction is much more sensitive than X-Ray as interaction as is many thousands times stronger in comparison. Again, in TEM, ED patterns can be obtained instantly with sufficient quality better than 0.1 nm resolution.

The electron diffraction structure analysis (EDSA) in materials was originally developed by the Russian School of structure analysis in the 50´s. Many different structures of organic and inorganic substances have been determined up to now and gave a great contribution in the crystallography and crystal chemistry of solids[1]. Recently, many structures of beam sensitive materials have been resolved successfully[2] through the use of direct phasing of ED intensities. The method has been demonstrated to work well[3] both in theory and practice – using maximum entropy approach -even in bulk inorganic structures (over a wide range of thickness) and with conventional TEM probe size (100 nm) instruments.

The key for reliable atomic structure analysis is linked to the increasing of reliability and precision of the electron diffraction data.

Speed of data acquisition is an issue, as radiation damage usually occurs (specially for organic compounds) for expositions larger than a minute. In fact, in order to resolve successfully a structure in electron crystallography we need to accurately (equal or better than 1 % precision) determine the intensity of all different spots (up to 200) present in an electron diffraction pattern and correct for dynamical diffraction contribution, specially for strong reflections.

In practice, in both light element detection and structure analysis is usually enough precise measurement of < 50-100 reflections to have accurate picture of the crystal structure, though in X-Ray several hundred of reflections have to be measured.[1]

The precision of measuring intensities (dynamic range from 1 to 10 exp 6) must be very accurate specially for the weakest intensities (at large diffraction angles), as they have big influence on the accurate determination of the light atoms in the structure (small Z like H, O, N).

In this work, we present a new device (electron diffractometer) able to collect efficiently ED intensities with high precision and resolve structures uniquely from ED data.

2.   STRATEGY FOR RECORDING ED INTENSITIES

Up to now most ED patterns are usually recorded on photographic films: such films are excellent for reproducing contrast detail in TEM images, however they seriously suffer from lack of dynamic range for recording intensity measurements from ED patterns, therefore they are not adequate for quantitative ED analysis. In fact , precision of measurement of ED from films can be as low as 30 %, specially for weak intensities.

Considerable improvement over film detection can be obtained with either a slow scan CCD camera interfaced to a TEM or with imaging plates. The CCD camera attached to a YAG crystal can have a good linearity in intensity detection, however can present several drawbacks : one is the presence of overflow artifacts when ED is recorded, but even more serious problem can be the potential damage of the CCD detector after continuous exposures to high diffracted ED beam intensities.

Even if in the near future those problems could be partially resolved in some commercial CCD cameras the main problem of precision measurement (up to 1 %)of all ED intensities -and specially the weakest ones - with a CCD camera will still remain in the future. For instance, accuracy in intensity estimation for high angle (> 2 A-1) reflections is very poor (up to 25 % accuracy).

Imaging plates are made with the same area as the photographic films and can be used together with films in a TEM. Exposed plates are read with a laser beam and the system is claimed to have a wide dynamic range over 100 times more than a commercial CCD camera.

Although the precision of measurement (up to 1 %)of all ED intensities - and specially the weakest ones – cannot be resolved with this technique, ED intensities have been used successfully recently to confirm structure of brucite ($Mg(OH)_2$) and reveal hydrogen atoms in the structure[4].

3.   ELECTRON DIFFRACTOMETRY DEVICE AND BEAM PRECESSION

For crystal structure determinations, which includes the determination of electrostatic potential and analysis of chemical bonding of crystals, the precision of measuring ED intensities should be high and preferably the same for all reflections. This is true also for the weakest intensities (i.e. those resulting from large diffraction angles), since they have a significant influence on the accurate determination of the light atoms in the structure.

The background to the electron diffractometry technique is that intensities of all ED reflections are measured point by point (better than 1% accuracy) during scanning of the diffraction pattern against a fixed detector[5]. This system has obvious advantages over CCD cameras used in TEM electron microscopy. Impressive results of old type electron diffractometers adapted in EDC- electron diffraction cameras (not TEM) show that complex structures can be refined accurately (like Lizardite 1T mineral $Mg_3Si_2O_5(OH)_4$) to a crystallographic residual up to R= 3.1%. Light atom (H, O) positions can be localized within the cell with high accuracy[6].

Other very interesting results show that even the splitting of hydrogen positions in brucite $Mg(OH)2$ can be clearly revealed in the electron diffraction structure analysis[7].

Although measurements with diffractometer interfaced with EDC cameras have been performed at 80-100 kv, however, this old-type system has a lot of limitations linked to the extremely long time (several hours) to scan ED patterns and the beam size (from microns to mm) of the electron diffraction cameras. Again ,the problem of correcting intensities from dynamical contribution has not been addressed satisfactory, as primary extinction (dynamical) corrections have been proposed for known structures using the Blackman formula[6,7].

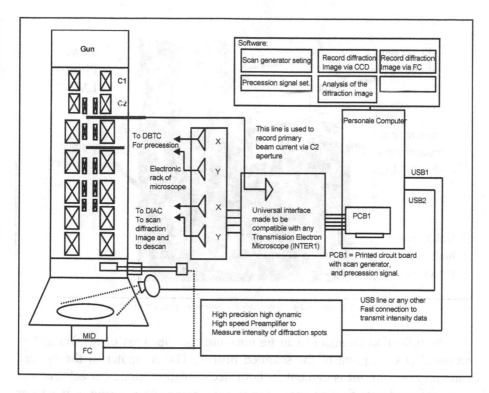

Figure 1 Schematics of the electron diffractometer + precession interface device adapted to a TEM. Diffractometer can be adapted either below the camera plates, or in the 35mm port.

In order to propose a satisfactory solution for ED structure analysis, we adapted a newly designed electron diffractometry device to a commercial TEM microscope (model EM 400 Philips working at 120 kv) in the Lab of Electrochemistry, ULB, Brussels.

With the aid of this prototype, an adequate scanning unit –interfaced to the TEM –scans with pre-determined step size resolution- a part or whole of ED pattern against a fixed detector. Electron beam is deflected by means of deflector coils in the TEM which are situated after the sample. Fixed detector can be either a combination of a scintillator and a photomultiplier or a Faraday cage (one or multiple). Detector is fitted at the bottom of the TEM column, but can also be adapted in the 35 mm port, if the port below the TEM column is occupied by e.g. a CCD camera (see fig. 1).

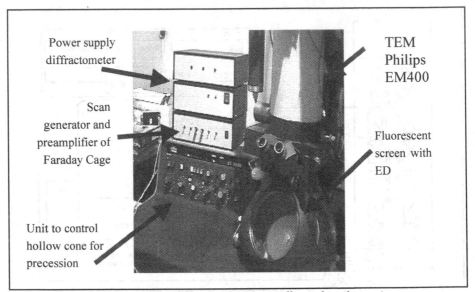

Figure 2. *Scheme of the precession controller and its electronics*

The reflection intensities can be measured as the sums of the intensities recorded at each point of the scanned profile. The computer of the system can carry out the measurement in both accumulation mode (to achieve the same required statistical accuracy for all reflections either strong or weak) or in constant time mode. Although in accumulation mode, precision of the order of 1% can be achieved for all reflections in the ED pattern (dynamical range of 10^6), measurement time is within the order of 10 min up to now. However, it is foreseen that in near future, measurement times of 1 min can be achieved for up to 50 reflections.

In order to properly address the problem of measuring intensities with minimum dynamical diffraction contributions we used the technique where the electron beam is precessed by means of beam tilt coils in the TEM. These coils are situated before the sample in combination with a similar precession of the ED pattern via deflection coils below the sample[8].

The measurement of ED intensities in a precession mode allows all ED intensities to be recorded under reduced influence from dynamical diffraction and secondary scattering contributions, therefore permitting structure analysis of improved precision. By precessing a incident beam at a constant angle around a zone axis in combination with a similar precession of the ED pattern below the specimen[8], the equivalent mechanism of the precession of the specimen is obtained.

The combination of the precession and scanning of the ED pattern reduces the sensitivity of ED intensities to crystal thickness, reduces the effect of Ewald sphere curvature and can also reduce a lot of multiple beam dynamical effects contribution to the reflections (see fig. 3).

This way all ED reflections are much more kinematical in character e.g. a quality of structure analysis can be greatly improved. as has been shown also in recent publications by several authors[9,10,12].

The use of precession can also be helpful to observe wider areas of reciprocal space than can be usually observed without precession , or can be helpful to check zone axis orientations (see fig. 4 as example).

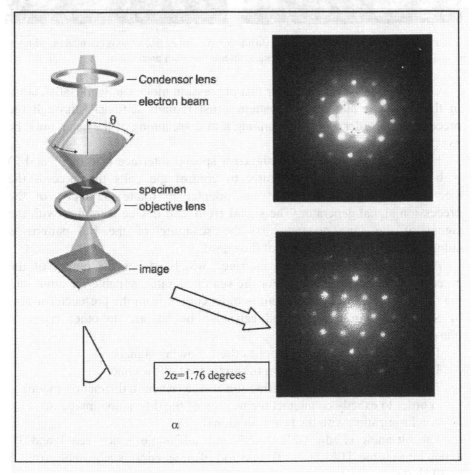

Figure 3. (a)Beam movement in a TEM after applying "conical dark field/precession" mode (b)electron diffraction (ED) pattern of a $Li_xMn_2O_4$ nanocrystal in (111) orientation, with no application and after application of "precession mode": as it can be observed dynamical contribution to all reflections is greatly reduced.

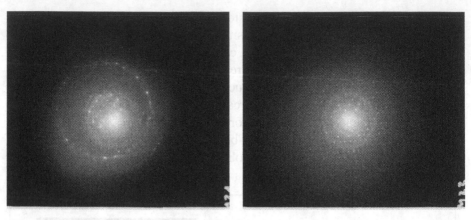

Figure 4. (left)Nd Al3(BO3)4 electron diffraction pattern near zone axis orientation, without precession (right) same orientation with precession

One important aspect in order that precession mode can work satisfactory in the TEM, is that the ED pattern must remain stationary through the precession. In order to do so, scanning and de-scanning of the beam must be exactly compensated.

For beam precession, we designed a special interface (see fig. 1 and 2) with a precession signal generator to control the coils that precess the electron beam and an inverter and adapter to rotate the phase of the precession signal generator. The signal from said device combines with the signal of the scan generator so the scanning of the ED pattern is compensated during precession of the beam.

Again, there is a universal interface which adjusts the output of the precession signal generator and/or the scan generator, suitable for input into the coils of the TEM. This system is independent from the presence (or not) of STEM unit in a TEM, and can also be adapted to other types of transmission microscopes.

The precession controller supplies the following signals:
− The signal to control the beam tilt angle for hollow cone.
− The signal to 'descan' the diffraction image (when in diffraction mode) in order to exactly counteract the motion of the diffraction image in synchronization with the beam tilt signal.

The tilt angle is adjustable. Maximum half angle is between 1 and 3°, depending on the TEM coils. The period of precession is adjustable from 5 to 30 Hertz.

The precession controller is connected to the microscope via some extra wires. When it is switched on, it is fully independent from the working mode of the microscope. The normal TEM mode of operation is the diffraction mode.

When it is switched off, the system is fully disconnected from the microscope by galvanic separation and no extra signal is left from the controller to the microscope.

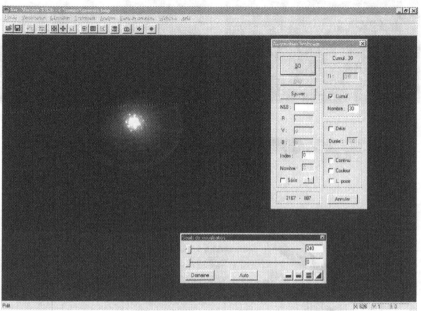

Figure. 5 Graphical user interface of the CCD camera that permits the user define interactively the area of interest to be scanned of ED pattern for precise measurement of intensities.

One important aspect of the electron diffractometer , is that any ED pattern can be captured before scanning by a CCD camera that is placed off-axis in relation with the ED pattern (can be any commercial CCD, even a webcam see fig.1). A dedicated software corrects for any type of optical distortions that may be produced due to the position of viewing angle of the CCD in relation with the ED pattern. This has the big advantage that the user that may observe and define interactively the reflections or the area of the ED pattern that will be scanned, allowing at the same time the main beam to be blanked, in order to avoid radiation damage.

In fig .6a.b we can observe as examples how the user can define different areas of the ED pattern (lines, or regions around spots) can select in order to speed up time of measurement and avoid unnecessary beam damage, or concentrate the study in selected group of reflections (e.g symmetry-related reflections or superstructure reflections).

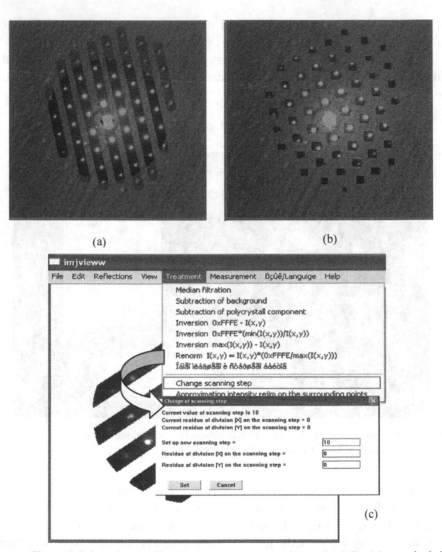

(a)   (b)

(c)

Figure 6. Selected areas like lines (a) or areas around spots(b) defined interactively by the user in the CCD image of the ED pattern before scanning in a Si crystal.(c) example of GUI for scanning step selection. Beam is blanked during scanning data selection.

Once the region to be scanned is selected, the pattern can be automatically indexed; with a separate window in the software interface (Fig. 6c) permits the user to define the scanning step resolution or the time that the reflection can be measured in accumulation mode .After measurement, all reflections are given with intensities corrected from background.

4.   CASE STUDY: REVEALING LIGHT LITIUM ATOMS IN A SPINEL STRUCTURE BY ELECTRON DIFFRACTOMETRY

In order to evaluate the sensibility of the technique to light atom detection in the structure we applied it to the purpose of localising light Lithium atoms in a known spinel structure compound.

This constitutes an interesting industrial problem, since compounds with chemical formula $Li_xMn_2O_4$ (crystal cell: cubic, 0.82 nm; space group: Fd3m) are found to have interesting electrochemical properties depending on the Li stoichiometry in the structure[11]. This material has a wide potential and it is used for Lithium –based batteries in cell phone applications.

Mechanism of energy storage is based on lithium insertion and extraction from the host structure , therefore atomic arrangements of lithium atoms have a direct effect on electrochemical properties.

Although X-Ray powder diffraction does confirm the spinel structure of these compounds, it is impossible to evaluate with this technique the Li content neither to confirm the Li existence in the structure.

According to Vainstein[1] the relative detectability of light atoms in presence of heavy ones in a 2-dimensional projection can be expressed, in the electron diffraction case, as:

$$W_{2d\ ED} = (Z_{light}/Z_{heavy})* e^{0,7} \tag{1}$$

In case of X-Ray diffraction the corresponding formula becomes:

$$W_{2d\ RX} = (Z_{light}/Z_{heavy})* e^{1,2} \tag{2}$$

The above formulas simply show that electron diffraction is most valuable for detecting light atoms in presence of heavy ones, even when there are important differences between the atomic numbers of compared elements.

This relative "advantage" is expressed by:

$$W_{ED}/W_{RX} = (Z_{light}/Z_{heavy})* e^{0,5} \tag{3}$$

e.g. the relative detectability of Li atoms in presence of heavier Mn is 2.88 times easier to detect by electron diffraction than by X-Ray diffraction .

We examined spinel type $Li_xMn_2O_4$ samples in an EM 400 Philips TEM microscope where we have adjusted the prototype diffractometer in order to measure all zone axis reflections in accumulation mode with the same accuracy (less than 3%).

According to bibliography and under the assumption that the atoms are in the standard tetrahedral and octahedral positions imposed by the spinel structure (see table I), we anticipate the zone axis (110) to be one of the most interesting. This is so because it may reveal the position of Lithium atoms well separated from Mn and oxygen atoms (see fig. 7).

Although beam damage is an issue during the collection of the data (time is depending on the required scan resolution and section of reciprocal space examined), we are able to collect several reflections: as many as 51 reflections in (110) zone axis. We corrected these reflections from background and measured them with a high dynamic range and the same accuracy (< 3%).

Table I. [Positions for Li, Mn and O in the spinel structure]

Atom	Wyckoff Pos	X	Y	Z
Li1	8[a]	0	0	0
Mn	16d	5/8	5/8	5/8
O1	32e	0.386	0.386	0.386

Misorientation can be an issue during the time of collection of ED patterns as sometimes this can exceed 60 min in accumulation mode, and dynamical diffraction contribution is observed (we may anticipate its presence due to the appearence of "forbidden" kinematically reflections in the pattern like +- 002). However, is important to note that misorientation effects become less critical and intensity of such "forbidden" reflections is lowered after applying "precession mode" to the ED pattern. Similar results have also been observed by M.Gemmi[9] with Si samples.

At this stage of the experiments we would like to examine if Lithium atoms could be localized experimentally, as already theory indicates (see fig.7). Again, we compared our results regarding Lithium localisation using intensity values taken with and without using precession mode.

In order to estimate the presence of the atomic density of light Li atoms into the spinel structure it is necessary of first of all to estimate the effect of the cut off the Fourier- series on the view of the required (110) projections. It is necessary to do this because the projection of the potential on some plane can be influenced due to the limited amount of the reflections which forms the projection. In fig. 6 the theoretical projection has been calculated for enlarged set F_{hkl} (up to $\sin \theta\lambda \sim 1.6$ Å^{-1}) below. All atoms can be distinguished and are shown by the arrows. The heights of Li-atoms are small but are seen on the projection.

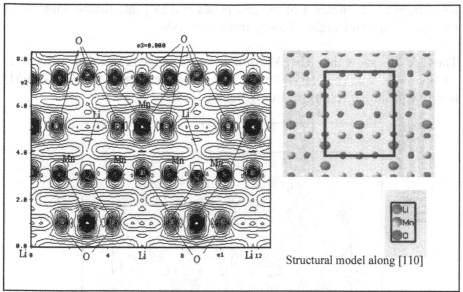

Structural model along [110]

Figure 7. *Projection of the electrostatic potential along (110)-direction for spinel, using the set of theoretical structure amplitudes in the angle range (sinΘ/λ) up to 1.6Å (h=k=l=1). The atomic positions are signed by names of atoms. It can be noted that Li-position intensities are too weak, but distinguishable. The projections were calculated by means of JANA 2000 program (and also by means of AREN, PROMETHEUS).*

Use of the precession technique should promote a decrease of multi-beam effects and an improvement of the situation with the inaccuracy of orientation. In this case we will have full set of orientations along the surface of cone and the diffraction pattern will be the sum of all patterns for every direction of the beam.

The experimental projection (Figure 8) calculated on the diffraction pattern with precession mode reveals clearly Lithium and oxygen positions with atomic density .positive where Lithium and oxygen atoms have ideal positions. Although those results can be considered as preliminary and further study is needed with much less measurement time for the ED pattern (1-2 min per 50 reflections), those results look very promising.
Again, it is important to note that in data taken without precession mode light atoms like lithium and oxygen do not appear well generally.Small crystal misorientations due to time of measurement and dynamical diffraction contributions to many reflections in o precession mode may well explain such results.

Recent study[12] performed on a HREM images taken with 300 kv FEG TEM and using exit-wave reconstruction software, also reveal presence of light lithium atoms in a LiCoO2 compound at 0.1 nm resolution . Again, is also observed a smearing of oxygen peaks and Li peak displacement from ideal positions with slight tilt away from zone axis.

Those results show that HREM combing with image processing software is an alternative way (but depending on sofisticted equipment like FEG TEM) to arrive to similar results regarding light atom localization.

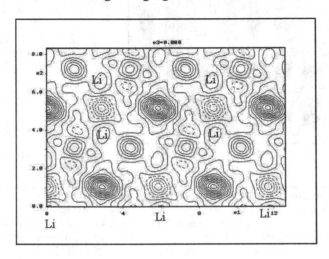

Figure 8. The experimental projection of the potential along <110> -direction, obtained with the precession mode.

5.   CONCLUSION

Electron diffractometry system with the combination of the precession technique can be very perspective experimental instrumentation for precise structural investigations. The technique can now be adapted in a commercial TEM (previously applied uniquely to electron diffraction cameras) taking advantage of the small beam size and can measure reflections in the ED pattern with same required precision for structure analysis.

The potential of the technique can be high , as any commercial TEM (< 100 KV) could be useful for structure determination via ED crystallography without the need (or in combination with) HREM images.

References

1.   Vainstein (1964), Structural Analysis by Electron Diffraction, Oxford Pergamon
2.   D.L Dorset, Structural Electron Crystallography, Plenum Press 1995

3. W. Sinkler, L. D. Marks, Ultramicroscopy 75 (1999) 251-168
4. M.S.Nickolsky, S.Nicolopoulos, A.P.Zhuklistov, B.B.Zvyagin, R.Ochs, Acta Cryst (2002) A58 (supplement)C173
5. A.S. Avilov et al, J.Appl. Cryst. 1999, 32, 1033-38
6. A.P. Zhuklistov, B.B. Zvyagin, Crystallography Reports, vol. 43, n° 6,, 1998, p.950
7. A.P. Zhuklistov et al, Crystallography Reports, vol. 42, n°5, 1997, 841-845.
8. R Vincent, P.A Midgley, Ultramicroscopy 53 (1994) 271
9. M.Gemmi Precession electron diffraction in EUROSUMMER SCHOOL: "Electron Crystalloggraphy and cryoelectron microscopy on inorganic materials and organic and biological molecules", Barcelona (Spain), July 23-27, 2001
10. C.S.Own, A.K Subramanian ,L.D.Marks L. Marks ,Microscopy and Microanalysis 10 , 96-104, (2004)
11. K. Dokko, M. Nishizawa, M. Mohamedi, M. Umeda, L. Uchida, J. Akimoto, Y. Takahashi, Y. Gotoh, S. Mizuta, Electrochem. and Solid State Lett:: "Electrochemical studies of Li-ion extraction and insertion of LiMn2O4 single crystal"
12. J.Gjonnes, V.Hansen , A. Krerneland Microscopy and Microanalysis 10 , 16-20, (2004)
13. Michael O´Keefe and Y.Shao Horn Microscopy and Microanalysis 10 , 86-95, (2004)

ROLE OF ELECTRON POWDER DIFFRACTION IN SOLVING STRUCTURES

JÁNOS L. LÁBÁR

Research Institute for Technical Physics and Materials Science, Budapest, Hungary and Department of General Physics, Eötvös Lóránd University, Budapest, Hungary

Key words: electron diffraction, structure determination, Rietveld refinement

1. INTRODUCTION

The process of solving an unknown structure consists of two main parts: structure determination (unit cell dimensions, space group symmetry, unit cell content, i.e. the approximate atomic positions) and refinement of this structure (reaching higher accuracy in atomic positions). Although the different steps are better elaborated in other papers in this volume, a brief account of the procedure is given here to elucidate the difference between single crystal and powder samples on the one hand, and between X-ray (neutron) diffraction and electron diffraction on the other hand. Our view of presentation is determined by that target in mind.

Since the variation of any physical property in a three dimensional crystal is a periodic function of the three space coordinates, it can be expanded into a Fourier series and the determination of the structure is equivalent to the determination of the complex Fourier coefficients. The coefficients are indexed with the vectors of the reciprocal lattice (one-to-one relationship). In principle the expansion contains an infinite number of coefficients. However, the series is convergent and determination of more and more coefficients (corresponding to all reciprocal lattice points within a sphere, whose radius is given by the length of a reciprocal lattice vector) results in a determination of the structure with better and better spatial resolution. Both the amplitude and the phase of the complex number must be determined for any Fourier coefficient. The amplitudes are determined from diffraction

185

T.E. Weirich et al. (eds.), Electron Crystallography, 185–195.
© 2006 Springer. Printed in the Netherlands.

experiments. The missing phases present the so-called phase problem. In an attempt to determine the missing phases, we can use information from either real-space or reciprocal space. Real space information about the phases appears in high resolution transmission electron microscope (HRTEM) pictures or can be presented by the knowledge that a given molecular fragment is present (and that latter can be used in simulation calculations). Reciprocal space information manifests itself in the form of mathematical considerations that the phases of Fourier coefficients (whose corresponding reciprocal lattice vectors form a closed polygon) are related by the so called structure invariants. The structure invariants are sum of phases for the reflections forming a polygon. This sum of phases is independent of the choice of the origin (although the individual phases appearing in the sum do depend on that choice). The simplest invariants are the phase triplets (from reflections forming a triangle) and the phase quartets (from reflections forming a rectangle). The so called direct methods rely on probabilistic formulas predicting the invariants from the known sizes of the corresponding amplitudes. Fourier synthesis with a finite number of coefficients provides an approximation to the structure. A model (a rough estimate) of the structure is obtained from this approximation and contains some refinable parameters (in atomic coordinates). The role of the final refinement step is not only to make the atomic coordinates more accurate, but also to validate the model (rough estimate). If the model is good, the final refinement can produce a very good match between the measured diffraction pattern and the pattern, calculated from the model (by refining its free parameters). The final refinement is generally done on a powder pattern using the Rietveld method (Rietveld 1967 and 1969; Young 1993).

As we shall see, electron diffraction from powder samples mainly found its (up to now limited) application in the refinement step and no unknown structures has been determined from scratch relying solely on electron diffraction from powder samples. The two main reasons for that are the collapse of the three dimensional information into one dimension and the poor resolving power of electron diffraction (in terms of $\Delta Q/Q$, the peak width).

2. STRUCTURE DETERMINATION

2.1 Structure determination from single crystal diffraction

Single crystal diffraction data are easier to study than powder data, if single crystal samples of good quality and proper size can be produced. That is because the separate Fourier coefficients (the separate structure factors) can be studied without the overlapping effect of the others (at the separate points of the reciprocal lattice). Measurement of the individual structure factor amplitudes reduces to the accurate integration of separate peak intensities over the background. The background in electron diffraction is significantly higher than for the other two methods, especially if energy filtering is not available. Identification and integration of individual reflections over this background can be done either with commercial programs, like ELD (Zou et al. 1993), of free programs, like QED (Belletti et al. 2000). Cell parameters can be routinely determined from (a combination of) single crystal diffraction patterns using either X-ray or neutron or electron diffraction (although the accuracy of the cell parameters obtained from electron diffraction is inferior to that from x-rays.) Symmetry determination relies on a systematic observation of the possible extinctions. Mainly kinematic nature of X-ray and neutron diffraction prevents one from an unambiguous identification of the space group, since kinematic diffraction obeys Friedel law (reflections with +g and –g have the same intensity, i.e. the recorded pattern is always centrosymmetric). Dynamic nature of electron diffraction is well utilized in convergent beam electron diffraction (CBED). With the help of CBED from thicker (100-200 nm) samples of good quality, the space group can either be determined unambiguously, or in the worst case, it can be reduced to a few (generally two) possibilities (out of the 230).

Application of Direct methods to X-ray data from a single crystal can solve the structure of structures with up to 2000 atoms in the asymmetric unit of the unit cell (Weeks at al. 1994, Burla et al. 2000). Several ten thousands of reflections are measured from a long series of tilting experiment for such a solution. A generally accepted rule of thumb in Direct methods that an order of magnitude higher number of indexed strong reflections are needed for a successful structure solution than the number of atoms in the asymmetric unit (Peschar et al. 2002). If the number of atoms in the asymmetric unit does not exceed 100, the success is almost guaranteed for single crystal X-ray data.

Direct methods are also applicable in electron diffraction, however, the dynamic nature of the recorded structure factors needs to be taken into account. When structure factors from strong reflections were determined with the dynamic approximation ($|F|\sim I$ instead of the kinematic $F|\sim I^{1/2}$), application of the Direct methods (SIR97 program) led to a determination of atomic coordinates with better than 0.2 Å accuracy (Weirich et al. 1999). The SIR97 program (Giacovazzo 1998; Altomare et al. 1999) was also fed with electron scattering factors (Jiang & Li 1984) in that study. Recently a freely distributed program, called EDM appeared, which is dedicated to the processing of electron diffraction data with Direct methods (Kilaas R and Marks L 2004). Structures with less than 30 atoms in the asymmetric unit were solved from single crystal electron diffraction data up to now. On the one hand this complexity is inferior as compared to X-ray determinations. On the other hand there is also an advantage that tiny crystals can be successfully used with electron diffraction. Their size can be several thousand times smaller than what is needed for X-ray diffraction. Selection (of grains for diffraction) aided by visual observation in the TEM is also an advantage.

2.2 Structure determination from powder diffraction

In powder diffraction the information in the direction of the scattering (=reciprocal lattice) vector is lost and all reflection with the same length of the scattering vector appear superposed. This collapse of the three dimensional reciprocal space into one dimension results in a multitude of peak overlaps. For application of the Direct methods, separate intensities must be assigned to all the individual reflections, so a partitioning of the measured intensity (which is a superposition of overlapping peaks) into the separate intensities of the components is needed first. The simplest of these coincidences is the superposition of all the symmetrically equivalent reflections of the same type of reflections (for example, the superposition of the eight equivalent general reflections in an orthorhombic system: $(hkl), (\overline{h}kl), (h\overline{k}l), (hk\overline{l}), (\overline{h}\overline{k}l), (\overline{h}k\overline{l}), (h\overline{k}\overline{l}), (\overline{h}\overline{k}\overline{l})$). Partitioning of them is simply done by applying the multiplicity factor (in the above example the measured intensity is equipartitioned into the 8 reflections listed). The next class is the exact overlap, when two (principally) different reflection types correspond to reciprocal vectors of same length. Examples include *(333)* and *(511)* in cubic systems or *(431)* and *(501)* in tetragonal systems. Since the coincidence of peak positions is exact in such overlaps, there is no way to separate the contributing intensity components experimentally. The last class of peak overlap is accidental, when the length of the corresponding reciprocal vectors differs less than the spectral resolution of the diffraction method in use. Contributions of such accidentally overlapping peaks can be more or less separated by using spectrum decomposition methods, which rely on applying predetermined peak shapes for the individual peaks. Using these peak shapes, the contribution of each of the components to any of the measured channels (diffraction angle value) can be calculated. Peak areas of the components are deduced from the over-determined boundary condition that the sum of the contributions in each channel must result the measured total intensity. (Without the predetermined peak shapes solution to the problem would not exist.) Two main methods became accepted for such intensity partitioning in X-ray and neutron diffraction, the La Bail (La Bail et al. 1988) and the Pawley (Pawley 1981) methods. Both of them resemble a Rietveld fit in which the peak intensities themselves are treated as fitting parameters. La Bail introduced an iterative procedure, in which the measured sum peak is partitioned according to the ratio of the components, determined as a result of the previous step of iteration. The procedure converges to the same values, irrespective of the starting proportions. Pawley's method is a least square fitting, supplemented with a constraint, in an attempt to keep positivity of the results. Although both methods are rather stable if large peaks with not-too-heavy overlap are present, instabilities can appear if small peaks are present over a large background and if the peaks overlap seriously. In such cases wild oscillations and unphysical (negative) intensity values might appear. At large scattering angles multiple overlapping of peaks is unavoidable and there is no peak-free region for reliable background determination. In this region all peak-extractions methods must be considered unreliable. (The same remark dos NOT apply to the Rietveld refinement, in which the peak intensities come from the structure model, as shown below.)

The complexity of the solvable structure strongly depends on the spectral resolution of the diffraction method in use. Structures with about 60 atoms in the asymmetric unit were solved from powder data combining synchrotron X-ray diffraction with refinement from neutron diffraction data from the same material (Morris et al. 1994; Admans 2000). About half of that complexity can be achieved with good laboratory X-ray diffractometers (Masciocchi et al. 1996; Kariuki et al. 1999). Neutron diffraction data can better be used for structure refinement than for structure determination, for the same reason.

2.3 The specialties of electron powder diffraction

The above remark about resolution also applies to electron diffraction, and in fact, resolution is one of the most serious limitations in structure determination based on electron diffraction from powder samples. The poor resolution in ED is mainly the consequence of the application of electromagnetic lenses with non negligible spherical aberration. (Improvement in instrumentation may promise increased power of the ED method (Nicolopoulos).) The application of a not perfectly parallel electron beam only aggravates the problem. If we apply the same rule of thumb considering the needed number of reflections as above, we see that with the present resolution, structure determination based on ED from powder samples alone can only be expected for structures with a very few atoms in the asymmetric unit. (The situation is better if structure *refinement* is considered.) The presence of dynamic effects in electron diffraction appears as an additional complication, which requires special treatment (Jansen).

In contrast to XRD, the background is rather high in electron diffraction experiments. The size of the background can partially be attributed to the poorer spectral resolution. Presence of electrons scattered on the apertures can not be completely neglected either.

Las but not least, sample preparation is also an important issue. If we want to examine nanocrystalline "powder" samples, The grain size must be just a few nanometers, the layer, formed by these nanocrystals must be as thin as possible (to minimize dynamic diffraction), continuous and self-supporting. In many cases not all these requirements are fulfilled simultaneously. The nanocrystalline material to be studied is frequently present on a thin supporting carbon layer. In such cases peak decomposition can not yield an acceptable fit unless the presence of the amorphous material (in the form of a few diffuse rings) is taken explicitly into account in the model to be fitted. The size of the background is also affected by scattering in such a carbon support.

2.4 Pattern decomposition with the Process Diffraction program

Extraction of individual line intensities without using an explicit structure model can be done in the ProcessDiffraction program (from version 4.2 upward), which is available free from the internet: http://www.mfa.kfki.hu/~labar/ProcDif . Part of the distortions introduced by lenses is compensated by averaging the intensity along ellipses in contrast to perfect circles. The efficiency of such a correction in case of 2% eccentricity is shown in **Figure. 1**. Eccentricity of that size is not uncommon in electron diffraction and it is difficult even to detect by the naked eye. It is generally identified by comparing the measured rings to a reference circle on the screen, or alternatively, by making measurements of ring diameters in different directions. An automatic peak search (using first and second derivatives) locates the (overlapping) peaks. The list of automatically detected peaks can be edited manually. In case of preliminary identified phases, peak positions can also be taken from a list of peaks, calculated for the identified phases.

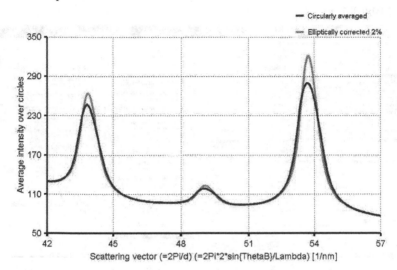

Figure 1. Effect of elliptical distortion on the resolution of the measured SAED pattern.

Quantitative pattern decomposition uses an empirical model for the background and the peak positions taken from the above list. Presence of broad, diffuse peaks under the interesting (narrower) peaks can also be identified manually. Parameters in the model are determined from a least square fit of the model pattern to the measured pattern. The primary parameters to optimize during the least square fit are the peak shape parameters and the peak area for each individual peaks. Optionally, background parameters and the parameters describing the amorphous components can also be optimized within the same iteration. At present, four different shape functions can be selected for the background (Gaussian, called "normal", log-normal, polynomial and spline) and three functions (Gaussian, Lorentzian and Pseudo-Voigt) can be selected for peak shapes. Peak width parameters can (optionally) be optimized individually or connected through a simple formula as a function of peak position. This is similar, but not identical to the approach introduced for XRD (Caglioti et al., 1958). In the absence of moving spectrometer components, the resolution of SAED patterns is constant over the scattering angle (as a starting estimate). Small higher order corrections to this estimate can also be taken into account. Parameters can be switched on and off during consecutive steps of iteration, while tolerance for detecting convergence can also be adjusted, as shown in **Figure 2**, which shows a part of the user interface. Probing of the parameter-space for the global minimum is done by using a downhill

SIMPLEX method (Nelder and Mead, 1995). Alternative functions for the background and for the peak shapes, furthermore alternative approaches for locating the global minimum are planned to be introduced in later versions of the program.

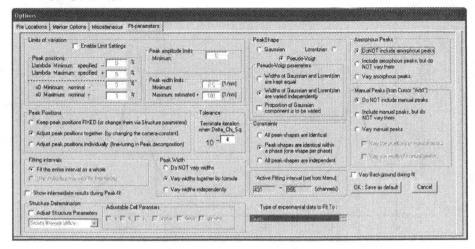

Figure 2. User interface for setting pattern decomposition parameters.

3.   STRUCTURE REFINEMENT

3.1   The Rietveld method

Structure refinement is not simply the last step of solving a structure, but it is also a whole independent world on its own. On the one hand, refinement can be done on a structure which is assumed (known) from a source different from that outlined above. On the other hand, refinement is not simply a procedure resulting more accurate atomic coordinates, but also a way to verify the starting model. The model can only be accepted with high probability if it can be refined to a good match between measured and computed patterns.

Most of the unknown structures is determined from single crystal diffraction and refined from powder diffraction. Refinement is done with the Rietveld method, which is a least square fitting of the computed pattern to the measured one, while structure parameters are treated as the primary fitting parameters. This is in contrast to the procedure in pattern decomposition, which is outlined above (where not the structure parameters, but the peak intensities were the primary fitting parameters). Beside the structure parameters, other parameters, like a scaling factor, parameters defining the background and the peak shapes are also adjusted during the fit. The same procedure can also be used for quantitative phase analysis, when a mixture of phases is refined and the relative proportion of the scaling factors obtained for the component phases characterizes the proportion of phases present.

Since scattering factors for neutron diffraction are independent of the scattering angle (in contrast to the case of XRD and ED), neutron diffraction data are more frequently used for refinement than the others.

3.2   Structure refinement from electron powder diffraction

As mentioned above, resolution is less of a problem during refinement than it is in pattern decomposition, since the contributions of the overlapping peaks to any measured channel can be calculated from the structural model during refinement, even if their separation would be a problem. The public domain program FULLPROF (Rodrigez-Carvajal) can also accept scattering factors as an input (beside measured intensities), so it can be also used for electron diffraction if proper input data are generated for it. As an example, two groups examined nanocrystalline $TiO2$, anatase applying the Rietveld method for refining electron diffraction data. Weirich and coworkers (Weirich et al. 2000a) integrated the ring intensities with the ELD program (Zou et al. 1993) and supplied it together with the electron scattering factors of Jiang and Li (1984) to the FullProf program for refining the atomic coordinates. They obtained close agreement with data deduced from previous neutron diffraction. Tonejc and coworkers (Tonejc et al. 2002) in a similar study supplied the intensity distribution integrated with the ProcessDiffraction program into the same FullProf program, also using Jiang's electron scattering factors. In this study, they also examined the variation of the cell parameters as a function of grain size in nanocrystalline anatase.

From version 4.2 upward, ProcessDiffraction can generate both types of data needed as an input for FullProf. The integrated intensities can be presented as a function of scattering angle (a small transformation from the previous form of presentation). Parameters of Jiang's model can be also output for the separate refinement with FullProf. As an alternative, ProcessDiffraction can also perform the refinement within itself. Electron scattering factors are calculated in ProcessDiffraction using the Weickenmeyer and Kohl (1991) model. Systematic testing has just started to compare refinement results from XRD and ED on the one hand and comparing ED result evaluated by FullProf and ProcessDiffraction on the other hand. Consistency of the results is also examined by comparing the pair correlation function obtained directly from the measured data (by Fourier-transformation) to the one calculated from the structure model obtained from the refinement.

4.   CONCLUSION

Information content in a powder diffraction pattern is reduced as compared to that in single crystal diffraction, due to the collapse of the three dimensional reciprocal space into a one dimensional space where the only independent variable is the scattering angle. The poorer the resolution of the diffraction method, the less the information content in the pattern (Altomare et al. 1995; David 1999). As a consequence, structure of less complex phases can be determined from power diffraction alone (fewer atoms in the asymmetric unit of the unit cell). However, refinement of the structure is not limited so seriously with resolution issues, so powder diffraction data are used in Rietveld refinement more frequently than in structure determination. Electron powder diffraction patterns can be processed and refined using public domain computer programs. The first successful applications of electron diffraction in this field were demonstrated on fairly simple structures.

Acknowledgement
    Financial support of the Hungarian National Science Fund (contract number OTKA T043437) is acknowledged.

Further reading
Admans G. ed. (2000), ESRF Highlights 1999, ESRF, Grenoble
Altomare A, Cascarano G, Giacovazzo C, Guadliardi A and Moliterni GG, (1995) J. Appl. Cryst. 28, 738-744
Altomare A, Burla MC, Camalli M, Cascarano G, Giacovazzo C, Guagliardi A, Moliterni AGG, Polidori G and Spagna R (1999) J. Appl. Cryst. 32, 115-118
Belletti D, Calestani G, Gemmi M and Migliori A (2000), Ultramicroscopy 57-65
Burla MC, Camalli M, Carrozzini B, Cascarano GL, Giacovazzo C, Polidori G, and Spagna R (2000), Acta Crystallogr. A, 56, 451-457
Caglioti G, Paoletti A and Ricci FP, (1958), Nucl. Instrum. Methods, 35, 223-228
David WIF (1999) J. Appl. Cryst. 32, 654-663

David WIF, Shankland K, McCusker LB and Baerlocher Ch, Structure Determination from Powder Diffraction Data, IUCr Monographs on Crystallography 13, Oxford University Press, 2002

Giacovazzo C (1998) Direct Phasing in Crystallography. Fundamentals and Applications, Oxford University Press

Jansen, MSLS, This volume

Jiang JS and Li FH (1984) Acta Phys. Sin. 33, 845-849

Kariuki BM, Calcagno P, Harris KDM, Philip D and Johnston RL (1999), Angew. Chem. 38, 831-835

Kilaas R and Marks L (2004), http://www.numis.northwestern.edu/edm

Le Bail A, Duroy H, and Fourquet JL (1988), Mater. Res. Bull. 23, 447-452

Lábár JL, Proc. EUREM 12, Brno (Frank L and F Ciampor , eds.) (2000), Vol III., I379-380

Masciocchi N, Cairati P, Carlucci L and Mezza G (1996), J. Chem. Soc., Dalton Trans. 13, 2739-2746

Morris RE, Owen JJ, Stalick JK and Cheetam AK (1994), J. Solid State Chem. 111, 52-57

Nelder JA and Mead R (1995) Computer Journal 7, 308-313

Nicolopoulos, This volume

Pawley GS (1981), J. Appl. Cryst. 14, 357-361

Peschar R, Etz A, Jansen J and Schenk H (2002), in Structure Determination from Powder Diffraction Data (Eds: David WIF, Shankland K, McCusker LB and Baerlocher Ch), 179-189

Rietveld HM (1967), Acta Crystallogr. 22, 151-152

Rietveld HM (1969), J. Appl. Cryst. 2, 65-71

Rodrigez-Carvajal J, Laboratoire Léon Brillouim, CEA Saclay, France, http://www-llb.cea.fr/fullweb/fp2k/fp2k.htm

Tonejc AM, Djerdj I and Tonejc A, (2002) Mater. Sci. and Eng. C 19, 85-89

Weeks CM, De Titta GT, Hauptman HA, Thuman P, and Miller R (1994), Acta Crystallogr. A, 50, 210-220

Weickenmeyer A and Kohl H (1991), Acta Cryst. A 47, 590-597

Weirich T.E, Zou X, Ramlau R, Simon A, Cascarano GL, Giacovazzo C and Hovmöller S (2000), Acta Crystallogr. A, 56, 29-23

Weirich T.E., Winterer M, Seifried S, Hahn H and Fuess H, (2000a), Ultramicroscopy 81, 263-270,

Young RA, Ed. (1993), The Rietveld Method, IUCr Monography 5, Oxford Science Publication.

Zou XD, Shukarev Y and Hovmöller S (1993), Ultramicroscopy, 52, 436-444.

GAS-PHASE ELECTRON DIFFRACTION FOR MOLECULAR STRUCTURE DETERMINATION

ISTVÁN HARGITTAI

Institute of General and Analytical Chemistry, Budapest University of Technology and Economics and Structural Chemistry Research Group of the Hungarian Academy of Sciences at Eötvös University, PO Box 91, H-1521 Budapest, Hungary

Abstract: Gas-phase electron diffraction is a unique source of structural information on free molecules. The present introduction gives a brief overview of the technique.

Key words: gas diffraction, molecular geometry, conformational equilibria, electron diffraction

1.   INTRODUCTION

Gas-phase electron diffraction is 75 years old and going strong. It is a remarkably interdisciplinary technique that bridges the heroic beginnings of structural chemistry and present-day sophisticated applications of multi-hyphenated techniques in the determination of molecular structure. It also bridges accurate structural information with our understanding of the nature of chemical bonding and a wealth of molecular properties. Linus Pauling made ample use of gas-phase electron diffraction data in his eternal classic *The Nature of the Chemical Bond*.[1] He did not just incorporate the information from the literature, he himself collected data from a rudimentary technique to which he had added important improvements. Today gas-phase electron diffraction is a source of structural information on a par with the most advanced quantum chemical computations and high-resolution spectroscopy. For electron crystallography and the structure determination of nanosized materials, gas-phase electron diffraction may seem to be on the periphery. Nonetheless, it is an integral part of both. It has fundamental

197

T.E. Weirich et al. (eds.), Electron Crystallography, 197–206.
© 2006 *Springer. Printed in the Netherlands.*

importance in the utilization of electron scattering in structure elucidation.[2] It also had pioneering contributions to the structure elucidation of nanosized materials.[3]

Gas-phase electron diffraction has several roots. It goes back to Peter Debye[4] who examined the diffraction of X-rays by randomly oriented rigid systems of electrons and found that the interference effects do not cancel in the scattered intensity distribution. It goes back to de Broglie's discovery about the wave nature of moving electrons. And it goes back to the first gas-phase electron diffraction experiments by Herman Mark and Raimund Wierl[5] from which the geometries of a set of simple molecules emerged.

The technique is based on the phenomenon that a beam of fast electrons is scattered by the potential from the charge distribution in the molecule. The molecular geometry can be extracted from the interference pattern and this geometry is characterized by the relative positions of the atomic nuclei and their motion. In terms of internal coordinates – the chemist's language – this is bond lengths, bond angles, and angles of torsion, augmented by mean vibrational amplitudes. In the best cases, the accuracy is in the region of a few thousandths of an angstrom for bond lengths and a few tenths of a degree for bond angles. For conformational equilibria, the presence of more than one form of internal rotation may be detected, their relative abundance may be determined, and their energy differences may be estimated. In fact, it was gas-phase electron diffraction that led to the discovery of conformational equilibria in the first place.

The present chapter provides an overview of the technique and its application, its purpose is to give an impression for the newcomer, but is not meant to be a guide for work with the technique. A two-volume monograph[6] by experts is available for advanced study of gas-phase electron diffraction and its stereochemical applications. Gas-phase electron diffraction has also been described in the context of other techniques[7] of structure determination with emphases of producing accurate structural information.

2. INTENSITY DISTRIBUTION OF ELECTRON SCATTERING

This expression is given here, without derivation, for the molecular contribution to the total electron scattering intensities (molecular intensities):

$$I_{\mathrm{m}}(s) = \frac{K^2 I_0}{R^2} \sum_{\substack{i=1 \\ i \neq j}}^{N} \sum_{j=1}^{N} g_{ij}(s) \exp(-\tfrac{1}{2} l_m^2 s^2) \frac{\sin\left[s\left(r_a - \kappa s^2\right)\right]}{s r_e} \qquad (1)$$

Here, K is a constant, I_0 is the incoming intensity, R is the distance of the scattered wave from the molecule (in practical terms, it is the distance between the scattering center and the point of observation), i and j are the labels of atoms in the N-atomic molecule, g contains the electron scattering amplitudes and phases of atoms, s is a simple function of the scattering angle and the electron wavelength, l is the mean vibrational amplitude of a pair of nuclei, r is the internuclear distance (r_e is the equilibrium internuclear distance and r_a is an effective internuclear distance), and κ is an asymmetry parameter related to anharmonicity of the vibration of a pair of nuclei.

The expression refers to vibrating molecules in random orientation in the gas phase, using the independent-atom model. The latter means that the atoms are supposed to retain their spherical symmetry of their electron-density distribution. This model thus ignores the deformation of the electron-density distribution due to chemical bonding. When such an effect is important, this model gives poor results. The complex atomic electron scattering amplitudes take account of intra-atomic multiple scattering; however, they ignore intra-molecular multiple scattering and may be inadequate for molecules containing one or more heavy atoms. Relativistic effects, polarization of atoms by scattered electrons, and the effect of electron exchange are usually ignored, but they also may have to be included in special investigations.

The g functions come from theory and three kinds of parameters are determined from an electron diffraction experiment, r (geometry), l (motion), and κ (anharmonicity of vibrations).

3. EXPERIMENT

The general requirements include a high-energy (typically 50 keV) monochromatic electron beam focused on the plane of registration, a minimal (but optimal) scattering volume where the electron beam crosses the beam of molecules, and the detection of the maximal amount of scattered electrons. The scheme of the traditional experiment is given in Figure 1 in which a photographic plate is used for registration and a rotating sector compensates for the steeply decreasing intensities; however, more sophisticated technologies may be used for both purposes.

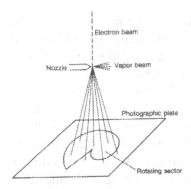

Figure 1. Scheme of traditional experiment.

An adequate vapor pressure of the compound to be studied is a prerequisite for the experiment; a typical vapor pressure is about 15 torrs; for heavier targets lower, for molecules of light atoms higher vapor pressure is needed. If there is uncertainty in the vapor composition or it is necessary to optimize it, a simultaneous mass spectrometric investigation, ultimately, a combined mass spectrometric-electron diffraction experiment[8] may be useful. Such a combined scheme is shown in Figure 2.

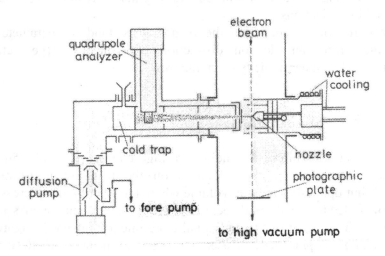

Figure 2. Scheme of a combined electron diffraction and quadrupole mass spectrometric experiment.

4.   STRUCTURE ANALYSIS

If a photographic plate is used for registration, first the optical density distribution of the electron diffraction pattern is determined, followed by various corrections, and the total electron diffraction intensity distribution is produced. The next step is the elimination of the contribution of atomic scattering and extraneous scattering, collectively called the background. The molecular contribution to the electron diffraction intensities – in short, molecular intensities – are considered to be the experimental data on which the determination of structural parameters are based, more often than not using a least-square refinement technique. Fourier transformation of the molecular intensities yields the so-called radial distribution, *f(r)*, this name is a misnomer because in reality it is related to the probability density distribution of the intramolecular internuclear distances. The total intensities, molecular intensities, and radial distribution for tetramethylsilane are displayed in Figures 3, 4, and 5.[9]

Figures 4 and 5 show not only the experimental distributions but also the distributions calculated for the best model of tetramethylsilane, which is characterized by the following bond lengths and bond angle and T_d symmetry and staggered methyl conformation, Si-C 1.877(4)Å, Si-H 1.110(3)Å, and Si-C-H 111.0(2)°. These are so-called average parameters (*vide infra*). The radial distribution is convenient to visually inspect the validity of a model and to read off some principal internuclear distances, but the quantitative determination of all the parameters is done on the basis of the molecular intensities. The refinement of parameters usually starts from an initial set of parameters. The expression of the molecular intensities is a non-linear relationship, a good choice of the initial parameters will ensure that the calculation reaches the global rather than a local minimum.

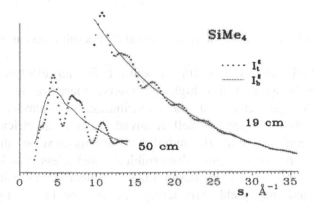

Figure 3. Total experimental intensities and background lines for tetramethylsilane. The two curves refer to two different scattering point to registration plane distances.[9]

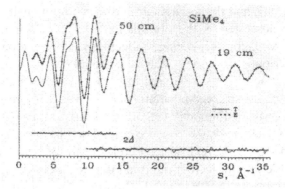

Figure 4. Molecular intensities (E – experimental and T – theoretical) for tetramethylsilane.[9]

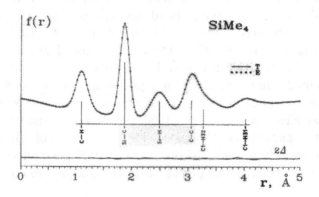

Figure 5. Radial distributions (E – experimental and T – theoretical) for tetramethylsilane. [9]

Tetramethylsilane is a fortunate subject for an electron diffraction investigation because it has high symmetry, therefore few parameters determine the geometry and the contributions of almost all of the internuclear distances appear well resolved. Less symmetrical and more complex molecules have usually radial distributions on which the individual contributions are not so easily discernible. Nonetheless, the least-squares refinement of the parameters based on the molecular intensities may be carried out and may yield satisfactory results. For increasing molecular complexity decreasing accuracy of the parameters may be the price to pay, but different investigations may be aiming at different goals. In any case, the gas-phase data yield – in the usual terminology of X-ray crystallography – a one-dimensional Patterson function, which indicates the limitations of the technique towards higher complexity of structures.

The vapor sample under investigation may not contain only one kind of species. It is desirable to learn as much as possible about the vapor composition from independent sources, but here the different experimental conditions need to be taken into account. For this reason, the vapor composition is yet another unknown to be determined in the electron diffraction analysis. Impurities may hinder the analysis in varying degrees depending on their own ability to scatter electrons and on the distribution of their own internuclear distances. In case of a conformational equilibrium of, say, two conformers of the same molecule may make the analysis more difficult but the results more rewarding at the same time. The analysis of ethane-1,2-dithiol data collected at the temperature of 343 kelvin revealed the presence of 62% of the anti form and 38% of the gauche form as far as the S-C-C-S framework was concerned.[10] The radial distributions calculated for a set of models and the experimental distribution in Figure 6 serve as illustration.

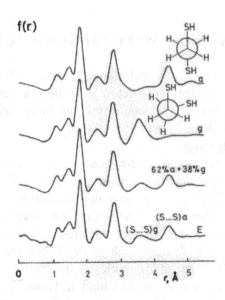

Figure 6. Ethane-1,2-dithiol has a mixture of the anti (a) and gauche (g) forms at the experimental temperature of 343 K, with respect to the S-C-C-S framework. The top three curves were calculated for models, the bottom curve is the experimental radial distribution.[10]

5. AVERAGE STRUCTURES

The r_a internuclear distance in the expression of molecular intensities (*vide supra*) is an effective parameter without rigorous physical meaning. It is, however, related in a good approximation to the thermal average internuclear distance ($r_g = r_a + l^2/r_a$). The thermal average distance (r_g) is different from the distance between average nuclear positions (r_α) due to the effects of perpendicular vibrations (perpendicular, that is, to the distance connecting the two respective nuclei). This is still a thermal average, and it can be extrapolated to zero kelvin, r_α^0. The latter differs from the equilibrium distance, r_e, by the consequences of anharmonicity of the vibrations along the distance connecting the two nuclei.

The following table summarizes the consequences of various vibrational effects on the internuclear distance types enumerated above:

Table 1. Representations of average structures

	r_g	r_α	r_α^0	r_e
Perpendicular vibrations	yes	no	no	no
Temperature	yes	yes	no	no
Isotope	yes	yes	yes	no

6.   COMPARISON WITH OTHER TECHNIQUES

In this section, we give a rudimentary survey of the capabilities of various techniques of molecular structure determination and the potentials of their concerted application, concentrating on three principal techniques. More comprehensive discussions are available. [7,11]

Microwave spectroscopy

It uses the pure rotational spectra and more generally, high-resolution rotational spectroscopy may be considered that is not restricted to the microwave region. Microwave spectroscopy is capable of investigating only molecules that have a permanent electric dipole moment, which excludes the most symmetrical molecules. In this, electron diffraction and microwave spectroscopy nicely complement each other. There are even more severe limitations as to the size and complexity of molecules amenable to a microwave spectroscopic structure elucidation. However, often, the patterns of correlation among the parameters are complementary that makes the joint application of the two techniques particularly advantageous. The internuclear distances determined from microwave spectroscopy may not have a rigorous physical meaning unless the row data had been corrected for vibrational effects. There is one representation that can be reached both in microwave spectroscopy and electron diffraction, viz., the distance between average nuclear positions extrapolated to zero kelvin, labeled r_α^0 in electron diffraction and r_z in microwave spectroscopy.

X-ray Crystallography

X-ray diffraction provides distances between the centroids of the electron density distribution and if the effects of its deformation are eliminated, they correspond to a good approximation to the distances between average nuclear positions. Gas-phase studies are not practicable, but crystallographic studies are the most widespread among molecular structure determinations. X-ray crystallography is well suited for investigating molecular systems of ever-increasing size, including large biological molecules. One of the caveats in the comparison of gas-phase electron diffraction and X-ray crystallographic data is that the consequences of intermolecular interactions need to be considered in the crystal. However, this difference between the two sets of data may also be a source of evaluating the impact of intermolecular interactions on the structure of individual molecules.

Quantum Chemical Calculations

This technique in many respects is becoming the most natural partner for gas-phase electron diffraction studies. For the time being, the overwhelming majority of quantum chemical calculations refer to the isolated, that is, gas-phase molecule. The representation of geometry is well defined and it is the equilibrium geometry (r_e distances). The capabilities of quantum chemical calculations are rapidly expanding, they are fast and they are relatively inexpensive.

Further techniques that are in close interaction with gas-phase electron diffraction are vibrational spectroscopy, mass spectrometry, NMR spectroscopy, and to a lesser extent some other techniques.

7. CONCLUDING REMARKS

Gas-phase electron diffraction is the technique of choice for many special problems of molecular structure determination. However, it has not become a mass-producing technique like X-ray crystallography or the quantum chemical calculations. With the proliferation of quantum chemical calculations some of the problems, namely, the accurate determination of relatively simple organic molecules that used to be solved by gas-phase electron diffraction have moved to the realm of these calculations. There are a wealth of other problems, mainly in inorganic chemistry,[12,13] that still necessitate the application of this rather demanding but instructive and amazing approach.

Acknowledgements

Our research is being supported by the Hungarian National Scientific Funds (OTKA T037978 and T046183).

References

1 Pauling, L., The Nature of the Chemical Bond and the Structure of Molecules and Crystals: an Introduction to Modern Structural Chemistry, Third Edition. Cornell University Press, Ithaca, New York, 1962.
2 Cowley, J. M., Ed., Electron Diffraction Techniques, Vols. 1 and 2. Oxford University Press (International Union of Crystallography Monographs on Crystallography Vol. 1), 1992.
3 Hargittai, I., "Lessons of a discovery. Fullerenes and other clusters in chemistry." In: Large Clusters of Atoms and Molecules. T.P. Martin, ed. pp. 423-435. NATO ASI Series E: Applied Sciences, Vol. 313. Kluwer Academic Publishers: Dordrecht, Boston, London, 1996.
4 Debye, P., Ann. Phys. **1915**, 46, 809.
5 Mark, H. F., Wierl, R., Z. Phys. **1930**, 60, 741.
6 Hargittai, I., Hargittai, M., Eds., Stereochemical Applications of Gas-Phase Electron Diffraction. Part A: The Electron Diffraction Technique; Part B: Structural Information for Selected Classes of Compounds. VCH, New York, 1988.
7 Domenicano, A., Hargittai, I., Eds., Accurate Molecular Structures: Their Determination and Importance. Oxford University Press (International Union of Crystallography Monographs on Crystallography Vol. 1), 1992.
8 Hargittai, I., Bohátka, S., Tremmel, J., Berecz, I., Hung. Sci. Instrum. **1980**, 50, 51.
9 Campanelli, A. R., Ramondo, F., Domenicano, A., Hargittai, I., Struct. Chem. **2000**, 11, 155.
10 Schultz, G., Hargittai, I., Acta Chim. Acad. Sci. Hung. **1973**, 75, 381.
11 Domenicano, A., Hargittai, I., Eds., Strength from Weakness: Structural Consequences of Weak Interactions in Molecules, Supermolecules, and Crystals. NATO Science Series. II. Mathematics, Physics and Chemistry – Vol. 68. Kluwer Academic Publishers, Dordrecht, Boston, London, 2002.
12 Hargittai, M., "Metal Halide Molecular Structures." In: Domenicano, A., Hargittai, I., Eds., Strength from Weakness: Structural Consequences of Weak Interactions in Molecules, Supermolecules, and Crystals. NATO Science Series. II. Mathematics, Physics and Chemistry – Vol. 68. Kluwer Academic Publishers, Dordrecht, Boston, London, 2002, pp. 191-211.
13 Hargittai, M., Chem. Rev. **2000**, 100, 2233.

PHASE IDENTIFICATION BY COMBINING LOCAL COMPOSITION FROM EDX WITH INFORMATION FROM DIFFRACTION DATABASE

JÁNOS L. LÁBÁR

Research Institute for Technical Physics and Materials Science, Budapest, Hungary and Department of General Physics, Eötvös Lóránd University, Budapest, Hungary

Key words: electron diffraction, phase identification, EDX

1. INTRODUCTION

Local composition is very useful supplementary information that can be obtained in many of the transmission electron microscopes (TEM). The two main methods to measure local composition are electron energy loss spectrometry (EELS), which is a topic of a separate paper in this volume (Mayer 2004) and x-ray emission spectrometry, which is named EDS or EDX after the energy dispersive spectrometer, because this type of x-ray detection became ubiquitous in the TEM. Present paper introduces this latter method, which measures the X-rays produced by the fast electrons of the TEM, bombarding the sample, to determine the local composition. As an independent topic, information content and usage of the popular X-ray powder diffraction database is also introduced here. Combination of information from these two sources results in an efficient phase identification. Identification of known phases is contrasted to solving unknown structures, the latter being the topic of the largest fraction of this school.

T.E. Weirich et al. (eds.), Electron Crystallography, 207–218.

2.   THE EDX METHOD

2.1   Production of X-rays

The primary process, leading to the production of element specific compositional information is the ionization of an inner shell in any of the atoms of the sample by the fast electrons of the TEM. The primary process of ionization is studied by EELS. Since an atom, ionized in an inner shell is unstable, de-excitation takes place within 10^{-14} s after the ionization. There are two alternative ways of de-excitation.

The first of them is a single electron process, in which an electron from a less-bound outer shell occupies the more-bound state that was emptied by the ionization. All the other electron states within the atom can be regarded unaltered to a good approximation. The difference between the energies of the initial and final states in this radiative transition is emitted in the form of a photon. Due to the large binding energies of the inner shells involved, the energy of the photon is in the X-ray range. Since energies of the atomic electron shells are quantized, the emitted photon has a well-defined energy, which is characteristic of the emitting atom.

The alternative way of de-excitation is a two-electron process. The energy difference in a transition, from the outer shell state to the inner shell electron state, emptied by the ionization, is carried away by emitting one of the loosely bound electrons of the same atom. All the other ("observer") electron states within the atom can be regarded as unaltered. The kinetic energy of the emitted (so called Auger) electron is also characteristic of the emitting atom and is used for chemical analysis in separate instruments. Since the application of Auger-electron spectroscopy (AES) is rare in the TEM, we are not elaborating this non-radiative process. The fraction of radiative de-excitations is a material-constant for a given atom and it is called the fluorescence yield (ω). $\omega = N_{radiative\ de\text{-}excitation}/N_{Total\ ionization}$. The atomic shells corresponding to principal quantum numbers 1, 2, 3, etc. are designated K, L, M, etc. All shells, except K, have several subshells, with slightly differing energies and the fluorescence yield is related to the subshell (ωK, ωL_{III}, ωM_V, etc.). Less bound electrons (light elements, outer shells) are more prone to non-radiative de-excitation, so their fluorescence yields are low (Figure 1). This fact presents the first problem with EDX analysis of the light elements.

Due to the presence of the several electron shells, a series of different X-ray lines are produced in cascades following inner shell ionizations. Notation of the lines corresponds to historical reasons (α is the strongest, β is the next strongest, etc.) and (in case of some minor lines) it is not always logical. The whole of this multitude of lines acts as a fingerprint, identifying the emitting atom. One of the lines is selected as the analytical line, the intensity of which is used to deduce the concentration of the element.

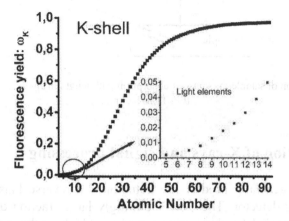

Figure 1. K-shell fluorescence yields as a function of the atomic number. Light element region is magnified. It can be seen that only a few photons are emitted out of one thousand ionizations in case of the lightest elements. As a consequence, analysis of light elements with EDX is less efficient than with EELS or AES.

2.2 Self-absorption of X-rays

Not all the photons propagating in the direction of the detector will reach the detector, since the radiation is attenuated by self-absorption of the emitted X-ray photons within the emitting sample itself. The attenuation is described by Beer's law: $I \sim I_0 * exp(-\mu * L)$, where μ is the absorption coefficient (material constant) and L is the absorption path length. The absorption path-length within the sample depends on the depth of generation and the direction to the detector, as shown in Figure 2. For a regular thin TEM sample with planparallel surfaces both of these quantities are related to local sample thickness in a simple way. From the analytical point of view, the sample is called thin if the effect of self-absorption on the measured number of photons is negligible. The fulfillment of this thin-film criterion is not only a function of the materials constants of the sample (absorption coefficients), but also of the geometry of the measurement (position of the detector, tilt of the sample). The two criterions, regarding the sample thin from the point of view of image formation on the one hand, and regarding the sample thin from the point of view of X-ray absorption, are unrelated.

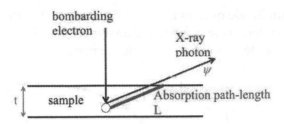

Figure 2. Relation of sample thickness to absorption path length in an idealistic, plan-parallel thin TEM sample.

2.3   Detection of X-rays and signal processing

The most generally used X-ray detector is a reverse biased p-i-n type semiconductor detector. The overwhelmingly large fraction of the detector volume is occupied by the intrinsic region, which is the active detecting element. The p-type and the n-type layers together with their metallic contacts are only needed to complete the device, but are only partially active and inactive in the detection process itself. The intrinsic region of the reverse biased diode is depleted of charge carriers. If an ionizing radiation (e.g. a fast electron) enters that region, a large number of electron-hole pairs are generated in the semiconductor and they are separated by the electric field of the reverse bias (Figure 3). Since the cross sections of the different scattering processes are well defined for a given detector material (e.g. Si), the amount of energy [that is spent in the different scattering processes while one electron-hole pair is generated] is (statistically) well defined. Consequently, the number of electron-hole pairs will be proportional to the total energy lost by the fast electron. A charge sensitive preamplifier, connected to the diode collects this charge and passes it to the main amplifier. The main amplifier measures the quantity of collected charge (corresponding to the energy of the detected particle) and stores the results of the separate measurements in a multi-channel analyzer (MCA), whose channels correspond to pre-selected energy intervals. This process assumes that the particles arrive at time intervals larger than the system's processing time (which is typically in the microseconds range). To ensure separate detection of particles, the electronic input is blocked until the processing of the charge from the previous particle is finished. If a second particle arrives within the processing time of the previous one, this "dead-time" of the system is elongated and particle-detection is not recorded. A too high input count rate can block the system completely, resulting in 100% dead-time and producing zero output counts. Recorded counts can be compared on the basis of the "live-time" (=real-time − "dead-time") spent while they were recorded. The sequence of particles with different energy is random, so the complete energy dispersive spectrum builds up parallel when different channel contents increase in random sequence.

When the particle to be detected is not an electron, but an X-ray photon, the detection is a two-step process. The absorbed photon creates a photo-electron, whose energy is only less than that of the photon by the binding energy of the electron. Since the de-excitation of the Si atom in the detector takes place within 10^{-14} s, the energy lost by the Auger electron (emitted in the most probable of the consecutive de-excitation processes) will also generate electron-hole pairs within the time of detection, so the two sets of electron-hole pairs are counted together, making the recorded total signal proportional to the original photon energy. In the less frequent case when the de-excitation of the detector's atom takes place with the emission of an Si-K photon, this photon is most probably absorbed with the creation of another photoelectron (from a less bound shell of the Si) within the detection time, so the detection of the total charge (corresponding to the energy of the originally detected photon) is ensured. In the very rare case (about 1 from a thousand) when this Si-K photon escapes the detector (in contrast to being absorbed), the charge recorded will be less than the energy of the original photon by the energy of an Si-K, i.e. 1.74 keV. It means that a tiny "escape peak" appears 1.74 keV left to the original peak in the MCA, as an artifact. Its proportion to the "mother-peak" is constant for a given detector. Its presence is easily recognized and can be corrected for. Escape peaks are more significant for Ge detectors than for Si detectors.

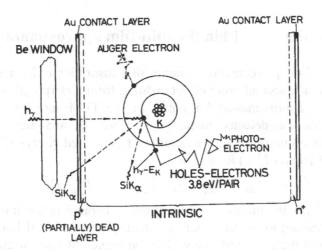

Figure 3. The EDX detector is a reverse biased p-i-n diode, with a charge-sensitive preamplifier attached to its contacts.

The width of the measured X-ray line is broader than the natural line width by about two orders of magnitude. The measured line width (so the spectral resolution of the detector) is determined by the statistical fluctuation of the number of electron-hole pair produced together with an additional electronic noise from the processing. In order to reduce the number of thermally generated electron-hole pairs and to reduce electronic noise, the detector and the charge-sensitive preamplifier are cooled to liquid nitrogen temperature. (Cooling is also needed by additional technical aspects, like immobilizing Li atoms in certain types of detectors, or preventing Al-contact from being oxidized by residual water-vapor in the vacuum in other types of detectors. The mentioned Li atoms are introduced into older types of Si-detectors to produce the intrinsic region by artificially compensating the B-atoms present due to the deficiencies in the cleanliness of the Si material used.)

3.   QUALITATIVE AND QUANTITATIVE ANALYSIS

Qualitative analysis is manifested in the identification of the elements present. It is based on Moseley's law, which points out that the energies of a pre-selected line-type (e.g. $K_{\alpha 1}$) lie on a monotonic, smooth curve as a function of the atomic number. Simultaneous read-out of the positions of the many lines present in the EDS spectrum acts as identifying fingerprints and results in a list of the element present in the excited volume.

3.1   Quantitation within the thin-film approximation

The thin film approximation assumes that absorption of X-rays within the sample (and any second order effect ensuing from absorption) is negligible. It is a good approximation for many of the TEM samples. Within this approximation the detected intensity (I_{Aa}) for the analytical line (a) of element A is proportional to the number of generated X-rays (G_{Aa}) and the detection efficiency (P_{Aa}) for this line.

$$I_{Aa} \propto G_{Aa} \cdot P_{Aa} = V_{Aa} \cdot \omega_{Aa} \cdot R_{Aa} \cdot P_{Aa},$$

where V_{Aa} is the number of ionizations generated in the initial electron (sub)shell leading to the transition, producing the analytical line and R_{Aa} is the fraction of the analytical line within the branch of lines originating from the same initial state. The detection efficiency of the detector is only a function of the energy of the detected photon. As the simplest example, if the analytical line is $K\alpha$, and the element is Zn, $V_{ZnK\alpha}$ is the total number of ionizations in the K-shell of any Zn atom, taking into account all the Zn atoms within the excited volume, which is proportional to the fraction of Zn atoms in the excited volume and the local thickness (and the illuminated area of the sample). Since the two main lines originating from this K-shell are the $K\alpha$ and $K\beta$, $R_{ZnK\alpha}=I_{ZnK\alpha}/(I_{ZnK\alpha}+I_{ZnK\beta})$ and the fluorescence yield is ω_{ZnK}. (For the shake of simplicity Zn $K\alpha1$ and Zn $K\alpha2$ lines were treated here as a single $K\alpha$ line, in accordance with the resolution of a simple detector.) The energy of the Zn $K\alpha$ line is 8.638 keV and most detectors have 100% efficiency at that energy, $P_{ZnK\alpha}=P(8.638 \text{ keV})\cong1$.

In summary, the intensity of the analytical line of any of the elements present in the excited volume to the sample is determined by four groups of data. The first ("*Sample*") is only dependent on the sample as a whole (density, local thickness), so being common for all elements in the excited volume. The second ("*Factor$_{Aa}$*") only depends on atomic data, the third is the fraction of the given element (c_A) and the fourth is the detection efficiency at the energy of the detected line. (Mass fractions appear in these formulas, as explained in the Appendix.)

$$I_{Aa} = Sample \cdot c_A \cdot Factor_{Aa} \cdot P(E_{Aa})$$

The sample-dependent factor cancels out when a ratio of two similar intensity formulas (for two elements) is formed, leaving a simple formula connecting the ratio of measured intensities

$$\frac{I_A}{I_B} = \frac{c_A}{c_B} \cdot \frac{Factor_{Aa}}{Factor_{Bb}} \cdot \frac{P(E_{Aa})}{P(E_{Bb})} = \frac{c_A}{c_B} \cdot k_{AB}$$

For a given detector and a given pair of elements the last two factors give a single constant (k_{AB}) that can be treated as a relative sensitivity factor. Both that factor and the method obtained their names after the two people who introduced them, Cliff and Lorimer (1975). The simplicity originates from the fact that the k_{AB} factor does not depend either on the rest of elements also present in the sample or on the other parameters of the sample (thickness, density), as far as the thin film criterion is fulfilled. The Cliff-Lorimer factors can either be calculated using the known parameters of the detector or can be measured if a well-characterized thin film sample (standard) is available. In the first case the method is standardless. In the second case the known weight fractions and the measured intensity ratio provides the Cliff-Lorimer factor for the pair of elements.

3.2   Detector efficiency

Surface of the cooled detector would attract contamination from the residual gases of the vacuum. To prevent the detector from being contaminated, the vacuum space of the detector is separated from that of the microscope by a thin window. The window itself is thermally isolated from the cold detector, so it does not attract contamination. An unwanted by-product of the presence of this window is the absorption of the photons (to be detected) by the window. Softer radiation (of lighter elements) is affected more. This is the second problem with the analysis of light elements with EDS.

3.3   Beyond the thin-film approximation

If the effect of absorption can not be neglected in the sample, the ratio of measured intensities contains the ratio of two exponentials from Beer's law. It corresponds to a difference inside the exponential. In practice, it is the mass absorption coefficients (μ/ρ) and the absorption path lengths expressed in mass thickness (ρL) that are used.

$$\frac{I_A}{I_B} = \frac{c_A}{c_B} \cdot k_{AB} \cdot \exp\left\{-\left[\left(\frac{\mu}{\rho}\right)_B - \left(\frac{\mu}{\rho}\right)_A\right] \cdot (\rho L)\right\}$$

It can be seen that calculation of the absorption correction also needs the values of the sample density and the linear value of the absorption path length beside the mass absorption coefficients. It is also important to note that it is the difference of the absorption coefficients, what counts in determining if the thin film criterion is fulfilled for a given sample (geometry). It frequently happens that the sample is thin for one pair of its components and thick if another pair of elements is considered from its components.

4.   THE POWDER DIFFRACTION DATABASE

The first collection of the powder diffraction data appeared in 1938, known as ASTM cards. Later, the International Centre for Diffraction Data (ICDD) issued the data in electronic form, which became known as the Powder Diffraction File (PDF). Majority of the users currently use the so called Pdf-2 version, updated in different years. It contains both measured and computed diffraction data in a flat-file structure. Beside some additional data, a list of d-values, Miller indices and relative intensities (for XRD) is stored in these files. While updating and extension of the Pdf-2 database continued, ICDD introduced a new product in the year 2000 (Faber and Fawcett 2002). The new, Pdf-4 database is in a relational database format and also contains atomic coordinates for the structures (beside the previously present data).

4.1 Qualitative phase identification with the ProcessDiffraction program

The ProcessDiffraction program (Lábár, 2000) helps phase identification by comparing the diffracted intensity data corresponding to known phases to the measured diffracted intensity-versus-scattering angle distribution, obtained by integrating the experimental diffraction ring patterns circularly (or over ellipses), as shown in Fig. 4. The known data are presented graphically (as "Markers"), overlaid against the measured distribution. Identification is done by visual comparison. Data, relevant to the known phases can be taken from two alternative sources. On the one hand, it can be calculated within the ProcessDiffraction program for all phases, which are completely known (including composition and the atomic coordinates). On the other hand, data can be searched directly in the Pdf2 database of ICDD (ICDD, 1998) for partially known phases. The database is not supplied together with the ProcessDiffraction program, however, legal users of the database can use it from within ProcessDiffraction without any additional requirement. Search can be done on the basis of elemental composition, which can be taken from EDS or EELS analysis. Logic filters help locating the candidate phases. Reduction of the candidate list can also apply restrictions on the largest d-values or to the three strongest lines.

4.2 Quantitative phase analysis with ProcessDiffraction

When all the phases present were identified, we can quantify their volume fraction in the analyzed volume similarly to the way the Rietveld-method is used for phase analysis in XRD. A whole profile fitting is used in ProcessDiffraction, modeling background and peak-shapes, and fitting the shape parameters, thermal parameters and volume fractions. Since the kinematic approximation is used for calculating the electron diffraction intensities, the grain size of both phases should be below 10 nm (as a rule of thumb), to avoid serious dynamic effects. The total layer thickness must also be as low as possible (preferably below 20 nm).

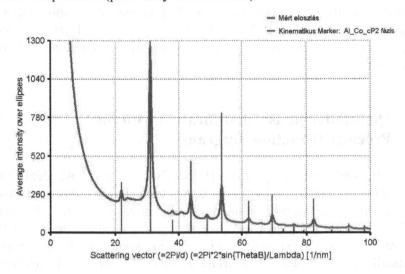

Figure 4. Usage of known information as "Marker" in the process of phase identification.

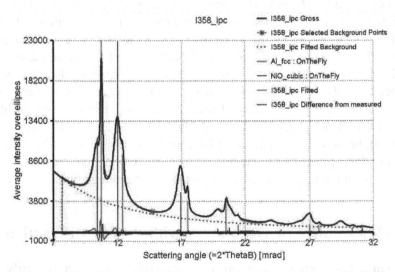

Figure 6. Measured distribution with fitted background and fitted whole-profile. Difference between measured and fitted curves is also shown. The agreement of measured volume fractions with the true values is within 2% relative.

As an example, result from a test sample is presented in Figure 6. A nano-crystalline thin film sample was prepared by vacuum evaporation of 100 Å NiO and 100 Å Al. The thicknesses were controlled by quartz thickness measurement. The proportion of phases was also checked by EDS. TEM BF and DF pictures show that the grain size of NiO is around 4 nm, while that of Al is around 10 nm. Quantitative phase analysis form the electron diffraction ring patterns (recorded in a TEM at 200 kV) was performed by ProcessDiffraction. The good fit of the profiles is shown in Fig. 6. The volume fractions were measured as 51 volume % of NiO and 49 % Al. The agreement is better than what can be expected in general. Extensive testing of the method is in progress. Texture-effects and dynamic corrections are to be incorporated in later versions.

5. APPENDIX: WHY INTENSITY IS PROPORTIONAL TO THE MASS FRACTION

There is a regularly returning confusion, why the generated X-ray intensity is proportional to the mass fraction of the elements present, in contrast to the atomic fraction. This fact is contradictory to our first physical anticipation, since the ionization cross section is defined for one atom, so it is the number of atoms that should be important (irrespective of their masses). It is a correct assumption and we recall it below, how the mass fraction results from the starting number of atoms.

Avogadro's number, N^0 (=6.02*10^{23}) gives the number of molecules in that many gramms of a chemical compound as the value of its molecular weight. If, for the sake of simplicity, we assume that the chemical formula of the molecule is X_nY_m than its molecular weight is nA_X+mA_Y and there are N^0n atoms of the X element and N^0m atoms of the Y element in the nA_X+mA_Y gramms of that compound. The important feature of the molecule here is the constant proportion of its constituents (and not the type of bond between them).

Let's take the excited volume in a thin layer first, for which we want to calculate the generated intensity during our X-ray analysis. The excited volume is assumed to be homogeneous, so it can be characterized by a constant composition (that is what we want to determine). Let the atomic fractions be c_i (i=1, *n* for the n elements present in the layer). Than, the material can be thought of being similar to a chemical compound (irrespective of the chemical bounds) made up of fixed "compositional blocks" with "block weight" $\Sigma c_i A_i$. If we take $\Sigma c_i A_i$ gramms of that material, it will contain N^0 "compositional blocks " [and (within that) $c_i N^0$ atoms of the ith element]. There will be $N^0/\Sigma c_i A_i$ "compositional blocks" in 1 gramms of this material and $\rho N^0/\Sigma c_i A_i$ "compositional blocks" in 1 cm^3. If we take a layer of thickness *t*, the number of "compositional blocks" is $t\rho N^0/\Sigma c_i A_i$ and within it, the number of the ith atoms is

$$N^0 \cdot \rho \cdot t \frac{c_i}{\sum_i c_i \cdot A_i} \qquad (3a)$$

per unit surface area. You can see that linear dependence on the number of atoms of an element in a given piece of material does not mean linear dependence on the atomic fraction of that element. Since the mass fraction can be calculated from the atomic fractions as $c^w_i = c_i A_i / \sum c_i A_i$, the number of the ith atoms in unit surface area of this layer can be rewritten as $t\rho N^0 c^w_i / A_i$.

The effective area of one atom from the point of view of ionization (as seen by one bombarding electron) is given by the ionization cross section, $Q(E)$. By multiplying one atom's area (cross section) with the number of atoms (in a unit surface area) we obtain the effective area (per unit surface area) of all the atoms seen by the bombarding electron:

$$Q(E) \cdot \left\{ N^0 \cdot \rho \cdot t \frac{c^w_i}{A_i} \right\} \qquad (3b)$$

The probability of ionization is given by the geometrical probability, i.e. by the ratio of that effective area to the total area considered, which is unit. That is why (3b) gives the ionization probability of the ith element per incident electron for the excited volume of our thin layer. Although the concentration dependence described in (3a) is identical to that given in (3b), it is expressed simpler in (3b), which describes a linear dependence on its variable (c^w_i) than how it is expressed in (3a) where both the numerator and the denominator depends on the variable (c_i).

So, we can see that linear dependence of the number of ionizations on the area density of the atoms (corresponding to a complicated dependence on the atomic fractions) translates into a simple linear dependence on the mass fraction for a thin layer.

Acknowledgment

Financial support of the Hungarian National Research Fund (contract number OTKA T043437) is acknowledged.

Further reading

Cliff G and Lorimer GW (1975) J. Microsc. 103, 203
Faber J and Fawcett T (2002) Acta Cryst. B 58, 325-332
Lábár JL, Proc. EUREM 12, Brno (Frank L and F Ciampor, eds.) (2000), Vol III., I379-380
Mayer 2004, this volume
S.J.B. Reed, Electron microprobe analysis, Cambridge University Press, 1993
D.B. Williams and C.B. Carter, Transmission electron microscopy, Plenum Press, 1996

ANALYSIS OF LOCAL STRUCTURE, CHEMISTRY AND BONDING BY ELECTRON ENERGY LOSS SPECTROSCOPY

Joachim Mayer
Gemeinschaftslabor für Elektronenmikroskopie
RWTH Aachen
Ahornstr.55, D-52074 Aachen
Germany

Abstract: In the present chapter, the reader will first be introduced briefly to the basic principles of analytical transmission electron microscopy (ATEM) with special emphasis on electron energy-loss spectroscopy (EELS) and energy-filtering TEM. The quantification of spectra to obtain chemical information and the origin and interpretation of near-edge fine structures in EELS (ELNES) are discussed. Special attention will be given to the characterization of internal interfaces and the literature in this area will be reviewed. Selected examples of the application of ATEM in the investigation of internal interfaces will be given. These examples include both EELS in the energy-filtering TEM and in the scanning transmission electron microscope (STEM).

Key words: Electron energy loss spectroscopy, EELS, ELNES, TEM

1. INTRODUCTION

In transmission electron microscopy, inner-shell excitation of the atoms by the beam electrons leads to characteristic edges in the electron energy-loss (EEL) spectrum. The onset energies of each edge can be used to identify the presence of individual chemical elements. The concentration of an element can be determined from an EEL spectrum if the pre-edge background is extrapolated and subtracted from the signal obtained above the edge. The near edge fine structures above the edge threshold reflect the

219

T.E. Weirich et al. (eds.), Electron Crystallography, 219–232.

densities of states in the conduction band. Using an energy filtering TEM, the information about the distribution of electrons which suffered a specific energy loss can be gathered in a two-dimensional way, a technique which is referred to as electron spectroscopic imaging (ESI). A higher spatial resolution, at the expense of serial spectra acquisition, can be achieved in the scanning transmission electron microscope (STEM). Using a high angle annular dark field detector, image information and spectra can be recorded at the same time. This makes it possible to perform atomic column resolved electron energy loss spectroscopy.

Over the past decade, analytical transmission electron microscopy (ATEM) has experienced a more rapid growth than any other major TEM technique. The main reasons for this development are the growing interest in the wealth of information that can be revealed by electron energy loss spectroscopy and the rapid spread of new instrumental developments, in particular field-emission guns and imaging energy filters. One of the trends in transmission electron microscopy is to consider a microscope not primarily as an instrument to obtain micrographs but as an experimental tool on which information from a sample can be obtained via various channels in parallel [1]. The channels are defined by the available detectors, such as two-dimensional detectors for imaging and diffraction, electron counting devices for STEM bright- and dark-field imaging, an electron energy-loss spectro-meter and an energy-dispersive X-ray spectrometer (EDS).

In solid state physics, the sensitivity of the EELS spectrum to the density of unoccupied states, reflected in the near-edge fine structure, makes it possible to study bonding, local coordination and local electronic properties of materials. One recent trend in ATEM is to compare ELNES data quantitatively with the results of band structure calculations. Furthermore, the ELNES data can directly be compared to X-ray absorption near edge structures (XANES) or to data obtained with other spectroscopic techniques. However, TEM offers by far the highest spatial resolution in the study of the densities of states (DOS).

2.   BASICS OF ELECTRON ENERGY-LOSS SPECTROSCOPY

2.1   Inelastic Scattering Processes

All analytical techniques in the TEM are based on the inelastic scattering of the fast beam electrons by the electrons of the atoms in the material investigated. The primary event in each case is the transfer of energy and momentum from the fast electron to a sample atom, thereby exciting the latter from its ground state to an excited state. The primary electron looses energy in this inelastic scattering event. The energy losses are characteristic of the element and the local coordination and bonding of the atoms in the sample. Measuring the energy-loss spectrum with appropriate detectors therefore allows the chemical composition and the bonding of materials to be analysed.

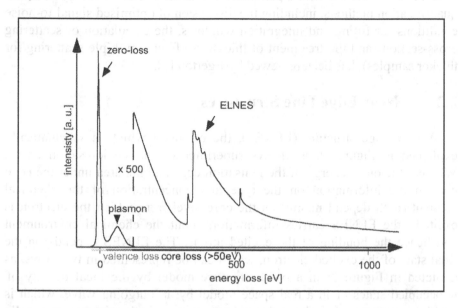

Figure 1. Schematic EELS spectrum for a wide range of energy losses covering the peak of the elastically scattered electrons (zero-loss peak), the range of the valence losses (plasmons, inter- and intraband transitions), and the core-ionization edges. EELS spectra have a high dynamic range. Note that the ionization edges are superimposed on a high background signal.

A schematic EELS spectrum for a wide range of energy-losses ΔE is shown in Figure 1, covering the peak of the elastically scattered electrons (zero-loss peak), the range of the valence losses (plasmons, inter- and intraband transitions), and the core ionisation edges for higher energy losses. The high dynamic range of the spectra makes it difficult to record low- and core-loss spectra (i.e. the spectra containing the zero-loss up to the valence losses and the core ionisation edges) in a single spectrum with good signal-to-noise ratio for the whole range of energy losses. Measurements with different acquisition parameters are therefore performed to obtain optimised low-loss and core-loss spectra. The ionisation edges are always superimposed on a background signal resulting from the tails of inelastic scattering processes with onset energies at lower energy losses. In order to use the ionisation edges for chemical quantification, this background has to be removed by fitting an analytical function to a pre-edge region and extrapolating it to the energy-loss range containing the ionisation edge. In most practical cases a function of the type

$$I(\Delta E, A, r) = A \exp(-r\Delta E) \tag{1}$$

where A and r are the fitting parameters, is used. The basic theory of EELS quantification methods, including the discussion of optimised signal-to-noise conditions for fitting and integration windows, the calculation of scattering cross-sections and the treatment of thickness effects (multiple scattering for thicker samples), has been reviewed by Egerton [2].

2.2   Near Edge Fine Structures

A near edge structure (ELNES), the origin of which is schematically explained in Figure. 2, is always superimposed on the ionisation edges. Whereas the onset energy of the ionisation edge and the area under the edge contain the information on the type and concentration of the chemical element (both depend mainly on the core level from which the electron is excited), the ELNES carries information about the chemical environment and hence the bonding of the excited atoms. The ELNES depends on the final state of the excited electron, which can be represented in two ways, as depicted in Figure 2: in a band structure model by the local density of unoccupied states or in a real space model by an outgoing wave, which is multiply scattered by the surrounding atoms. Both pictures are fully equivalent within the appropriate limits and may be used alternatively in trying to understand the observed ELNES.

According to Fermi's golden rule, the intensity variation $I(E)$ in the near-edge region is proportional to the density of unoccupied (i.e. final) states $N_u(E)$ and the square of the transition matrix element $M(E)$:

$$I(E) \propto N_u(E)|M(E)|^2 \tag{2}$$

(see e.g. ref. [3]). The transition matrix element is usually assumed to vary only slowly with energy and the ELNES hence provides a direct measure of the local density of unoccupied states. In addition, under normal experimental conditions, only allowed dipole transitions can be strongly excited, so that the actual measurement probes a site- and angular-momentum-resolved partial density of states. In these conditions the ELNES is fully equivalent to the near-edge structure in X-ray absorption spectra (XANES), which has been exploited for the experimental and theoretical understanding of the

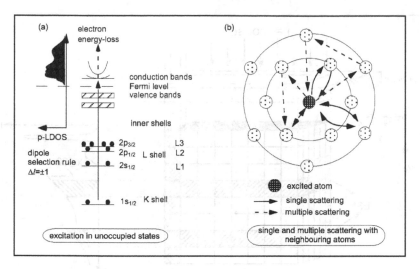

Figure 2. Schematic representation of two models for the origin of electron energy-loss near-edge structures (ELNES) for the core-ionization edges. (a) Transition of strongly bound core electrons into unoccupied states (b) multiple scattering description of ELNES.

observed ELNES in many different systems. The ELNES features of interest are usually found in the region between the edge onset and about 20-30 eV above the edge onset. Starting at about 50 eV above the edge onset, the fine structures are named EXELFS (extended energy-loss fine structures) and are again fully equivalent to the extended X-ray absorption fine structure (EXAFS). Both reflect the local arrangement of atomic scatterers and a quantitative analysis reveals information on the local pair correlation function. Since the ELNES has received far more interest than EXELFS in the last few years, only the former will be considered in the examples discussed in this chapter.

2.3   Data Acquisition: STEM and Energy Filtering TEM

Figure 3 depicts the three-dimensional data space that has to be evaluated in a spatially resolved quantitative ATEM analysis [4]. The graphical representation suggests that the intensity distribution $(x, y, \Delta E)$ can equally well be obtained by recording individual spectra or by employing the technique of electron spectroscopic imaging [5]. In the former case, which is illustrated in the examples of EELS in the dedicated STEM in chapter 4, only a small spot on the specimen is illuminated and EEL spectra are recorded, whereas in the latter case a large specimen area is illuminated and filtered images are recorded with an energy-selecting window of finite width δE. The theory and the application of ESI will be discussed in the next section.

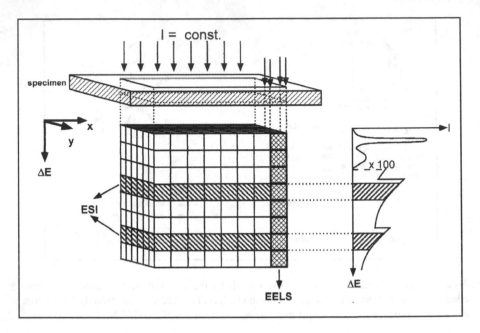

Figure 3. Three-dimensional data space constituting the basis of a spatially resolved quantitative TEM analysis.

The different approaches of ATEM are closely related to the available instrumentation. We distinguish serial and parallel EELS detectors, which allow individual spectra to be registered, and imaging filters, which allow images to be formed with electrons from a specific energy-loss window. Standard EELS detectors are fitted to the microscope below the viewing screen. Imaging filters can be placed either in-column or post-column. The most prominent examples of in-column filters are the Henry-Castaing electrostatic and the omega-type magnetic filters [6]-[8]. Several designs of post-column energy filters as well as parallel EELS spectrometers have been dveloped by Krivanek [5], [9]. The use of in-column and post-column imaging filters for ESI will be discussed in the next section.

3.   ENERGY FILTERING TEM

3.1   Elemental Mapping

In EELS analysis, the unspecific background under an edge has to be removed before the signal of the given element can be analyzed. In ESI images this has to be done for each individual pixel. The easiest method for background subtraction, the three window technique, is illustrated in Figure 4

[10]. Two ESI images are acquired in the background region before the edge and the extrapolated background is then subtracted from the ESI image containing the signal above the edge. As a result, the difference image only contains intensity in the areas where the corresponding element is present in the sample and one thus obtains a map of the distribution of this element. In the three window technique, however, the intensity in the difference image is not only depending on the concentration of the element but may also vary with thickness or Bragg orientation of the crystalline grains. Owing to the low intensity of the individual ESI images (up to a factor of 100 - 10 000x less than the corresponding bright field image) the difference image will also contain considerable noise which makes it impossible to detect elements in very small concentrations (below 1 at%). The noise in the elemental distribution images can be reduced by special image processing techniques, which may result in a loss of resolution. The detection limits in elemental distribution images are as low as one monolayer, e.g. for segregants at grain boundaries. The resolution limits are in the order of 1 nm, depending on the element and its concentration.

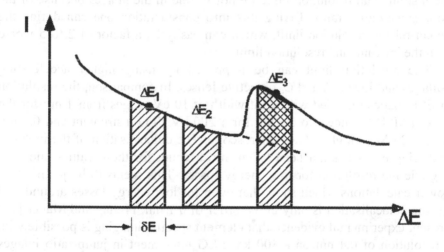

Figure 4. Three window technique used to obtain elemental distribution images.

In both techniques, the optimum position of the energy windows and their width depends on several parameters: 1) the intensity in the energy loss spectrum which decreases strongly with increasing energy loss, 2) the shape of the edge which only shows a sharp onset for the light elements, and 3) the width of the unstructured background region before the edge.

In most of the cases the distribution of several elements will be studied in one sample area. The resulting elemental distribution images can be combined in one image by using different colors for each element and overlaying the individual images. If two or more of the elements under investigation are present in one sample area, mixed colors will occur. Mixed colors thus reveal important information on the occurrence of phases which contain more than one of the elements under investigation [11].

3.2   Resolution Limits

The resolution limits in ESI images, or the elemental distribution images derived from them, is controlled by a number of factors: The ultimate limit is defined by the aberrations of the electron optical elements of the instrument and it is referred to as the instrumental resolution limit. However, there is also a degradation of the resolution by the delocalisation of the inelastic scattering process. Newer calculations show that this contribution is small and can be ignored for inner shell loss edges with energy losses of 100 eV and higher. In many cases, the dominating factor emerges from the statistical nature of the inelastic scattering processes and the weak signal resulting from the small inelastic scattering cross sections. Thus, structures close to the instrumental resolution limit are not visible in the images because of the poor signal/noise ratio. Taking this into consideration one can define the object-related resolution limit, which can easily be a factor of 2 to 5 worse than the instrumental resolution limit.

The resolution limit can be improved by using higher accelerating voltages and lower C_s and C_c objective lenses. In comparison, the resolution in ESI images recorded with a slit width of 10 eV ranges from 1 nm for the Zeiss EM 912 Omega, to 0.5 nm for a 200 kV FEG instrument and 0.3 nm for a 1.25 MeV high voltage microscope. The delocalisation of the inelastic scattering event has not been taken into account for these values and may degrade the resolution for low energy losses. This seems to be justified by newer calculations which show that even for low energy losses around 100 eV the delocalisation is only in the order of 0.1 nm. Freitag and Mader [12] present experimental evidence that element specific imaging is possible with a resolution of 0.4 nm on a 300 kV FEG instrument in jump-ratio images obtained with the B-K edge.

3.3   Quantitative Analysis of ESI Series

The energy loss spectra extracted from an ESI series with n images can graphically be visualized in several different ways. The data are obtained as intensities $I(\Delta E)$ integrated over the energy window width δE defined by the slit aperture. A simple plot would consist of a series of n data points which give the integrated intensities at the center positions ΔE_i ($i = 1 \dots n$) of the corresponding energy windows (Figure 5). Most analysis programs use a bar representation showing the intensities in steps with a width which corresponds to the energy increment. We modify this type of representation for the spectra in the core loss region by using linear interpolation between the individual data points (Figure 5). The resulting spectra resemble very closely the spectra which would be obtained with a parallel energy loss (PEELS) detector with much higher sampling frequency. At the present stage, we also use data produced by this linear interpolation for the quantitative analysis [13]. It can easily be shown that summing (or integrating) over all data points produced by linear interpolation exactly reproduces the original intensities, as long as the integration extends from one original data point to any other. Graphically this can also be seen from the equality of the two hatched triangles in Figure 5.

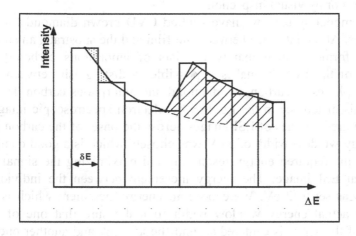

Figure 5. Spectrum extracted from a series of electron spectroscopic (ESI) images around a core loss edge. For a quantitative analysis a slit width is dE = 10...20 eV.

The background extrapolation and subtraction is performed via a power-law background fit. We have found that, using the spectra obtained by linear interpolation, very accurate background fits can be obtained. However, it should be kept in mind that the linear interpolation is only an approximation. To become more accurate, a modeling of the exact functional dependence of the intensity variation for windows with a finite width δE is required. As an experimental example, we want to report on investigations performed in the materials system Al_2O_3–Ti–Cu [14]. A thin interlayer of titanium is introduced between an Al_2O_3 substrate and a Cu metallisation layer to enhance the adhesion of the copper on the sapphire substrate [15]. Both, the titanium interlayer and the Cu film were grown by molecular beam epitaxy. The results show, that the thickness of the titanium layer can be controlled with an accuracy of better than one monolayer during the deposition process in the MBE machine [15].

3.4   Mapping of ELNES

From a whole series of ESI images, information on the ELNES can be retrieved for any given area in the image. The energy loss spectrum is obtained by simply extracting the intensity from the same area in the series of ESI images and plotting it as a function of the corresponding energy loss. Basically, this can be performed for each individual pixel in the images. However, the resulting spectra would be very noisy. In order to reduce the noise, the intensities have to be integrated over a certain area in the images. Prior to this, drift correction can be applied to the individual ESI images in order to align the corresponding areas properly in the series of images. The magnitude of the drift correction can either be determined by cross correlation or by visual inspection.

As a model system, we have studied CVD grown diamond films on Si substrates. At the interface between the film and the substrate, an amorphous layer is formed which mainly consists of amorphous carbon [16]. An analysis of the ELNES makes it possible to distinguish between the two different phases of carbon, i.e. diamond and amorphous carbon. In order to reveal this difference in the ELNES by electron spectroscopic imaging, we have acquired a series of ESI images across the onset of the carbon K-edge. An energy window width of 5 eV was chosen, which is a good compromise between the required energy resolution and maximizing the signal in each individual ESI image. The energy increment between the individual ESI images was set to 2 eV. We choose an energy increment which is smaller than the actual energy window width to make sure that one of the ESI images of the series is centered around the π^*-peak and another one around the σ^*-excitations. In the studies, we were able to distinguish between the two modifications of carbon with high spatial resolution [16].

4.   SPATIALLY RESOLVED EELS WITH FOCUSED PROBE

The major experimental difficulty in spatially resolved studies e.g. of interfaces is to record the very weak signal reliably from the interface with good signal-to-noise ratio and subsequently to distinguish clearly the interface signal from signals arising from the adjacent bulk phases. Various techniques to address this problem have been discussed in the literature. Using a focused probe in a STEM, the three main methods currently in use are:
−   Line scans of equidistant spot measurements across the interface.

- The spatial difference technique, which makes it possible to extract the interfacial signal from measurement areas with nm-size dimensions.
- Spot measurements with a probe focused on the interface. Using a probe with a diameter of 2 Å or less in combination with Z-contrast imaging makes it possible to perform atomic column-resolved EELS.

Each of these methods has distinct advantages and disadvantages and the best way to record the spectra will depend on the experimental situation and especially on the stability of the interface at the atomic level. A brief survey of the main applications of the three techniques will be given in the following.

In the line-scan method, pioneered by Colliex and co-workers [17,18] the electron probe is scanned across the region of interest (for example the interface) and a series of spectra recorded. Acquisition of line scans is the method of choice if gradual changes across an interface, resulting from inter-diffusion for example, are to be investigated [19]. The method has mainly been used to obtain profiles of the chemical composition of materials. However, the data can also be analysed with respect to changes of the ELNES at interfaces or to study valence losses (interface plasmons) at interfaces [20]. If interfacial ELNES is the aim of the analysis, the length of the line scan and the number of spectra recorded should be chosen such that the successive measuring spots (defined by the position increments and the effective probe diameter) overlap. A good spatial sampling should be reached. Even if only information on the interfacial layer itself is sought, experimental recording of a line scan has some advantages over a spot measurement: If the interface drifts during the measurement the line scan will nevertheless cross the interface in general. Inspection of the data then helps to identify the spectra containing ELNES components arising from the interface from among the larger data set.

Spectra can also be recorded not only along one line scan but by scanning a whole region of interest with the focused electron probe in a raster pattern. This data set is usually called a *spectrum image* [4],[21] and now represents the spectral intensity $I(\Delta E, x, y)$ with two spatial coordinates x and y. The method of recording this data cube (see Figure 3) sequentially, spectrum by spectrum (spectrum imaging), is complementary to the method of electron spectroscopic imaging, which has been introduced in Sect. 3 of this chapter. The advantage of the spectrum image approach to the filling of the data cube is that a complete spectrum is available for every pixel of the image. This can be useful to improve the background fitting or for the spatially resolved analysis of ELNES changes.

In spatial difference measurements, the beam is scanned within a rectangular area (box) during acquisition of the spectra. These box measurements are repeated on the interface and in the bulk material on both sides. Early applications of the spatial difference method can be found in [22]. Müllejans and Bruley [23] gave a systematic explanation of the method, which is presented schematically in Figure 6. The rectangular scanning area is used to reduce electron irradiation damage and to ensure that the interface is in fact included and will remain within the measuring area during acquisition. The disadvantage is that the contribution of the atoms i at the interface to the total signal I_{meas} measured while scanning the beam across the interface can become very small if larger box sizes are chosen. Even for boxes of a few nm width and length, the interfacial contribution can be as low as 10%. If we consider the interface signal as the useful signal and the bulk contributions which have to be subtracted as the background, the box size should be carefully chosen in such a way that the signal-to-background ratio will be high enough to ensure that a meaningful signal is obtained, compared to the statistical and systematic errors.

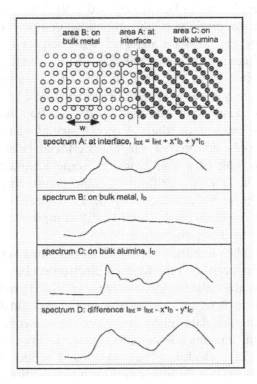

Figure 6. Schematic explanation of the spatial difference method using rectangular boxes as measuring areas for the interface and the two adjacent bulk phases.

When the electron probe of a STEM can reliably be positioned on the interface for the whole acquisition time, the spot measurement will necessarily contain a high contribution from the interface atoms. A measurement with sub-nanometre electron probes in a STEM in combination with high-angle annular dark-field imaging (Z-contrast imaging) constitutes the ultimate local interfacial analysis [24]. A prerequisite for such a measurement is that the atomic structure can be resolved while scanning the probe being used for the analysis across the interface.

5. CONCLUSION

In the present contribution we have presented two alternative approaches for a spatially resolved EELS analysis: ESI in an energy filtering TEM and the standard EELS method, in which a fine probe is stepped across a sample and EELS spectra are recorded consecutively. ESI makes it possible to obtain two-dimensional information in much shorter time than in the scanning approach. In comparison, the main advantages of a dedicated STEM are the higher energy resolution of ~ 0.5 eV and the better spatial resolution in the range of 0.2 - 0.5 nm. However, using an EFTEM with an FEG emitter a similar spatial and energy resolution can be reached in ESI studies which means that the ESI approach is clearly advantageous if two-dimensionally resolved information is sought. Or in other words, the use of a FEG-EFTEM will make it possible to select the most appropriate way to analyse the energy loss space in each case - via PEELS acquisition in spot mode or via ESI series in the TEM imaging mode.

References

[1] C. Jeanguillaume and C. Colliex, Ultramicroscopy 45, 205 (1992).

[2] R. Egerton, *Electron Energy-loss Spectroscopy in the Transmission Electron Microscope.* 2nd edition, Plenum Press, New York/London 1996.

[3] M. Inokuti,. Rev Mod Phys 43, 297 (1971).

[4] C. Jeanguillaume and C. Colliex, Ultramicroscopy 28, 252 (1989).

[5] O.L. Krivanek, A.J. Gubbens, and N. Dellby, Microsc Microanal Microstr 2, 315 (1991).

[6] S. Lanio, Optik 73, 99 (1986).

[7] W. Probst, G. Benner, J. Bihr, and E. Weimer, Adv Mater 5, 297 (1993).

[8] M. Tanaka, K.Tsuda, M. Terauchi, K. Tsuno, T. Kaneyama, T. Honda, and M. Ishida, J Microscopy 194, 219 (1999).

[9] O.L. Krivanek, A.J. Gubbens, N. Dellby, and C.E. Meyer, Microsc Microanal Microstruct 3, 187 (1992).

[10] C. Jeanguillaume, P. Trebbia, and C.Colliex, Ultramicroscopy 3, 237 (1978).

[11] J.Mayer, D.V. Szabo, M. Rühle, M. Seher, and R. Riedel, J. Eur. Cer. Soc. 15, 703 (1995).

[12] B.Freitag, and W. Mader, J. Microscopy 194, 42 (1999).

[13] J.Mayer, U. Eigenthaler, J.M. Plitzko, and F. Dettenwanger, Micron 28, 361 (1997).

[14] J.M.Plitzko, and J. Mayer, Ultramicroscopy 78, 207 (1999).

[15] G. Dehm, C. Scheu, G. Möbus, R. Brydson, and M. Rühle, Ultramicroscopy 67, 207 (1997).

[16]   J. Mayer, and J.M. Plitzko,. J. Microsc. **183,** 2 (1996).

[17]   M. Tence, M. Quartuccio, C.Colliex, Ultramicroscopy **58**, 42 (1995).

[18]   C. Colliex, M. Tence, E. Lefevre, C. Mory, H. Gu, D. Bouchet, and C. Jeanguillaume, Microchim Acta **112**, 71 (1994).

[19]   C. Gatts, G. Duscher, H. Müllejans, and M. Rühle, Ultramicroscopy **59**, 229 (1995).

[20]   P. Moreau, N. Brun, C.A. Walsh, C. Colliex, and A. Howie, Phys Rev B **56,** 6774 (1997).

[21]   J.A. Hunt and D.B.Williams, Ultramicroscopy **38**, 47 (1991).

[22]   S.D. Berger and S.J. Pennycook, Nature **298**, 635 (1982).

[23]   H. Müllejans and J. Bruley, J Microsc **180**, 12 (1995).

[24]   G. Duscher, N.D. Browning, and S.J. Pennycook, phys stat sol (a) **166**, 327 (1999).

Crystal structure determination from electron microscopy data

FROM FOURIER SERIES TOWARDS CRYSTAL STRUCTURES
A survey of conventional methods for solving the phase problem

Thomas E. Weirich
Gemeinschaftslabor für Elektronenmikroskopie, Rheinisch-Westfälische Technische Hochschule Aachen, Ahornstraße 55, D-52074 Aachen (Germany)

Abstract: This introduction to the phase-problem in crystallography is addressed to those who are not familiar with the standard techniques used for crystal structure determination. Whereas the basic concepts of Fourier synthesis and the related phase-problem are introduced in the first section, the second part focuses on some methods that have successfully been used to solve structures from electron data, i.e. crystallographic image processing (CIP) of high-resolution electron microscopy images, Patterson vector-maps and conventional direct methods.

Key words: phase-problem, crystallographic image processing, Patterson method, direct methods, structure determination, electron diffraction

1. INTRODUCTION

About 1915 W.H. Bragg suggested to use Fourier series to describe the arrangement of the atoms in a crystal [1]. The proposed technique was somewhat later extended by W. Duane [2] and W.H. Zachariasen was the first who used a two-dimensional Fourier map in 1929 for structure determination [3]. Since then Fourier synthesis became a standard method in almost in every structure determination from diffraction data.

1.1 Fourier synthesis in one and two dimensions

In order to understand how Fourier synthesis works let us examine the pattern in Figure 1. We can easily see that the pattern is a repetition of a

T.E. Weirich et al. (eds.), Electron Crystallography, 235–257.

black square – thus we can say that the pattern is periodic in two dimensions. If our task would be to reproduce the pattern, we can use a rubber stamp to make copies of the square by printing one by one until the whole pattern is complete. The motif on our stamp - the black square - is the basic unit of the pattern, or as a crystallographer would say, this is a *unit cell*. In principle we can choose any part from the pattern as unit cell, provided it fits seamlessly with its neighbouring copies. One possible example for a legal unit cell is shown in Figure 2. However, reproduction of a pattern by hand is a laborious and time consuming work. For this reason it appears much smarter if a computer program would do this work for us. Since standard computers have usually no eyes to see the pattern we must tell the computer in his language (which is mathematics) how large the unit cell is and at which position black squares appear in the unit cell.

Figure 1. A two-dimensional periodic pattern composed of black squares

Since we deal with a periodic pattern, it is possible to apply a technique that was originally invented by the French physicist and mathematician *Jean Baptiste Joseph FOURIER* (1768–1830). Fourier was the first who showed that every periodic process (or an object like in our case) can be described as the sum (a superposition) of an infinite number of individual periodic events (e.g. waves). This process is known as *Fourier synthesis*. The inverse process, the decomposition of the periodic event or object yields the individual components and is called *Fourier analysis*. How Fourier synthesis works in practice is shown in Figure 4. To keep the example most simple, we will first consider only the projection (a shadow image) of the black squares onto the horizontal *a*-axis in the beginning (Figure 3).

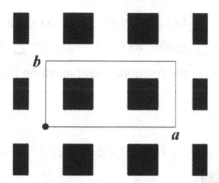

Figure 2. Enlarged view of a legal repeating unit (unit cell) that can be used by continuous repetition to produce the pattern shown in Figure 1. The boundaries of the two-dimensional unit cell are indicated by the outlined box. The origin of the unit cell at the lower left corner is marked by the black bullet, the two principal axes of the unit cell are labelled by the letters a (x direction) and b (y direction).

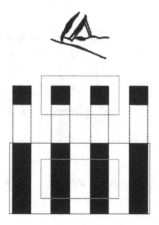

Figure 3. Projection of the unit cell contents onto the a-axis.

In this demonstration of a Fourier series we will use only cosine waves to reproduce the shadow image of the black squares. The procedure itself is rather straightforward, we just need to know the proper values for the amplitude A and the index h for each wave. The index h determines the frequency, i.e. the number of full waves trains per unit cell along the a-axis, and the amplitude determines the intensity of the areas with high (black) potential. As outlined in Figure 4, the Fourier synthesis for the present case is the sum of the following terms:

$$-1000 \cdot cos\ (2 \cdot \pi \cdot \underline{2} \cdot X) \underline{+78} \cdot cos\ (2 \cdot \pi \cdot \underline{4} \cdot X) \underline{+326} \cdot cos\ (2 \cdot \pi \cdot \underline{6} \cdot X)$$
$$\underline{-84} \cdot cos\ (2 \cdot \pi \cdot \underline{8} \cdot X)\ \underline{-180} \cdot cos\ (2 \cdot \pi \cdot \underline{10} \cdot X) = \sum A \cdot cos\ (2 \cdot \pi \cdot h \cdot X)$$

Herein denotes $X = x\ /\ a$ the fractional position along the a-axis (X varies between zero and 1).

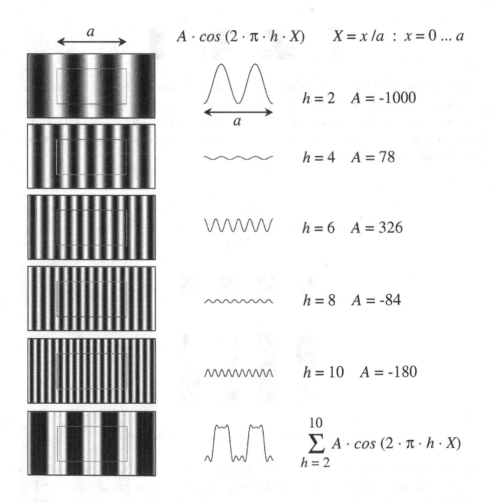

Figure 4. Principle of Fourier synthesis in one dimension. In this simple example of a Fourier series with cosine waves we need to know the amplitude A and the index h for each wave. The index h gives the frequency, *i.e.* the number of full wave trains per unit cell along the a-axis. The left row of images shows how the intensity within the unit cell changes for each Fourier component. The last image at the bottom gives the result after superposition of the waves with index $h = 2$ to 10 (areas with high potential are shown in black, brighter areas in the map indicate low potential). The corresponding intensity profiles along the a-axis for one unit cell are shown in the middle row. The ripples in the profile of the Fourier sum arise from the limited number of components that have been used in the synthesis (termination errors). If the

number of components would be very large (increasing frequency), these ripples will disappear and the image becomes identical to that shown in Figure 2. Note that the components with odd indices ($h = 1, 3, 5, ...$) were omitted in the series since their amplitude is zero.

1.2 The role of the phases

The picture of the projected unit cell in Figure 3 shows that the black squares have their centre along the *a*-axis at $X = 1/4$ and 3/4. In agreement with this finding the wave with the strongest amplitude ($A = -1000$) and frequency *h* = 2 (= two full wave trains) accumulates high potential at exactly these coordinates (top image in Figure 5). However, if we change the sign of the amplitude from minus to plus, the appearance of the map is quite different (image at the bottom). This wave starts at the origin of the unit cell with high potential, it generates another maximum at $X = 1/2$ and reaches the end of the unit cell again with a maximum. Inspection of the corresponding intensity profiles along the *a*-axis shows that changing the sign from minus to plus causes a phase shift ϕ between the two waves that is equal to 180 degrees or π (recall that a full wave train is 2π or 360 degrees).

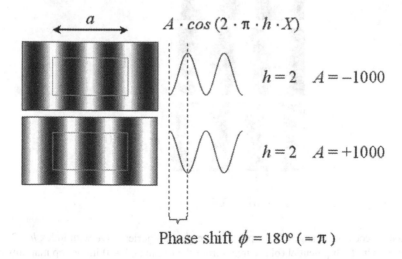

Figure 5. Changing the sign of the amplitude from minus to plus causes a phase shift of 180° in the Fourier map. Every cosine wave with a positive amplitude starts at the origin of the unit cell with a maximum (high potential); cosine waves with negative amplitudes on the other hand produce low (zero) potential at the origin.

Since cosine waves naturally start at the origin ($X = 0$) with a maximum, we can generally say that changing the sign of the amplitude from plus to minus causes a 180 degree ($\Delta X = a / 2h$) shift of the wave maxima. If we want to change the wave not only by fixed values of 180 degrees but by any angle ϕ, we have to rewrite the cosine term for a single Fourier component in the following form: $|A| \cdot cos\,(\,2 \cdot \pi \cdot h \cdot X - \phi\,)$

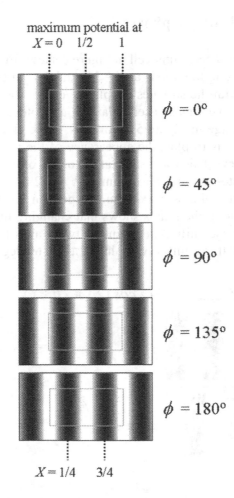

Figure 6. Successive changes of the phase value ϕ of a Fourier wave with index $h = 2$ moves the region with high potential (black areas) from the origin at $X = 0$ in the top map towards $X = 1/4$ in the map at the bottom. This shows that the value of the phase ϕ determines the positions with high potential within the unit cell, whereas the amplitude $|A|$ just affects the intensity. Note, that the maps with a phase shift of $\phi = 0°$ and $\phi = 180°$ have a centre of symmetry at the origin of the unit cell, whereas the other maps have <u>no</u> symmetry centre. From this we can draw another important conclusion: if we put the origin of the unit cell on a centre of symmetry we have <u>only</u> two choices for the phase value, $\phi = 0°$ or $\phi = 180°$. As we will see later, this feature is of great importance for solving centrosymmetric crystal structures.

What happens if we successively change the phase value ϕ from zero to 180 degrees is illustrated in Figure 6. We can see that the black areas with high potential move continuously from the origin of the cell at $X = 0$ to the right until the maximum is located at $X = 1/4$ for $\phi = 180°$. From this we can draw the following important conclusion: *the phase values of the Fourier components determine where the regions of high potential appear in the unit cell, whereas the amplitudes just affects the intensity of the maxima.*

With this background we are now able to generate a two-dimensional picture of our unit cell. For this task we have to superimpose the Fourier components along the a-axis and b-axis. This procedure of course requires to take care that all phases have their correct values according to the common unit cell origin (see Figure 7).

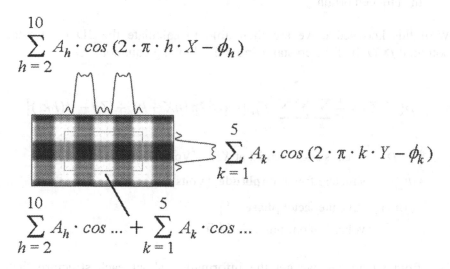

$$\sum_{h=2}^{10} A_h \cdot \cos (2 \cdot \pi \cdot h \cdot X - \phi_h)$$

$$\sum_{k=1}^{5} A_k \cdot \cos (2 \cdot \pi \cdot k \cdot Y - \phi_k)$$

$$\sum_{h=2}^{10} A_h \cdot \cos \ldots + \sum_{k=1}^{5} A_k \cdot \cos \ldots$$

Figure 7. By superposition of only five Fourier components along the two principal axes ($h = 2, 4, 6, 8, 10$; $k = 1, 2, 3, 4, 5$) it is already possible to obtain the highest potential in the map at the positions of the black squares. If we begin to add also Fourier components with mixed indices hk (these correspond to waves which travel diagonal through the unit cell) we would see, that the light grey areas which link the black squares will start to disappear.

1.3   The problem of the missing structure factor phases

Now let us assume that the black squares in the above example are the atoms in a real crystal structure and we want to locate these atoms by help of a Fourier synthesis. As we have seen in the previous paragraph we then need to know

- ... the size of the 3D unit cell, *i.e.* the length of the *a*, *b* and *c*-axis (and angles between them, if different from 90 degrees)
- ... the location of the unit cell origin
- ... the amplitude and phase for each Fourier component with respect to the unit cell origin

With this knowledge we are then able to calculate the 3D electrostatic potential $\rho(XYZ)$ of the crystal using the following equation:

$$\rho(XYZ) = \frac{1}{V} \sum_h \sum_k \sum_l |F_{hkl}| \cdot \cos\left[2\pi(hX + kY + lZ) - \phi(hkl)\right]$$

ρ = electrostatic potential at XYZ [Volts $\cdot$ Å^{-3}]

$|F_{hkl}|$ = structure factor **amplitude** [Volts]

$\phi(hkl)$ = structure factor **phase** [$°$]

V = volume of one unit cell [Å^3]

But from where can we get the information about each structure factor amplitude $|F_{hkl}|$ and phase ϕ_{hkl} ? If the crystal structure is known, the required parameters can be calculated by using the following equation:

$$F_{hkl} = |F_{hkl}| \cdot e^{i\phi_{hkl}} = |F_{hkl}| \cdot \cos\phi_{hkl} + i|F_{hkl}| \cdot \sin\phi_{hkl} = A + iB$$

$$A = \sum_j f_j \cdot \cos 2\pi(hx_j + ky_j + lz_j)$$

$$B = \sum_j f_j \cdot \sin 2\pi(hx_j + ky_j + lz_j)$$

$$\phi_{hkl} = \arctan\frac{B}{A}$$

Herein denotes f_j the atomic scattering factor of atom j in the unit cell, the x_j y_j z_j are the corresponding fractional coordinates of the atom j and the *hkl* are the (*Miller*) indices of the Fourier component (see below). If the structure is unknown, we can obtain some part of the required information from a *diffraction experiment*, since diffraction works in some way quite similar as Fourier analysis. If a thin crystal is exposed to a parallel beam of high-energy electrons, it can be observed that the crystal diffracts small amounts of the incident beam in certain directions (see Figure 9). Each beam in the diffraction pattern arises from lattice planes in the crystal which are of the same type, *i.e.* which have the same distance (d-spacing) to each other. If the incident beam with wavelength λ strikes a pile of equivalent lattice planes under the angle θ and the term $\lambda / \sin \theta$ is equal to twice the d-spacing, then the beam is quasi "reflected" at these lattice planes and a diffracted beam leaves the crystal at the angle θ (BRAGGS-law). Each of the diffracting lattice planes virtually cuts the crystal and hence also each unit cell into small slices. In order to define how the unit cell is cut by these lattice planes we assign an index triple *hkl* (the *hkl* are integer numbers!) to each set of equivalent planes. According to this nomenclature a lattice plane with indices *hkl* is constructed by connecting the points *a/h*, *b/k* and *c/l* in the unit cell (recall that *a*, *b* and *c* are the length of the axes of the unit cell). The corresponding d-spacings are given by the length of the normal (90°) vector which points from the origin of the unit cell to the lattice plane (see figure 8). It turns out that these d-spacings are identical to the length of the individual wave trains that we have used in our earlier example for Fourier synthesis. Thus we can associate every diffracted spot in a diffraction pattern with an individual Fourier component that is defined by a lattice plane with index *hkl*. In order to obtain a real space image of the atoms in the unit cell we then need to collect a sufficient number of Fourier components from diffraction and perform a Fourier synthesis. Unfortunately this is not that straightforward in practice as it might appears. In a standard diffraction experiment with electrons (or X-rays) we usually obtain information about the size of the unit cell (axes, angles) and we can measure the square of the (structure factor) amplitudes for each Bragg reflection. However, the most important parameter we need to know for locating the atoms in the unit cell – the phase value – is lost due to the lack of phase-sensitive detectors that can measure the phase shift between the incoming diffracted beams. This severe lack of information is the so called ***phase-problem*** of crystallography.

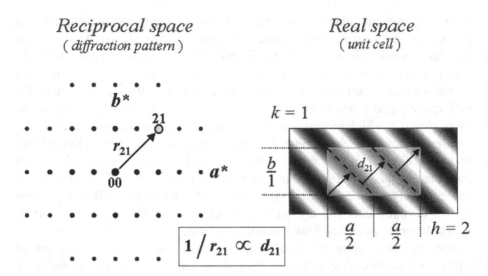

Figure 8 Each spot in a diffraction pattern can be considered as a Fourier component, *i.e.* a wave that travels through the crystal with a certain amplitude and frequency. The square of the amplitude of each component can be measured directly in the diffraction experiment and the frequency is determined by the d-spacing of the particular set of lattice planes *hkl* which have caused "reflection" of the incident beam. The above example in 2D shows a Fourier wave with indices *hk* = 21 and the corresponding d-spacing between the parallel lattice planes of the same type. Note, that there are three full wave trains (= three times the d-spacing) in this case that travel through one unit cell. The direction of the waves across the unit cell is determined by the indices *hk* , *i.e.* the wave front is the straight line between the point *a/h* on the *a*-axis and the point *b/k* on the *b*-axis. The cosine wave starts in this case at the origin of the unit cell (lower left corner) with a maximum (phase shift of the wave = 0°) and leaves this unit cell after three full wave trains also with a maximum (upper right corner). Nevertheless, the wave could also start with any other phase value which would lead to high potential (black areas) at other places in the unit cell. As mentioned in the main text, the amplitude and the frequency (d-spacing) of the Fourier component can be measured in the diffraction experiment, but **not** the phase shift with respect to the phases of the other waves which cross the cell – this drawback is the phase problem of crystallography.

2.   METHODS FOR SOLVING THE PHASE PROBLEM

2.1   Crystallographic Image Processing (CIP)

In contrast to X-rays, electrons can be focussed by magnetic lenses to give images of the investigated objects. This is the basic principle behind every transmission electron microscope (TEM). As shown by the sketch in figure 9 the central part of every TEM is the objective lens. This lens "collects" all diffracted electron beams from the crystal and "sorts" them in the back focal plane (diffraction plane) of the lens according to their 2θ Bragg-angle (→ d-spacing) and relative orientation of the lattice planes in the crystal.

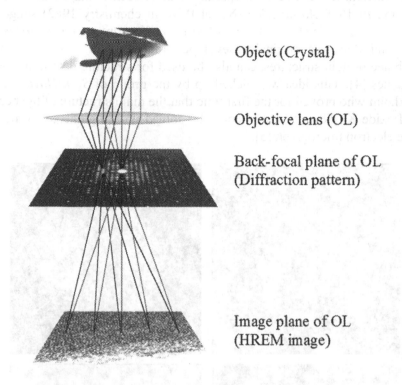

Object (Crystal)

Objective lens (OL)

Back-focal plane of OL
(Diffraction pattern)

Image plane of OL
(HREM image)

Figure 9. Simplified ray diagram (Abbe diagram) that shows simultaneous formation of the diffraction pattern and the corresponding real space image in a transmission electron microscope (TEM).

At the back focal plane we have to look if we want to record an electron diffraction pattern. However, the diffracted electrons normally don't stop at the back focal plane of the lens and if they are properly focussed they will recombine to give a (slightly blurred) image of the object in the image plane. Provided the investigated crystal is thin (less than about 100 Å) and it is well aligned along a prominent viewing direction (which is usually a low-order zone axis), then it is even possible to image single atom columns by a conventional TEM (figure 10a). If we manage to obtain such a high-resolution (HR) TEM image which shows the projected crystal structure up to a certain resolution, the task to solve the crystal structure in projection appears again at first sight straightforward. However, HRTEM images can only be interpreted in terms of projected structure if they were obtained under special conditions (e.g. *Scherzer* defocus which shows all atom columns in black). The main reason for this drawback are the imaging properties of the objective lens which depend on several parameters, e.g. sample thickness and defocus of the objective lens. This is the reason for the severe difficulties in direct interpretation of HRTEM images in terms of structure. In 1979 *Aaron Klug* (Nobel Prize in chemistry 1982) suggested that a technique which was later named *crystallographic image processing* (CIP) and that was originally developed for structure determination of membrane protein structures, can also be used for structure determination of inorganics [4]. This idea was picked up by the group of *Sven Hovmöller* in Stockholm who proved for the first time that the main structure of the heavy-metal oxide $K_{8-x}Nb_{16-x}W_{12+x}O_{80}$ could be solved with high accuracy from a single electron micrograph [5].

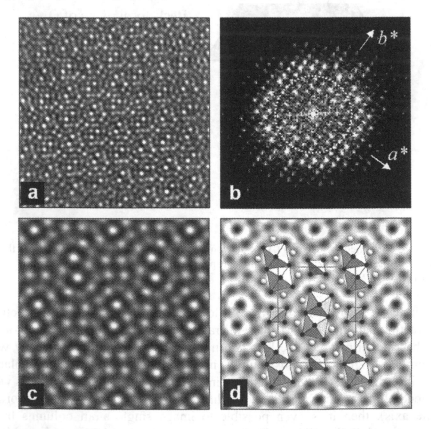

Figure 10. Crystallographic Image Processing (CIP) of high-resolution electron microscopy (HREM) images. (a) HREM image of α-Ti_2Se recorded with a 300 kV TEM (Jeol 3010UHR, point resolution 1.7 Å) along the [001] zone axis. (b) Fourier transform (power spectrum) of the HREM image (only the amplitudes are shown). The position of the white ring marks the first crossover of the Contrast Transfer Function (CTF) which is used to determine the defocus value ($\Delta f \approx$ -650 Å). The structure factor phases of all reflections outside the white ring have a phase difference of 180° compared to their true value. The latter causes inversion of image contrast. (c) Lattice averaged image ($p2$ symmetry) deduced from the amplitudes

and phases in the power spectrum before CTF-correction. (d) Projected pseudo-potential map (*p2gg* symmetry) after correction of the phase-shifts imposed by the CTF showing all columns of atoms in black. The average agreement of atomic co-ordinates determined from the pseudo-potential map and the superimposed model from X-ray diffraction is about 0.2 Å. (• = Ti; o = Se).

Since then this method has been used to solve numerous other complex crystal structures [6-13]. Because solving a structure from a single projection requires a short (3 to 5 Å) crystal axis, the method was later extended to combine the information from several orientations which allows also to uncover structures with pronounced overlap of the atom columns in projection. This technique was applied in 1990 to solve the 3D structure of the mineral staurolithe $HFe_2Al_9Si_4O_4$ [14, 15] and more recently to determine the structure of the huge quasicrystal approximant v-AlCrFe [16] which contains 129 atoms per asymmetric unit. How CIP works to solve a crystal structure from projected data is shown in figure 10 (for further details see [17]).

2.2 Patterson (interatomic vector) maps

Even without having the structure factor phases, e.g. from electron microscopy images, it is possible to get some insight into the atomic architecture of a crystal. A simple but powerful method to get this information was introduced by *A.L. Patterson* about 70 years ago. Following Patterson the Fourier synthesis is carried out using the squared structure factor amplitudes $F_{hkl}{}^2$ which are equal to the measured intensities for the reflections with index *hkl*. Moreover, all phase values must be set to zero, which leads to the following (auto-correlation) function:

$$P(UVW) = \frac{1}{V}\sum_h\sum_k\sum_l |F_{hkl}|^2 \cdot \cos 2\pi(hU + kV + lW)$$

P = intensity in Patterson-space at *UVW* [Volts$^2 \cdot$ Å^{-3}]

$|F_{hkl}|$ = structure factor amplitude [Volts]

V = volume of one unit cell [Å^3]

In contrast to Fourier synthesis, which yields with electron diffraction data high electrostatic potential at the positions of the atoms, the maps obtained from Patterson synthesis show peaks at the tips of vectors. The length of each vector (drawn from the origin of the Patterson map) corresponds always to the distances between pairs of atoms and the direction each vector points to is the same as for the particular couple of atoms in the crystal. Moreover, the peak heights allow to identify the type of the involved atoms. Thus a Patterson map is generated by translation of all interatomic vectors in the unit cell to a common origin. What this means becomes more clear by the following exercise. Copy the whole unit cell in figure 11a (including the three atoms) on a sheet of transparent paper. Then move each atom to the origin of the unit cell and draw straight lines (vectors) from the origin to the other two atoms. A vector map as shown in figure 11b is obtained if this procedure is repeated for all three atoms. From this we can derive that N atoms per unit cell generate N^2 peaks per cell in the Patterson map, whereof N peaks overlap in the origin (= strongest peak in the map). The latter peak represents the distance of each atom to itself (vectors with zero length). Although the peaks in a Patterson map correspond 1:1 with the direction and intensity of the interatomic vectors in the crystal structure, the major drawback of these maps is that the positions of the vectors within the unit cell remain undefined. Nevertheless, if there are only a few heavy atoms in a crystal structure – strong peaks in the Patterson map - it might be possible to derive their co-ordinates and so at least partially to solve the structure. Missing (light) atoms can then located by difference Fourier synthesis *F(observed)* minus *F(calculated)* using the tentative structure model with the heavy atoms only (*heavy atom method*).

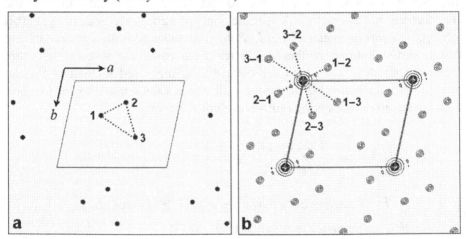

Figure 11. Patterson (vector) maps: The structure with three atoms per unit cell in (a) causes appearance of 6+1 peaks in the Patterson map in (b). The peaks in the Patterson map can be generated by shifting each atom to the origin of the unit cell and drawing vectors to the other two atoms (the tips of the vectors are marked by peaks), e.g. placing atom #2 at the origin generates the vectors 2-1 and 2-3. The strong peak at the origin corresponds to the distance of each atom to itself (vector with zero length). Note, that the Patterson map in (b) has a center of symmetry (2-fold axis at the origin), whereas the structure in (a) has no center of symmetry. This is one of the properties of Patterson maps that they always possess a center of symmetry.

Provided the structure under investigation contains a known structural fragment, we can screen the Patterson map for it. For this purpose the fragment must be converted into Patterson space at first. The received motif is then rotated around the origin until a reasonable match with the experimental Patterson map is obtained. Since this procedure allows to fix only the orientation of the real space fragment, the fragment must be located in the next step within the unit cell according to space group symmetry (e.g. on rotation or screw axes). This partial solution of the crystal structure can again be used as input for difference Fourier synthesis which assist to locate the missing atoms in the structure. Structures which have been solved from Patterson maps with electron diffraction data are for example Nd_2CuO_4 [18] and Al_mFe [19].

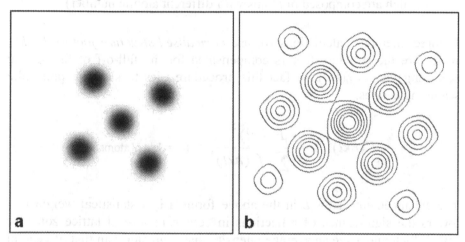

Figure 12. In order to check the experimental Patterson map for a known structural fragment it must be converted into Patterson space first. This is shown in the above sketch for the fragment of a body centered cube which is projected along the 4-fold axis (a). The corresponding motif in Patterson space is shown in (b). Note, that the atom in the center of the fragment defines a 2-fold axis what reduces the number of different peaks in the Patterson map.

2.3 Sayre equation and direct methods

If the Patterson method cannot be applied because the structure has no or too many heavy atoms, it is possible to use another approach for phase determination, the so-called *direct methods*. *"By the term direct methods is meant that class of methods which exploits relationships among the structure factors in order to go directly from the observed magnitudes* $|E|$ *to the needed phases* ϕ*"* (Herbert A. Hauptman, Nobel lecture, 9. Dec., 1985). The direct method approach for solving the phase problem uses probability methods to predict the phase relations within certain sets of reflections. The theory behind direct methods assumes that the following conditions hold:

- The electron density is everywhere positive (a probability can not be negative!). Nevertheless, that the positivity of the density function is not a necessary prerequisite for solving the phase problem via direct methods was recently shown [20]

- There are only discrete, non-overlapping point atoms in the structure (this requires diffraction data with resolution of about 1 Å)

- The crystal contains only one type of atoms (is seldom the case, but direct methods are in practice very robust and work also for structures which are composed of atoms with different atomic number)

In direct methods calculations we use *normalised structure factors E(hkl)*, which are the structure factors compensated for the fall-off of the atomic scattering factors $f_j(hkl)$. In fact this procedure tries to simulate point-like scattering centres.

$$E(hkl)^2 = \frac{|F(hkl)|^2}{\varepsilon \cdot \sum_j f_j(hkl)^2} \quad (j = 1 \ldots N \text{ atoms})$$

The enhancement factor ε in the above formula is a statistical weight that lowers the significance of reflections in certain reciprocal lattice zones or rows which may have an average intensity that is greater than that of general reflections. The ε factor depends upon the crystal class and is listed in the *International Tables for Crystallography*. Among other relations, *Herbert A. Hauptman & Jerome Karle* (both awarded for the Nobel prize in chemistry in 1985), derived the most important relation for _centrosymmetric_ space groups, the Σ_2 formula:

$$\Sigma_2: \quad S(E_H) \approx S\left(\sum_K E_K E_{H-K}\right)$$

This formula is also known as *Sayre equation*, since *David Sayre* discovered around 1952 a similar expression that was later confirmed by the theoretical derived Σ_2 formula. The practical meaning of the Σ_2 formula is that the sign S of the normalised structure factor amplitude E_H depends on the signs of structure factors E_K and E_{H-K}. For example, if E_K is positive and E_{H-K} is negative, the product of these two is negative ("+" · "-" $\Rightarrow$ "-") and hence the sign of E_H is negative. If E_K and E_{H-K} are both positive it follows that E_H is also positive ("+" · "+" ⇒ "+"). From the previous we know already that a structure with inversion center (a centrosymmetric structure) forces the structure factor phases ϕ_{hkl} to zero or 180 degrees (= "+" or "-"). The Σ_2 formula is commonly used to determine the phase values from *triplet relations* or *origin invariant sums*. These origin invariant sums are simply those, where the Miller indices *hkl* add up to a value that does <u>not</u> change if the origin of the unit cell is shifted (the sum of all *hkl* must be always equal to zero). Moreover, it was derived from theory that for large *E(hkl)* a high probability exists that the value of the phase sum is zero.

$$0 \approx \phi(h_1, k_1, l_1) + \phi(h_2, k_2, l_2) + \phi(-h_1 - h_2, -k_1 - k_2, -l_1 - l_2)$$

That this is a plausible assumption is demonstrated by the following two examples. In the first example we will consider a crystal that has all atoms on the corners of a 2D unit cell as shown in figure 13. Accordingly we have three structure factors whose phases are related by the above equation:

$$\phi(10) + \phi(01) + \phi(-1-1) = 0 \text{ resp. } \phi(10) + \phi(01) = \phi(11)$$

Friedels law for centrosymmetric structures states that $\phi(hkl) = \phi(-h-k-l)$

In the Fourier maps of the three components we see that the (10) amplitude with phase equal to zero generates high potential at the beginning and end of the unit cell along the *a*-axis. The same situation exists for the (01) structure factor along the *b*-axis. The question is now how the third component, the (11) amplitude is affected by the other two? If we assume that the phase of the latter structure factor is 180° (the cosine wave will start at the origin with a minimum!) high potential appears along the line which connects the points ½ *a* and ½ *b* – but obviously there are no atoms! In order to add to the high potential at the corners of the unit cell, this wave <u>must</u> start at the origin with a maximum as shown in the third graph and hence the phase shift has to be zero. Superposition of these three Fourier components yields now to high potential at the corners of the unit cell where the atoms are located.

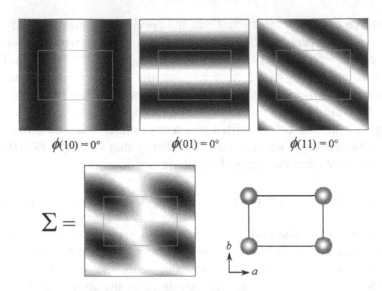

$\phi(10) = 0°$ $\phi(01) = 0°$ $\phi(11) = 0°$

$\Sigma =$

Figure 13. Principle of direct methods using triplet relations. As shown in the lower right-hand image the trial structure consists of atoms which are located at the corners of the unit cell. According to the Σ_2 formula (*Sayre* equation) a strict phase relation exists within a certain set of three reflections (a triplet) with large normalized structure factor amplitudes E_{hkl}. Such a triplet or origin invariant sum is defined as $h_1k_1l_1 + h_2k_2l_2 + h_3k_3l_3 = 0$ or $h_1k_1l_1 + h_2k_2l_2 = -h_3-k_3-l_3$ [(10) + (01) = (11) for the above example in 2D]. The practical meaning of the Σ_2 formula is that not only the indices *hkl* have to add up to zero, but also the phase values. To understand this point consider what would happen if the phase of the (11) wave in the above example is changed from zero to 180°, i.e. the wave which crosses the unit cell in diagonal direction will then begin with a minimum (white areas) at the unit cell origin instead with a maximum (black). This simple example shows that the (11) wave must have zero phase otherwise it would generate peaks half way along the *a* and *b*-axis.

Lets check what happens with the phases if we shift the origin of the unit cell towards ½ *a* and ½ *b* of the old cell (indicated by the arrow in figure 14). In order to gather high potential at the centre of the revised unit cell, the (10) wave, which travels along the *a*-axis, needs to start with a minimum (phase shift 180°). The same applies to the (01) component that travels along the *b*-axis. According to the Σ_2 formula we should be able to predict the phase shift for the third, the (11) reflection: The first reflection caused a phase shift by 180° ("-") and the second reflection imposes another phase shift by 180°. In agreement with the Σ_2 formula we need to end up with a total phase shift of 0° ("+") for all three reflections. Since the sum of the first two reflections is already 360° or 0° ("-" · "-" $\Rightarrow$ "+"), the third (11) component must cause a phase shift equal to zero (180° + 180° + 0° = 0° or expressed in sign notation "-" · "-" · "+" $\Rightarrow$ "+"). This can also be verified by inspection of the of images shown in figure 14. This proves the most important ideas of direct methods for centrosymmetric structures, namely that the phase value of a reflection within a triplet of large normalised structure factors can be predicted with high probability if the other two phase values are known. Moreover, the sum of the phases is always equal to zero, regardless which of the allowed unit cell origins was selected (remember, that the origin of the unit cell must be on a centre of symmetry to fix the phase values to zero and 180° !). As illustrated by the last example with just three reflections, it is not necessary to know all structure factor phases if we are already satisfied with a blurred "image" of the crystal structure which shows all atoms at their nearly correct position. As pointed out earlier, direct methods work most reliable for the strong normalised structure factors, so only these will be used to find a rough structure model (*trial structure*) with atom co-ordinates about 0.2 Å in error. Once a trial structure can be stablished, it can be refined to much higher precision using all available reflections. Nevertheless, in practice only 1/3 or less out of all measured reflections (usually those with $E_{hkl} > 1$) are used to find the trial structure. After fixing the origin of the unit cell by assigning phases to a certain triplet one has to seek for other triplets that contain structure factors with already known phases. In this way it is possible to expand the resolution of the trial structure step by step until individual atoms are resolved in the Fourier map. Since solving a structure by hand is a very time-consuming, computer programs have been developed which perform the phasing of the data within a few seconds. Such programs automatically select the reflections with large normalised structure factors from the whole data set and assign a number of random phases to some strong E-values.

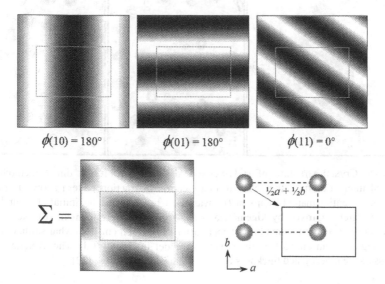

$\phi(10) = 180°$ $\qquad$ $\phi(01) = 180°$ $\qquad$ $\phi(11) = 0°$

Figure 14. Principle of direct methods using triplet relations (continued). In this case the origin of the unit cell was shifted half a unit cell along the *a* and *b*-axis. Whereas the indices of the three waves remain unchanged, the relative phase values must change in order to generate a peak maximum at the centre of the unit cell. However, the sum of the phases is still zero: the first (10) wave turns the phase by 180°, the second (01) reflection adds another 180°, which leads to a total change of 360° or 0°. Since we need to arrive at zero degrees after adding the third (11) wave, the latter can only be 0°.

After checking all phases for their consistency within the reduced data set, the program tries to expand the phasing also to reflections with lower E-values. This step is called *phase extension* and is done by exploiting phase relations between three (triplets) or sometimes four reflections (quartets). If this procedure fails and no reasonable structure model is found, the program starts again from the beginning and checks another set of random phases. In case of success, a residual (R-factor) is calculated that tells us how good the experimentally determined structure factor amplitudes fit the calculated structure factors from the trial structure. The trial structure with the lowest R-factor finally represents the most likely solution. One of the computer programs that can solve crystal structures from electron diffraction data is *SIR97* [21]. This program uses however, much more advanced strategies for phasing and improved reliability checks than described above (e.g. the trial structure is also checked for plausible atom distances). Until now several heavy-atom structures have been solved in a quasi-automatic manner from electron diffraction with *SIR97*. Among them are the previous unknown compounds $Bi_8Pb_5O_{17}$ [23], β-Ti_2Se [24], $Ti_{45}Se_{16}$ [25] and Ti_2P [26].

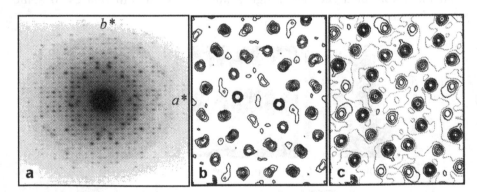

Figure 15. Crystal structure of α-Ti_2Se solved in projection via direct methods using quantified intensities from the selected area electron diffraction pattern shown in (a) [film data]. The potential map (E-map) in (b) was used to construct an initial structural model which was later improved by kinematical least-squares (LS) refinement (c). Note that the potential of the selenium atoms in (c) appear after LS-refinement somewhat stronger than the surrounding titanium atoms (see the structural model in figure 10d). The average effective thickness of the investigated thickness of the crystal is about 230 Å [22].

Another goal in this area is the recent *ab initio* determination of the framework structure of the heavy-metal oxide $Cs_xNb_{2.54}W_{2.46}O_{14}$ from 100 kV precession electron diffraction data using the same approach [27] (for the precession method see also the article of *Stavros Nicolopoulos* and coworkers in this book).

2.4   Other approaches

Aside the above outlined methods for solving the phase problem there exist also other methods or sometimes even mixed approaches which exploit information from several sources. Most important and a true competitor to traditional direct methods is the *maximum entropy method* [28, 29] which has been applied for solving several molecular structures from electron data (see also the article of *Chris Gilmore* in this book). The above mentioned mixed approaches are often helpful in cases with substantial dynamical diffraction. Structure factor phases are obtained from high-resolution electron microscopy images of thin crystal areas and used to fix the phases of the low-order reflections in order to guide the subsequent phasing process towards the correct solution [30-35]. Last but not least an alternative method to the here proposed crystallographic image processing of high-resolution electron microscopy images deserves to be mentioned. This method is known as *exit-wave reconstruction* and retrieves the real projected potential of a crystal from a through-focus series of images (see also the article of *Christian Kübel* in this book). This method was for example used to determine the crystal structure of Mg_5Si_6 precipitates in aluminium [36].

References

1.   Bragg, W. H. (1915) „X-rays and crystal structure", *Philos. Trans. Roy. Soc.* **215**, 253-274.
2.   Duane, W. (1925) „The calculation of the x-ray diffracting power at points in a crystal", *Proc. Nat. Acad. Sci. U.S.* **11**, 489-493.
3.   Zachariasen, W. H. (1929) „The crystal structure of potassium chlorate", *Z. Kristallogr.* **71**, 501-516.
4.   Klug, A. (1979) „Image Analysis and Reconstruction in the Electron Microscopy of Biological Macromolecules", *Chemica Scripta* **14**, 245-256.
5.   Hovmöller, S., Sjögren, A., Farrants, G., Sundberg, M., Marinder, B.O. (1984) "Accurate atomic positions from electron microscopy", *Nature* **311**, 238-241.
6.   Wang, D. N., Hovmöller, S., Kihlborg, L., Sundberg, M. (1988) "Structure Determination and Correction for Distortions in HREM by Crystallographic Image Processing", *Ultramicroscopy* **25**, 303-216.
7.   Li, D. X., Hovmöller, S. (1988) "The Crystal Structure of $Na_3Nb_{12}O_{31}$ Determined by HREM and Image Processing", *J. Solid State Chem.* **73**, 5-10.
8.   Hovmöller, S., Zou, X., Wang, D. N., Gonzalez-Calbet, J. M., Vallet-Regi, M. (1988) "Structure Determination of $Ca_4Fe_2Ti_2O_{11}$ by Electron Microscopy and Crystallographic image Processing", *J. Solid State Chem.* **77**, 316-321.

9. Zou, X. D., Hovmöller, S., Parras, M., Gonzalez-Calbet, J. M., Vallet-Regi, M., Grenier, J. C. (1993) "The Complex Perowskite-related Superstructure of $Ba_2Fe_2O_5$ solved by HREM and CIP", *Acta Cryst* **A49**, 27-35.

10. Sundberg M., Zakharov, N. D., Zibrov, I. P., Barabanenkov, Yu. A., Filonenko, V. P., Werner, P. (1993) "Two High-Pressure Tungsten Oxide Structures of W_3O_8 Stoichiometry deduced from High-Resolution Electron Microscopy Images", *Acta Cryst.* **B49**, 951-958.

11. Weirich, T. E., Ramlau, R., Simon, A., Hovmöller, S., Zou, X. (1996) "A crystal structure determined with 0.02 Å accuracy by electron microscopy", Nature 382, 144-146.

12. Weirich, T. E., Ramlau, R., Simon, A., Hovmöller, S. (1997) „Exact atom positions by electron microscopy? – A quantitative comparison to X-ray crystallography", In: Electron Crystallography, D. L. Dorset, S. Hovmöller and X. Zou Eds., Kluwer Academic Publishers, Dordrecht, 423–426.

13. Ezersky, V. I., Rochman A. D., Talianker, M. M. (2001) „Crystal Structure of New Approximant Phase in Al-Fe-V-Si System", *Mat. Res. Soc. Symp. Proc.* **643**, K4.2.1-K4.2.6.

14. Downing, K. H., Meisheng, H., Wenk, H. R., O'Keefe, M. A. (1990) "Resolution of oxygen atoms in staurolite by theee-dimensional transmission electron microscopy", *Nature* **348**, 525-528.

15. Wenk, H. R., Downing, K. H., Meisheng, H., O'Keefe, M. A. (1992) "3D Structure Dertermination from Electron-Microscope Images: Electron Crystallography of Staurolite", *Acta Cryst.* **A48**, 700-716.

16. Zou, X. D., Mo, Z. M., Hovmöller, S., Li, X. Z., Kuo, K. H. (2003) "Three-dimensional reconstruction of the v-AlCrFe phase by electron crystallography", *Acta Cryst* **A59**, 526-539.

17. Zou, X. D. (1999) „On the Phase Problem in Electron Microscopy: The Relationship Between Structure Factors, Exit Waves, and HREM Images", *Microscopy Research and Technique* **46**, 202-219.

18. Bougerol-Chaillout, C. (2001) „Structure Determination of Oxide Compounds by Electron Crystallography", *Micron* **32**, 473–479.

19. Gjønnes, J., Hansen, V., Berg, B. S., Runde, P., Cheng, Y. F., Gjønnes, K., Dorset D. L., Gilmore, C. J. (1998) „Structure Model for the Phase Al_mFe Derived from Three-Dimensional Electron Diffraction Intensity Data Collected by a Precession Technique. Comparison with Convergent-Beam Diffraction.", *Acta Cryst.* **A54**, 306–319.

20. Hauptman, H. A., Langs, D. A. (2003) „The phase problem in neutron crystallography", *Acta Cryst.* **A59**, 250-254.

21. Altomare, A., Burla, M. C., Camalli, M., Cascarano, G., Giacovazzo, C., Guagliardi, A., Moliterni, A. G. G., Polidori, G., Spagna, R. (1999) „SIR97: a new tool for crystal structure determination and refinement", *J. Appl. Cryst.* **32**, 115–119.

22. Weirich, T. E. (2003) „Electron Diffraction Structure Analysis: structural research with low-quality diffraction data", *Z. Kristallogr.* **218**, 269-278.

23. Gemmi, M., Righi, L., Calestani, G., Migliori, A., Speghini, A., Santarosa, M., Bettinelli, M. (2000) „Structure determination of $Bi_8Pb_5O_{17}$ by electron and powder X-ray diffraction", *Ultramicroscopy* **84**, 133–142.

24. Weirich, T. E., Zou, X. D., Ramlau, R., Simon, A., Cascarano, G. L., Giacovazzo, C., Hovmöller, S. (2000) „Structures of nanometre-size crystals determined from selected-area electron diffraction data", *Acta Cryst.* **A56**, 29–35.

25. Weirich, T. E. (2001) „Electron crystallography without limits? - Crystal structure of $Ti_{45}Se_{16}$ redetermined by electron diffraction structure analysis", *Acta Cryst.* **A57**, 183–191.

26. Gemmi, M., Zou, X. D., Hovmöller, S., Migliori, A., Vennström, M., Andersson, Y. (2003) „Structure of Ti_2P solved by three-dimensional electron diffraction data collected with the precession technique and high-resolution electron microscopy", *Acta Cryst.* **A59**, 117-126.

27. Weirich, T. E., Portillo, J., Cox, G., Hibst, H. Nicolopoulos, S. (2005) "*Ab initio* determination of the framework structure of the heavy-metal oxide $Cs_xNb_{2.54}W_{2.46}O_{14}$ from 100 kV precession electron diffraction data", *Ultramicroscopy* (accepted for publication).

28. Gilmore, C. J. (1999) „Maximum entropy methods in electron crystallography", *Microscopy Research and Technique* **46**, 117–129.

29. Voigt-Martin, I. G. (1999) „Electron crystallography and non-linear optics", *Microscopy Research and Technique* **46**, 178–201.

30. Huang, D. X., Liu, W., Gu, Y. X., Xiong, J. W., Fan, H. F., Li, F. H., (1996) „A Method of Electron Diffraction Intensity Correction in Combination with High-Resolution Electron Microscopy" *Acta Cryst.* **A52**, 152– 157.

31. Sinkler, W., Bengu, E., Marks, L. D. (1998) „Application of Direct Methods to Dynamical Electron Diffraction Data for Solving Bulk Crystal Structures", *Acta Cryst.* **A54**, 591–605.

32. Li, F. H. (1998) „Image processing based on the combination of high-resolution electron microscopy and electron diffraction", *Microscopy Research and Technique* **40**, 86–100.

33. Sinkler, W.; Marks, L. D. (1999) „Dynamical direct methods for everyone", *Ultramicroscopy* **75**, 251–268.

34. Fan, H. F. (1999) „Direct methods in electron crystallography: Image processing and solving incommensurate structures", *Microscopy Research and Technique* **46**, 104–116.

35. Chukhovskiia, F. N., Poliakovb, A. M. (2003) „Domino phase-retrieval algorithm for structure determination using electron diffraction and high-resolution transmission electron microscopy patterns", *Acta Cryst.* **A59**, 48-53.

36. Zandbergen, H. W., Andersen, S. J., Jansen, J. (1997) "Structure determination of Mg_5Si_6 particles in Al by dynamic electron diffraction studies", *Science* **277**, 1221-1225.

CRYSTAL STRUCTURE DETERMINATION BY IMAGE DECONVOLUTION AND RESOLUTION ENHANCEMENT

Hua Jiang[1,2], Fang-hua Li[2] and Esko. I. Kauppinen[1,3]

[1]*VTT Processes, Nanotechnology Group, P.O. Box 1602, 02044 VTT, Finland*;
[2]*Chinese Academy of Sciences, Institute of Physics & Center for Condensed Matter Physics, Beijing 100080, P.R. China*;
[3]*Helsinki University of Technology, New Materials Center, 02044 VTT, Finland.*

Abstract: Two techniques for crystallographic image processing are introduced aiming at determining the crystal structure and defect structure respectively at atomic resolution. The techniques are based on a simple image contrast theory that extends the weak object approximation to thicker crystals. Dynamical effect and other intensity distortions are properly corrected. Examples of applications to crystal structure and defect structure are given.

Key words: Electron Crystallography, Image Processing, HREM, Electron Diffraction

1.   INTRODUCTION

In a high-resolution transmission electron microscope (TEM) the electron wave mainly experiences two basic processes during its propagation from specimen to image: interaction with the specimen and modulation by the electron-optical lens system. During the first process, the electron wave is loaded with the structure information of the specimen. In the second process, the structure information is transferred. Thanks to the unique instrumental design in the TEM, the electron wave which carries the structure information of the specimen forms an electron diffraction pattern (EDP) on the back-focal plane of the objective lens and an image on the imaging plane. Therefore, the EDP and the image can simultaneously be available in the TEM. The image represents the specimen in real space, while the diffraction pattern is a depiction of the specimen in reciprocal space. They both

T.E. Weirich et al. (eds.), Electron Crystallography, 259–273.

characterize the crystal structure; hence both are extensively applied in crystal structure analysis.

In contrast to X-ray crystallography, electron diffraction is more suitable for studying crystal structures of surfaces, thin films, and crystals of small volumes. However, development of the electron diffraction technique is impeded by the phase problem and by strong dynamical effects. High-resolution electron microscopy (HREM) can be used for crystallographic studies as well. As it is well known, an HREM image is a complex function of the interaction between the high energy electrons with the electrostatic potential in the specimen and the magnetic fields of the objective lenses in the microscope. Accordingly, in HREM structure analysis, both the imaging process and the electron-specimen interaction process are virtually inverted.

In order to solve such inverse problems in HREM, the trial-and-error method based on image simulation has proved successful if the examined structure can be modeled beforehand. However this is often not possible. Another attractive idea is to solve the problems by image processing without relying on the prior knowledge of the structure. For this purpose many different techniques have been developed. One is known as exit wave reconstruction followed by structure derivation from the exit wave. The wave function at the exit face of the object can be reconstructed based on the linear or non-linear imaging theory from two or more images taken under various defocus conditions [1-4]. The information loss due to the zero crosses of the contrast transfer function (CTF) in one image can be compensated from the information contained in other images taken with different defocus conditions, and the information distortion in the high spatial-frequency region due to the strong oscillation of the CTF can be eliminated. Since the early 1990s, exit wave reconstruction was studied extensively and developed as a direct method for phase and structure retrieval in HREM [5-10]. After the exit wave is reconstructed, structure retrieval would be straightforward if the object is thin enough to act as a phase object, then the phase is proportional to the electrostatic potential of the structure projected along the beam direction. In the case of a thicker object, the problem becomes complicated due to dynamical diffraction. In this case more effort is needed to "invert" the electron scattering process in the object so as to retrieve the projected structure [10].

Another way of solving the inverse problems in HREM proceeds directly from the image to the structure, skipping the intermediate step of exit wave reconstruction [11-18]. In this way the electron diffraction data and some diffraction analysis methods that were originally developed in X-ray crystallography may be introduced into the image analysis so that the inverted processes of electron-optical imaging and that of electron dynamical diffraction can be treated simultaneously. This way can be characterized as a sort of electron crystallographic image processing (CIP) technique. The rationality of introducing diffraction data into image analysis relies on the mutual supplement of structure information between image and diffraction.

In this contribution, we introduce two kinds of electron crystallographic image processing techniques that have been proven useful in structure analysis of minute crystals and defects.

2. FUNDAMENTALS

2.1 Weak-phase-object approximation

It is well known that under the weak-phase-object approximation (WPOA) [19], the *image intensity function* is linear to the convolution of the *projected potential* distribution function $\varphi_t(x, y)$ and the inverse Fourier transform (FT) of the *contrast transfer function* (CTF) $T(\mathbf{u})$ of the electron microscope:

$$i(x, y) = 1 + 2\sigma\varphi_t(x, y) \otimes \mathsf{F}^{-1}[T(\mathbf{u})] \tag{1}$$

where the symbols $\otimes$ and F^{-1} are operator of convolution and inverse Fourier transform respectively. $\sigma = \pi/\lambda E$ is an interaction constant with λ the electron wavelength and E the accelerating voltage. $\varphi_t(x, y)$ is the *projected potential* of the crystal along the incident beam direction, which is not represented here as a function of the crystal thickness t, but obviously it is related to t. By Fourier transform, equation (1) yields

$$I_{img}(\mathbf{u}) = \delta(\mathbf{u}) + 2\sigma F(\mathbf{u}) T(\mathbf{u}) \tag{2}$$

Here $I_{img}(\mathbf{u})$ is the Fourier components of the *image intensity function*. $\mathbf{u}$ is the lattice vector in reciprocal space. $F(\mathbf{u})$ is the Fourier spectrum of the projected potential $\varphi_t(x, y)$. In crystallography, $F(\mathbf{u})$ is the *structure factor* of the crystal.

Equation (1) shows that **HREM images contain information of the crystal *projected potential* as well as all microscopic aberrations**, which is characterized by the CTF. According to Equation (2), under WPOA, **Fourier transform of the HREM image yields the crystallographic structure factors that are multiplied by the CTF**. Therefore, it is reasonable to derive the crystal structure from HREM images by crystallographic image processing techniques.

2.2   Effects of crystal thickness on the image contrast

Under the WPOA, equation (1) does not show the thickness influence on the image intensity (contrast). However, the WPOA only applies to very thin specimens of the order of a few nanometers or even thinner, depending on the types of the atoms in the crystal. In reality, the projected potential $\varphi_t(x, y)$ varies significantly with the crystal thickness t. To illustrate the dependence of the image contrast on the thickness t, a simple image intensity formula, named pseudo-Weak-Phase-Object-Approximation (pseudo-WPOA), was proposed by Li et al [20]. Referring to the multislice theory [21], the pseudo-WPOA assumes that all slices are identical to each other but ignores the secondary electron scattering between adjacent slices. As a consequence, the pseudo-WPOA supposes an artificial crystal structure which is isomorphic to the examined one. The projected potential of the artificial structure, hereafter named *pseudo potential*, can be expressed as

$$\varphi'(\mathbf{r},t) = \varphi_t(\mathbf{r}) + \Delta\varphi(\mathbf{r},t) \tag{3}$$

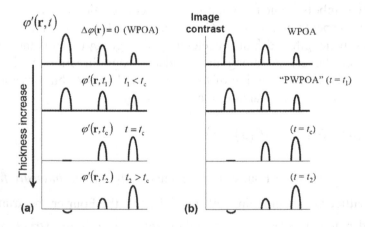

Figure 1. Schematic diagram showing change of (a) the pseudo potential; (b) the image contrast, with thickness t for three differently weighted atoms. The leftmost atom is the heaviest one.

The crystal thickness t is here included as a parameter in the increment potential $\Delta\varphi(\mathbf{r},t)$ (and hence the pseudo potential). For a weak-phase object, $\Delta\varphi(\mathbf{r},t)$ becomes zero and $\varphi'(\mathbf{r}, t)$ degenerates into $\varphi_t(\mathbf{r})$. The peak number and positions in $\varphi'(\mathbf{r},t)$ are the same as in $\varphi_t(\mathbf{r})$, but the peak height in $\varphi'(\mathbf{r}, t)$

is different from that of the corresponding peak in $\varphi_t(\mathbf{r})$. For a very small crystal thickness, the peak heights in $\varphi'(\mathbf{r},t)$ increase as the thickness becomes larger, and the increase speed is different for different type of atoms in the same crystal. They increase more rapidly for light atoms than for heavy ones. In Fig. 1a the peak positions and the relative peak heights of the pseudo potential for different thickness values are shown schematically. Corresponding to Fig. 1a, Fig. 1b indicates that the image contrast is linear to $\varphi_t(\mathbf{r})$ under the WPOA and linear to $\varphi'(\mathbf{r},t_1)$ as well under the pseudo-WPOA. In general, with the increase of crystal thickness, the peak height of the heaviest atom (the leftmost one) reaches a maximum value at a certain thickness, then decreases gradually to zero (see the bottom but one picture in Fig. 1a), and finally becomes negative at the peak center (see the bottom picture in Fig. 1a). For heavy atoms the peak heights become zero at a smaller crystal thickness than those for light atoms in the same crystal. The critical thickness, t_c, of a crystal is defined as the thickness, for which the peak height, or say, the contrast of the heaviest atom just becomes zero. Evidently the contrast of the heaviest atom becomes negative when the thickness is above the critical thickness.

The image contrast which is predicted by the pseudo-WPOA is in agreement with that given by multislice simulation with the crystal thickness below and around the critical value [22, 23]. The critical thickness of a crystal depends on the electron wavelength and the types of the atoms that constitute the crystal.

Therefore, if the specimen thickness is below the critical thickness, equation (2) can be replaced with

$$I_{\text{img}}(\mathbf{u}) = \delta(\mathbf{u}) + 2\sigma F'(\mathbf{u})T(\mathbf{u}) \tag{4}$$

where $F'(\mathbf{u})$ is the Fourier transform of $\varphi'(\mathbf{r},t)$, hereafter named the *pseudo structure factor*. Equation (4) implies that **when the crystal thickness is large but below a critical value, the *pseudo structure factor* $F'(\mathbf{u})$ can be extracted from Fourier transform of the HREM image.**

2.3 Electron diffraction data

Electron diffraction takes place as early as when the incident electrons interact with the specimen. The objective lens acts as a diffractometer to make the electron diffraction pattern observable at its back-focal plane. Electron diffraction is not affected by the CTF of the objective lens so that the resolution limit of the diffraction pattern is much higher than that in the image. Typically the EDP contains reflections up to resolution of 1 Å or better, depending on the thickness of the crystal and the electron wavelength. However, the electron diffraction data are affected by many other factors, such as dynamical diffraction, unflatness of the specimen, the crystal shape factor and Ewald sphere curvature effect etc. Other factors like crystal bending, crystal tilt, electron irradiation effects and so on also complicate the simple relationship between the intensities and the structure factor moduli. In order to apply electron diffraction data into structure analysis, the intensities must be corrected in suitable ways. A method of diffraction intensity correction will be discussed in section 3.2.

As discussed above, information extracted for crystal structure analysis from HREM images and from the corresponding EDP compensates for each other. In the image there exist both the amplitudes and phases of the crystal *structure factor* though the image resolution is limited by the resolution of the microscope. The EDP yields the structure factor amplitudes at a much higher resolution but all phases are lost. The advantage of combining HREM and electron diffraction is obvious. Structure analysis by electron diffraction when compensated with some initial phases that are retrieved from the HREM image would simplify the phase problem. On the other hand, the image analysis, if reinforced by diffraction data, offers the possibility of improving the image resolution.

3.   IMAGE PROCESSING TECHNIQUE FOR CRYSTAL STRUCTURE DETERMINATION

It is reported that the two-stage image processing technique based on combination of high resolution electron microscopy and electron diffraction is applicable to determination of crystal structures [24]. In the first stage, the image is deconvoluted to induce a structure image at the resolution of the microscope. Generally this resolution is not high enough to resolve all atoms in the structure. In the second stage, the image resolution is improved by phase extension using the corresponding electron diffraction data that have been properly corrected. The correction of the diffraction data is often performed in the second stage.

3.1   Image deconvolution

It is known from equation (4) that the *pseudo structure factor* can be derived by

$$F'(\mathbf{u}) = \frac{I_{\text{img}}(\mathbf{u})}{2\sigma T(\mathbf{u})} \qquad (\mathbf{u} \neq \mathbf{0}) \tag{5}$$

where $T(\mathbf{u})$ is the *contrast transfer function*. $I_{\text{img}}(\mathbf{u})$ is the Fourier component of the HREM image. The CTF $T(\mathbf{u})$ mainly depends on the defocus value under which the image was taken. The defocus value can be determined by deconvolution techniques based on the direct method [14] or on the principle of maximum entropy [25]. One may refer to the literatures for more technical details about the image deconvolution.

The image after deconvolution gives a pseudo potential map with a resolution of the microscope. In the deconvoluted image atoms appear at correct positions, but the darkness of atoms is not linear to the atomic weight. Usually, not all atom columns can be resolved if the image is taken with a medium-voltage microscope. For instance, only metallic atoms of oxides can be seen in the obtained deconvoluted images when using a 200 kV microscope. In order to see oxygen atoms, the resolution of the deconvoluted image needs to be enhanced with electron diffraction data.

3.2 Image resolution enhancement

As discussed in section 2.3, the electron diffraction intensities need to be corrected before being employed for structure analysis. An empirical method has been set up to correct simultaneously all kinds of distortions in the diffraction data by referring to the heavy atom method and the Wilson statistic technique in X-ray crystallography. After correction, the intensity of each diffraction beam can approximately lead to the modulus of the corresponding structure factor [26].

Firstly, coordinates of relatively heavy atoms in the examined crystal are read out from the deconvoluted image, hence the partial structure factor $F_p(\mathbf{u})$ containing only the contribution from those heavy atoms in one unit cell are calculated. Secondly, divide the reciprocal space into a number of circular zones and the intensity of each beam within the *i*-th zone is corrected as

$$I_c(\mathbf{u})_{H_i \pm \Delta H} = \frac{\left\langle \left| F_p(\mathbf{u}) \right|^2 \right\rangle_{H_i \pm \Delta H}}{\left\langle I_o(\mathbf{u}) \right\rangle_{H_i \pm \Delta H}} I_o(\mathbf{u}) \tag{6}$$

where I_c and I_o represent respectively the corrected and the observed diffraction intensities. H_i is the averaged H in the *i*-th zone, ΔH_i the half-width of the i-th zone and $<>$ denotes the averaging operation. The validity of the method was proved with the crystal $YBa_2Cu_3O_{7-\delta}$ [26].

Resolution enhancement is achieved by phase extension of the deconvoluted image combined with the moduli of the corresponding pseudo structure factor derived from the corrected diffraction data. Phase extension is performed by using the DIMS program [27] which is developed based on the direct method in X-ray crystallography. After carrying out the first cycle of phase extension, more atoms will appear in the resulting potential map than in the deconvoluted image. Then a new set of partial structure factors is calculated, by which the observed diffraction intensities can be scaled for the second cycle. This process is continued until all atoms appear in a stable potential map. Finally, the structure is refined by Fourier synthesis that is widely used in X-ray crystallography.

It is emphasized that the final result is the structure map of the examined crystal rather than a pseudo structure map. This is because the diffraction intensities have been pushed towards the corresponding kinematical values during the calculation of partial structure factor in each cycle of the correction. In addition, in the final step, structure refinement by Fourier synthesis modifies the peak heights towards the true values to some extent. It is obvious that all the missing structure information due to the CTF zero transfer is mended after phase extension. The amplitudes are provided by the electron diffraction data, and the phases are derived from the phase extension. As a result, the resolution of the structure analysis by this method is determined by the electron diffraction resolution limit.

3.3   An application to study on the boron positions in $(Y_{0.6}Ca_{0.4})(SrBa)(Cu_{2.5}B_{0.5})O_{7-\delta}$

It is found that the compound $(Y_{0.6}Ca_{0.4})(SrBa)(Cu_{2.5}B_{0.5})O_{7-\delta}$ with the ratio of Cu and B atoms equal to 5:1 is a high-temperature superconductor [28]. In principal, the crystal structure of this compound is isomorphic to that of $YBa_2Cu_3O_{7-\delta}$. It would be interesting to make clear how B atoms substitute for Cu atoms in the crystal.

Boron is a light atom. Compared with other atoms in the crystal, B has a relatively stronger scattering power for electrons than that for X-ray. Thus the electron diffraction technique has been successfully applied to study positions of boron atoms in several other crystals [10, 29]. We study the position of B atoms in the crystal of $(Y_{0.6}Ca_{0.4})(SrBa)(Cu_{2.5}B_{0.5})O_{7-\delta}$ by means of the image processing technique.

Electron diffraction study indicates that the unit cell is orthorhombic with lattice parameters $a = 3.85$ Å, $b = 3.86$ Å and $c = 11.5$ Å. The [010] plane group of the crystal is $p2mm$ [30].

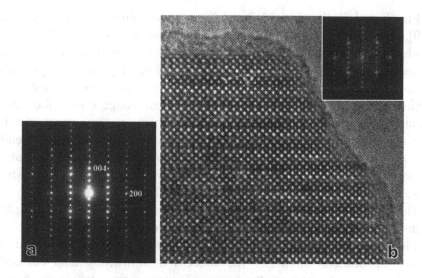

Figure 2. (a) EDP and (b) HREM image of the crystal taken along [010] direction.

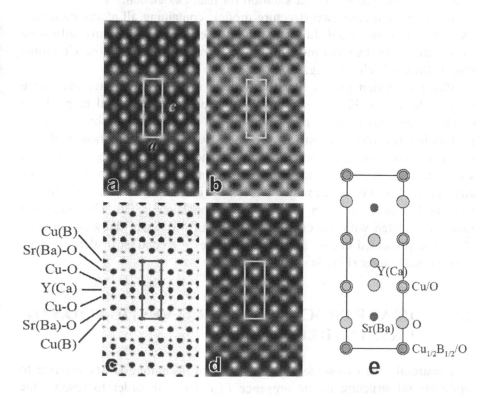

Figure 3. (a) Symmetry-averaged image; (b) deconvoluted image; (c) final projected potential map after image processing; (d) simulated image based on the structure model given in (e).

Fig. 2a and 2b show the EDP and the corresponding HREM image along [010], taken with a JEOL-2010 electron microscope operated at 200 kV, of which the point resolution is 1.94 Å. The noise filtering and symmetry averaging were performed on a thin area from Fig. 2b yielding a symmetry-averaged image (Fig. 3a). Image deconvolution is performed based on the maximum entropy principle. Due to the Fourier image effect, the correct deconvoluted image is carefully selected among several other possible solutions [31]. Fig. 3b shows the deconvoluted image where all metallic atoms are resolved as individual dark dots. Boron atoms and oxygen atoms are not visible in Fig. 3b because of the resolution limit of the microscope. Phase extension is expected to improve the resolution.

The success of electron diffraction data correction is vital to phase extension because, if they are not properly corrected, the projected potential map is not linear with the atomic weight. This makes it difficult to study the atomic substitution. Obviously the success of diffraction data correction depends, to a great extent, on the proposed structure from which the $F_p(\mathbf{u})$ is calculated. It is believed that a better match of the proposed model to the true structure would lead to a better solution for phase extension.

In the present case, two structure models containing all atoms including oxygen were constructed. In one of them, the B atoms randomly substitute for Cu atoms; in the other model, the B atoms substitute for those Cu atoms only in the Cu-O chains (Fig. 3e).

Phase extension proves that the second model gives better and more reasonable results. Fig. 3c shows the final projected potential map of the crystal along [010] with resolution up to 1 Å that is obtained after performing the phase extension for two cycles in combination with the diffraction data correction based on the second proposed mode. Hence, it is supposed that, in the examined structure, B atoms replace those Cu atoms sited in the Cu-O chains. Image simulations based on the multislice theory were performed to confirm the proposed model in Fig. 3e. The simulated image calculated with the crystal thickness of 46 Å and defocus value of -650 Å is presented in Fig. 3d, which matches the contrast of the averaged experimental image (Fig. 3a) pretty well.

4.   IMAGE PROCESSING FOR INVESTIGATION OF CRYSTAL DEFECTS

Apparently the two-stage image processing technique is not feasible to study crystal structure in the presence of defects. In order to resolve the defect structure at atomic level, Li et.al [32, 33] proposed an image processing method based on the high-resolution imaging technique with a medium-voltage FEG microscope, combined with image deconvolution and dynamical diffraction effect correction. This approach is successfully applied to investigate the core structure of a dislocation [34] and a misfit dislocation complex [35] in a SiGe/Si epilayer.

4.1 Methods

Due to the high coherence of electron source in an FEG electron microscope, the information resolution limit of an image can reach the atomic level, much higher than the point resolution. However, the structure information, especially those in the high-resolution region, is seriously distorted due to the strong oscillation of the CTF. The distortion can be removed by the image deconvolution technique.

The principle of image deconvolution is based on the same theory of the pseudo-WPOA but the procedure is different in the case of a defect structure. Replace $F'(\mathbf{u})$ in equation (5) with $F''(\mathbf{u})$, which is the structure factor corresponding to an artificial unit cell with a large size, where the examined defect exists.

In practice, the technique mainly contains three parts: defocus determination, image deconvolution and dynamical diffraction correction. Firstly, the defocus is roughly determined by means of the Thon map [36] from an amorphous region close to the interesting area. Then several trial defocus values close to that of the roughly determined are assigned with a small focus step, say 5 Å. So a set of $F''_{\text{trial}}(\mathbf{u})$ is obtained by equation (5) from the diffractogram of the image including the examined defect. After that, to minimize the dynamical effect, all $F''_{\text{trial}}(\mathbf{u})$ are corrected by forcing the integrated amplitudes of reflections to be equal to the amplitudes of corresponding structure factors $F(\mathbf{u})$ of the perfect crystal [33]. The corrected $F''_{\text{trial}}(\mathbf{u})$ for the i-th pixel is given by

$$\left| F''_{\text{trial}}(\mathbf{u}) \right|_i^{\text{cor}} = K_u \left| F''_{\text{trial}}(\mathbf{u}) \right|_i = \frac{|F(\mathbf{u})|}{\sum_i \left| F''_{\text{trial}}(\mathbf{u}) \right|_i} \left| F''_{\text{trial}}(\mathbf{u}) \right|_i \qquad (7)$$

where K_u is the correction coefficient that is a constant for all pixels within one reflection but different for other reflections.

Accordingly, a series of trial potential maps of the defect structure is achieved by inverse Fourier transform of the corrected $F''_{\text{trial}}(\mathbf{u})$. Finally, from among those trial maps a best one can be selected, in which atoms in both the perfect and the defect regions are best resolved.

4.2   An application to study of a dislocation structure in $Si_{0.76}Ge_{0.24}/Si$

The crystal structure of SiGe is isomorphic to that of Si. In the [110] projected structure of $Si_{0.76}Ge_{0.24}$, the distance between two adjacent atoms is about 1.4 Å, which is close to the information resolution limit of a 200 kV FEG high-resolution electron microscope. In order to reveal individual atom columns after image processing, the structure information of reflections 111, 220, 113 and 004 with the spatial frequency up to (1.4 Å)$^{-1}$ should all be included in the original image. This can be viewed from the diffractogram of the image.

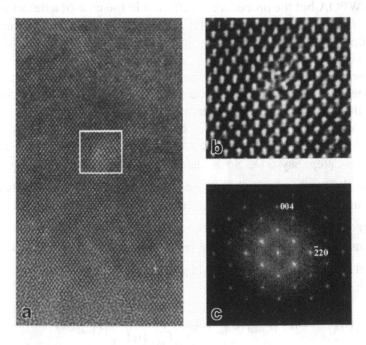

Figure 4. (a) [110] image with a Lomer dislocation in the framed region; (b) magnified image of the framed area in (a) and (c) diffractogram of a circular area in (a).

Figure. 4a is a [110] projected high-resolution image taken with a JEM-2010FEG electron microscope with a spherical aberration coefficient 0.5 mm. A dislocation can be seen in the framed area, of which a magnified photo is shown in Fig. 4b. It can be seen that two extra half {111} planes running from the top left to the bottom right and from the top right to the bottom left, respectively. Both of them end in the center of the picture.

Although it can be recognized that the dislocation is of the Lomer type, the atomic configuration is not clearly shown in the image.

Figure 4c is the diffractogram obtained from a circular area about 175 Å in diameter, with the dislocation at the center. All reflections with the spatial frequencies for resolving the two adjacent atoms appear with rather dominant intensity. This implies that no significant reflection is lost by CTF crossovers. Based on the Thon diffractogram [36] method, the defocus value is roughly determined to be -390 Å.

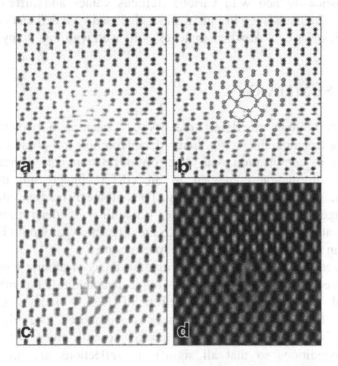

Figure 5. (a) Deconvoluted image after correction of the dynamical diffraction effect; (b) the same as (a) but showing the atomic arrangement schematically; (c) result of deconvolution without dynamical diffraction correction and (d) simulated image based on (b).

Seven trial $F''_{trial}(\mathbf{u})$ sets in all were obtained from the corresponding trial defocus values from -405 Å to -375 Å with a step of 5 Å, so seven corrected trial $F''_{trial}(\mathbf{u})$ sets were then calculated. Inverse Fourier transform of those $F''_{trial}(\mathbf{u})$ sets yields seven potential maps accordingly. Figure 5a shows the best map, in which all atoms are best resolved. The atomic configuration of the dislocation core structure is seen more clearly by linking the adjacent atoms to show the bonding situation (Figure 5b). It can be seen that atoms forming the core structure of Lomer dislocation are resolved individually. Each dark dot illustrates the position of an atomic column projected in the [110] direction. A five-member ring and a seven-member ring form a symmetric undissociated core without dangling bond. This is in agreement with the Hornstra model [37]. Fig. 5c is the potential map obtained with the same defocus value as Fig. 5a, but without dynamical diffraction correction. The differences between Fig. 5a and Fig. 5c show clearly the positive effect of dynamical diffraction correction.

Based on the model shown in Fig. 5b, image simulation is performed by the multislice method with various defocus values and different crystal thickness. The one shown in Fig.5d was calculated with defocus - 400 Å and crystal thickness 61.4 Å, which matches quite well with Fig. 5b by contrast.

5.   SUMMARY

For determining the defect-free crystal structure, a two-stage image processing technique including image deconvolution and phase extension is developed in combination with electron diffraction data correction. The structural information contained in an HREM image and that in the corresponding EDP are supplementary. To combine diffraction intensity data with image phase data affords the possibility of determining crystal structures at a resolution that is close to the diffraction resolution limit, much higher than the point-resolution of the microscope.

To reveal the core structure of a crystal defect at atomic level, another image processing method based on image deconvolution combined with dynamical diffraction effect correction of HREM images taken with medium-voltage FEG electron microscopes is proposed. The original image should be carefully selected from among several others with different defocus conditions so that all significant reflections are present in its diffractogram. The core structure of defects can be revealed in the final result with a resolution up to the information resolution limit of the FEG electron microscope.

The pseudo-WPOA theory proves the validity of introducing diffraction crystallographic methods based on the kinematical diffraction theory into HREM structure analysis.

References

1.   Schiske, P.: Proceedings 4th European Conference on Electron Microscopy, Rome, (1968) 145-146.
2.   Kirkland, E. J.: Ultramicroscopy 15 (1984) 151-172.

3. Kirkland, E. J.; Siegel, B. M.; Uyeda, N.; Fujiyoshi, Y.: Ultramicroscopy 17 (1985) 87-103.
4. Saxton, W. O.: Proc. 11th Intern. Congr. on Electron Microscopy, Kyoto, (1986) Post-deadline paper, 1.
5. Van Dyck, D.; Op de Beeck, M.: Proc. 11th Intern. Congr. on Electron Microscopy, Seattle, (1990) 26-27.
6. Van Dyck, D.; Lichte, H.; Van der Mast, K. D.: Ultramicroscopy 64 (1996) 1-15.
7. Coene, W.; Janssen, G.; Op de Beecl, M.; Van Dyck, D.: Phys. Rev. Lett. 69 (1992), 3743-3746.
8. Jia, C.L; Thust, A.: Phys. Rev. Lett. 82 (1999), 5052.
9. Tang, D.; Jannsen J.; Zandbergen, H. W.: Acta Cryst. A51 (1995) 188-197.
10. Jansen, J., Tang, D.; Zandbergen, W.; Schenk,H.: Acta Crystallogr. A 54 (1998) 91.
11. Li, F. H.: Acta Physica Sinica 26 (1977) 193-198 (In Chinese).
12. Ishizuka, K.; Miyazaki, M.; Uyeda, N.: Acta Cryst. A38 (1982) 408-413.
13. Fan, H. F.; Zhong, Z. Y.; Zheng, C. D.; Li, F. H.: Acta Cryst. A41 (1985) 163-165.
14. Han, F. S.; Fan, H. F; Li, F. H.: Acta Cryst. A42 (1986) 353-356.
15. Hovmöller, S.: Utramicroscopy 41 (1992) 121-136.
16. Zou, X. D.: Electron Crystallography of inorganic Structure – Theory and Practice, Chemical Communication 5 (1995).
17. Dong, W.; Baird, T.; Fryer, J. R.; Gilmore, C. J.; MacNicol, D. D.; Bricogne, G.; Smith, D. J.; O'Keefe, M. A.; Hovmöller, S.: Nature 355 (1992) 605.
18. Sinkler, W.; Marks, L. D.: Ultramicroscopy 75 (1999) 251-268.
19. Spence, J.C.H: Experimental high-resolution electron microscopy, Clarendon Press , Oxford (1981).
20. Li, F. H.; Tang, D.: Acta Cryst. A41 (1985) 376-382.
21. Cowley, J. M.; Moodie, A. F.: Acta Cryst. 10 (1957) 609-619.
22. Li, F. H.; Hashimoto, H.: Acta Cryst. B40 (1984) 454-461.
23. Tang, D.; Teng, C. M.; Zou, J.; Li, F. H.: Acta Cryst. A42 (1986) 340- 342.
24. Li, F. H.: Proc. 13th Int. Congr. on Electron Microscopy, Paris, (1994) 481-484.
25. Hu, J. J.; Li, F. H.: Ultramicroscopy 35 (1991) 339–350.
26. Huang, D. X.; Liu, W.; Gu, Y. X.; Xiong, J.W.; Fan, H. F.; Li, F. H.: Acta Cryst. A52 (1996) 152-157.
27. Fu, Z. Q.; Fan, H. F.: J. Appl. Cryst. 27 (1994) 124-127.
28. Che, G. C., Liu, G. D., Wu, F., Chen, H., Jia, S. L., Dong, C. & Zhao, Z. X.: Physica (Utrecht) C, 341-348 (2000) 391-394.
29. Cowley, J. M. Acta Cryst. 6 (1953), 522-529.
30. Wang, H. B.; Jiang, H.; Li, F. H.; Che, G. C.; Tang, D.: Acta Cryst. A58 (2002) 494-501.
31. Wang, H.B.; Wang, Y.M.; Li, F.H.: Ultramicroscopy, in press (2004).
32. He, W. Z.; Li, F. H.; Chen, H.; Kawasaki, H.; Oikawa, T.: Ultramicroscopy 70 (1997) 1-11.
33. Li, F. H.; Wang, D.; He, W. Z.; Jiang, H.: J. Electron Microscopy 49 (2000) 17-24.
34. Wang, D.; Chen, H.; Li, F. H.; Kawasaki, K.; Oikawa, T.: Ultramicroscopy 93 (2002) 139-146.
35. Wang, D; Zou, J; He, W.Z.; Chen, H., Li, F. H.; Kawasaki, K.; Oikawa, T.: Ultramicroscopy 98 (2004) 259-264.
36. Thon, F.: Phase contrast electron microscopy. In: Electron Microscopy in Material Science (Ed. U. Valdre), Academic press, (1971) 570-625.
37. Bourret, A.; Desseaux, J.; Renault, A.: Philosophical Magazine A 45 (1982) 1-20.

STRUCTURE DETERMINATION FROM HREM BY CRYSTALLOGRAPHIC IMAGE PROCESSING

Xiaodong Zou and Sven Hovmöller

Structural Chemistry, Arrhenius Laboratory, Stockholm University, SE-106 91 Stockholm, Sweden

Abstract: This chapter demonstrates that it is possible to perform *ab initio* crystal structure determination by HREM. The various steps in a crystal structure determination; recording and quantifying HREM images, analysis and processing of these data to retrieve the projected potential of the crystal and finally determine the atomic coordinates are described.

Key words: HREM, Structure factors, Crystallographic image processing, Contrast transfer function

1.   INTRODUCTION

Electron crystallography is an important technique for determination of unknown crystal structures, complementing X-ray and neutron diffraction. The birth of electron crystallography dates back to the discovery that electrons possessed both particle and wave properties. The crystallographers Pinsker, Vainshtein and Zvyagin, solved inorganic crystal structures from electron diffraction patterns, notably texture patterns [1,2,3]. They designed and used their own electron diffraction cameras and quantified electron diffraction intensities and treated them kinematically. In spite of this early start in 1947, electrons have not been used much for crystal structure determination outside Moscow until the last two decades. Unfortunately, the development of electron crystallography for the study of inorganic crystals was long hampered by an exaggerated fear of dynamical effects.

In the early days of high resolution electron microscopy (HREM), some special classes of structures were solved by recognising basic units of a

275

T.E. Weirich et al. (eds.), Electron Crystallography, 275–300.

projected structure, and determining their arrangement in larger unit cells. The extensive studies of so-called block oxides constituted the beginning of high resolution electron microscopy on inorganic compounds[4]. This meant that a model had to be proposed and verified by comparisons (usually qualitative) with extensive contrast calculations based on dynamical scattering theory. Typically a set of images was calculated, with a range of defocus and crystal thickness values[5]. Structure determination *ab initio* from HREM was not considered to be practicable.

Experience from a number of structure determinations in recent year have proved in practice that unknown crystal structures can be solved from HREM images, irrespective of whether the structures contain light or heavy elements, provided the image is taken from a thin crystal. There is no need to guess the experimental conditions, such as defocus and crystal thickness, since these can be determined experimentally from HREM images. Furthermore, the very important parameters astigmatism and crystal tilt, which in most cases of image simulations have been set to zero although they often cannot be neglected, can also be determined experimentally directly from HREM images. The distortions caused by the above mentioned factors can then largely be compensated for by crystallographic image processing. Random noise can also be eliminated by averaging over many unit cells. The projected crystal symmetry can be determined and imposed exactly to the data. In this way a projected potential map is reconstructed. For structures with one short unit cell axis (≤ 5 Å), atomic coordinates are read out directly from the map, with a precision of 0.2 Ångström or better. For more complex structures, several projections can be merged into a 3D structure. Finally, it is possible to improve the structure model by least squares refinement against accurately quantified SAED data. After refinement, the atoms are typically located within 0.02 Ångström of their correct positions determined from X-ray crystallography.

Here we will demonstrate that it is possible to perform *ab initio* crystal structure determination by HREM. The various steps in a crystal structure determination; recording and quantifying HREM images, analysis and processing of these data to retrieve the projected potential of the crystal and finally determine the atomic coordinates are described.

2. INTERACTIONS BETWEEN ELECTRONS AND MATTER

Electrons interact thousands of times more strongly with matter than X-rays do. This has the advantage that extremely small crystals can be studied, down to a size of a few nanometers in all directions. This is about a million times smaller than what is needed even for X-ray diffraction using a synchrotron. Many compounds form so small microcrystals that electrons are the only possible source for analysis of their structures.

On the other hand, the strong interaction between electrons and matter gives rise to dynamical effects[13] which complicate quantitative analysis of the experimental data. This has led to a pessimistic view of the possibility of direct crystal structure determination by electron crystallography[14], especially for compounds containing heavy elements. Already after penetrating a few nanometers into the sample, a considerable fraction of the incident beam has been scattered. These scattered electrons may well be scattered again as they propagate through the sample. This multiple scattering results in a diffraction pattern or an image which no longer can be interpreted as a linear function of the structure factor amplitudes or projected crystal potential. It has been widely assumed that this multiple scattering is so severe that not even the thinnest crystals that can be obtained experimentally can be treated as singly scattering (kinematical) objects and that a direct interpretation of HREM images in terms of structure in general would not be possible. Based on this argument, image simulations have been considered necessary for interpretation and validation of suggested structure models.

Plots of amplitudes and phases of the diffracted beams at the exit surface of a crystal calculated by image simulation seem to show rapid changes of both amplitudes and phases with increasing crystal thickness, so that inorganic crystals can not be treated by a simple linear kinematic model if they are thicker than about ten or twenty Ångströms. However, these rapidly changing phases are partly due to the electron wave propagation in the crystal. After the effects of propagation have been removed, phases of the strong diffracted beams are close to the crystallographic structure factor phases, even for crystals thicker than 100Å. Furthermore, the phases obtained from the Fourier transform of the HREM images are not the same as the phases of the diffracted beams at the exit surface of the crystal. The former are affected by the defocus and astigmatism of the objective lens. The relation between the different types of phases, is described by Zou[15]. It has been shown experimentally that the structure factor phases (which are the ones that are needed for a structure determination) can be correctly determined directly from HREM images of relatively thin crystals[10, 15, 16]. This is also supported by theoretical considerations[17,18].

3.   CRYSTAL POTENTIAL AND STRUCTURE FACTORS

Electrons are scattered by the electrostatic field generated by the electrostatic potential difference in a crystal. Atoms in a crystal give sharp and positive peaks to the potential. The relation between the potential $V(r)$ and the structure factors $F(u)$ are

$$V(r) = k\sum_{u} F(u)\exp[-2\pi i(u \cdot r)] \qquad (1)$$

where k is a constant. The potential at any point in the crystal can be calculated by adding the vectors $F(u)\exp[-2\pi i(u \cdot r)]$ for all the structure factors $F(u)$, i.e. by Fourier synthesis. In fact, each vector of reflection u, together with that of its Friedel mate $-u$, generate a cosine wave[19].

$$F(u)\exp[-2\pi i(u \cdot r)] + F(-u)\exp[-2\pi i(-u \cdot r)] = 2|F(u)|\cos[\phi(u)-2\pi(u \cdot r)] \quad (2)$$

The direction and the periodicity of each cosine wave are given by its index $u = (hkl)$, the amplitude of the cosine wave is $2|F(u)|$, proportional to the structure factor amplitude $|F(u)|$. More importantly, the positions of the maxima and minima of the cosine wave (in relation to the unit cell origin) are determined by the structure factor phase $\phi(u)$. If both the amplitudes $|F(u)|$ and the phases $\phi(u)$ of the structure factors for all reflections u are known, the potential $\varphi(r)$ can be obtained by adding a series of such cosine waves.

An example of this procedure is shown in Fig. 1. This example shows the build-up of the 2D potential of Ti_2S projected along the short c axis, but the principle is the same for creating a 3D potential. The potential is a continuous function in real space and can be described in a map (Fig. 1). On the other hand, the structure factors are discrete points in reciprocal space and can be represented by a list of amplitudes and phases (Table 1). In this Fourier synthesis we have used the structure factors calculated from the refined coordinates of Ti_2S[20].

If the Fourier synthesis is carried out by adding in the strong reflections first, we will see how fast the Fourier series converges to the projected potential. The positive potential contribution from the reflection is shown in white, whereas the negative potential contribution is shown in black. Most of the atoms are located in the white regions of each cosine wave, but the exact atomic positions will not become evident until a sufficiently large number of structure factors have been added up.

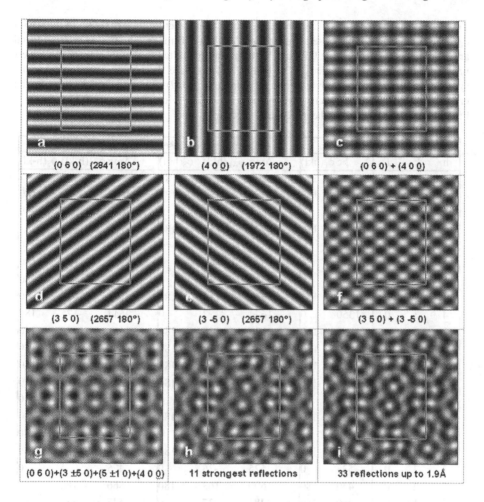

Figure 1 Fourier synthesis of the projected potential map of Ti_2S along the c-axis. Amplitudes and phases of the structure factors are calculated from the refined atomic coordinates of Ti_2S and listed in Table 1. The space group of Ti_2S is *Pnnm* and unit cell parameters a= 11.35, b=14.05 and c=3.32 Å.

Among all the 33 reflections up to 1.6 Å, the strongest one is (0 6 0). This reflection generates a cosine wave which cuts the a axis zero times per unit cell and the b axis 6 times. It is the phase of each reflection which determines where the maximum and minimum of the cosine wave are. The phase of (0 6 0) is 180° (Table 1) and thus $\cos 180° = -1$, the contribution to the potential at the origin is negative (black) (Fig. 1a). Similarly, the cosine wave given by (4 0 0) reflection cuts the a axis 4 times and the b axis zero times. The phase of (4 0 0) is also 180° so its contribution to the potential at the origin is also negative (Fig. 1b). The summation of these two cosine waves is shown in Fig. 1c.

Table 1   Amplitudes $F_c(hkl)$ and phases $\phi_F(hkl)$ of all structure factors $F(hkl)$ of Ti_2S with $d_{hkl} > 1.9$Å compared with those obtained from the HREM image. The phases $\phi_{ctf}(hkl)$ listed are after compensation for the CTF and with the origin shifted to the position of the lowest phase residual for pgg symmetry. The phases $\phi_{sym}(hkl)$ are the phases after imposing the symmetry. The experimentally determined phases $\phi_{sym}(hkl)$ are virtually identical to the crystallographic structure factor phases $\phi_F(hkl)$; only four (highlighted) out of 33 reflections have wrong phases and all of them are weak reflections.

h	k	l	Structure factor		Amplitudes and phases from experimental image					
			Fc	ϕ_F(hkl)	Amp	ϕ_{AF}(hkl)	ϕ_S(hkl)	Amp	ϕ_{AF}(h-kl)	ϕ_S(h-kl)
0	6	0	2841	180	2706	-133	180	-	0	-
3	5	0	2657	180	2371	174	180	4022	-179	180
5	1	0	2553	0	2609	49	0	3099	2	0
4	0	0	1972	180	1904	140	180	-	-	-
1	7	0	1935	0	613	-10	0	625	19	0
5	2	0	1841	0	1241	-62	0	1553	132	180
5	4	0	1560	0	255	15	0	417	-104	180
4	5	0	1504	180	516	115	180	586	-20	0
4	3	0	1350	0	1674	7	0	2544	-175	180
2	6	0	1308	180	1026	-140	180	1455	180	180
5	3	0	1262	180	530	110	180	642	160	180
4	4	0	1129	180	837	-140	180	1322	133	180
3	6	0	1030	180	274	176	180	336	-22	0
3	4	0	707	0	843	-39	0	1006	180	180
2	5	0	630	180	448	-122	180	887	-1	0
1	2	0	584	0	721	-6	0	1131	169	180
4	2	0	560	180	281	165	180	251	-137	180
3	3	0	533	0	525	-36	0	475	5	0
2	2	0	459	180	690	163	180	866	180	180
3	1	0	440	180	481	37	0	547	-33	0
1	5	0	439	0	431	-19	0	332	-38	0
1	6	0	416	0	242	36	0	304	-117	180
2	0	0	396	0	755	-164	180	-	-	-
4	1	0	392	180	312	179	180	186	45	0
3	2	0	258	180	-	-	-	-	-	-
0	2	0	236	180	721	-115	180	-	-	-
2	4	0	215	180	831	-178	180	406	119	180
2	3	0	193	180	636	5	0	436	126	180
1	3	0	192	180	1054	-1	0	745	-3	0
0	4	0	177	180	612	-175	180	-	-	-
1	1	0	147	180	1165	180	180	788	-138	180
1	4	0	123	180	308	-114	180	372	3	0
2	1	0	89	180	255	-38	180	360	-39	0

Among all the 33 reflections up to 1.6 Å, the strongest one is (0 6 0). This reflection generates a cosine wave which cuts the *a* axis zero times per unit cell and the *b* axis 6 times. It is the phase of each reflection which determines where the maximum and minimum of the cosine wave are. The phase of (0 6 0) is 180° (Table 1) and thus $\cos 180° = -1$, the contribution to the potential at the origin is negative (black) (Fig. 1a). Similarly, the cosine wave given by (4 0 0) reflection cuts the *a* axis 4 times and the *b* axis zero times. The phase of (4 0 0) is also 180° so its contribution to the potential at the origin is also negative (Fig. 1b). The summation of these two cosine waves is shown in Fig. 1c.

The cosine waves generated by the second strongest reflection (3 5 0) and its symmetry-related reflection (3 -5 0) are shown in Figs. 1d and e. Both cosine waves cut the *a* axis 3 times per unit cell and the *b* axis 5 times, however, they are oriented differently. The summation of the two symmetry-related cosine waves is shown in Fig. 1f).

When the 4 strongest independent reflections (in total 6 reflections including symmetry-related ones) are added, the map already shows some indication of where the atoms should be located within the unit cell (Fig. 1g). After the strongest 1/3 (11) of all the independent reflections has been included, all the atoms appear in the map (as white dots) (Fig. 1h). The map generated from all the 33 unique reflections (Fig. 1i) is only slightly better, because the 22 reflections further added in are weaker and so do not contribute very much to the Fourier map. The weak reflections are, however, equally important as the strong ones in the last step of a structure determination, the refinement.

In summary, as along as the crystallographic structure factor phases of the strongest reflections are correct, the reconstructed map represents the true (projected) potential distribution of the crystal. Once the potential distribution of the crystal is available, atomic positions can immediately be determined from the peaks of high potential in the map. Thus *to determine crystal structures is equivalent to determine crystallographic structure factors.*

4.   PHASES IN HREM IMAGES

Electron crystallography provides two major advantages over X-ray crystallography for determination of atomic positions in crystal structures; extremely small samples can be analysed and the crystallographic structure factor phases can be determined from images[21]. The crystallographic structure factor phases must be known in order to arrive at a structure model, but these phases cannot be measured experimentally from diffraction patterns.

It is the *raison d'être* of electron microscopy that the phase information is preserved in the EM images, such that they represent a magnified image of the object. DeRosier and Klug[21] recognised that the crystallographic structure factor phases could be extracted directly from the Fourier transforms of digitised images, under the assumption of weak scattering and linear imaging, i.e. for very thin crystals. This discovery, for which Klug was awarded the Nobel Prize in Chemistry in 1982, can be considered as the birth of structure determination from HREM images.

5.   HREM IMAGES AND PROJECTED POTENTIAL

The relation between an HREM image and the projected crystal potential is quite complex if the crystal is thick. To obtain an image which can be directly interpreted in terms of projected potential, crystals have to be well aligned, thin enough to be close to weak-phase-objects and the defocus value for the objective lens should be optimal, i.e. at the Scherzer defocus.

For a weak-phase-object, the Fourier transform of the HREM image $I_{im}(\mathbf{u})$ is related to the structure factor $F(\mathbf{u})$ by[22]:

$$I_{im}(\mathbf{u}) = \delta(\mathbf{u}) + k'T(\mathbf{u})F(\mathbf{u}) \tag{3}$$

where k' is a constant and $T(\mathbf{u})$ is called the contrast transfer function (CTF). The effects of the contrast transfer function will be discussed in section 8.

For an image taken at Scherzer defocus, where $T(\mathbf{u}) \approx -1$ over a large range of resolution, the structure factor $F(\mathbf{u})$ can be obtained from the Fourier transform of the HREM image $I_{im}(\mathbf{u})$:

$$F(u) \approx -\frac{1}{k'}I_{im}(u) = \frac{1}{k'}\exp(i\pi)I_{im}(u) \tag{4}$$

After fixing the unit cell origin (see section 10), the amplitudes and phases of the crystallographic structure factors are proportional to the amplitudes and phases of the corresponding Fourier components of the Fourier transform $I_{im}(\mathbf{u})$ of the image. All the phases in the Fourier transform $I_{im}(\mathbf{u})$ of the image, within the Scherzer resolution limit are shifted by 180° from the structure factor phases.

The projected potential can be obtained from the Fourier transform $I_{im}(\mathbf{u})$ of the image:

$$V(\mathbf{r}) \approx \frac{-1}{k'}\sum_{\mathbf{u}} I_{im}(\mathbf{u})\exp\left[-2\pi i(\mathbf{u}\cdot\mathbf{r})\right] = -\frac{I_{im}(\mathbf{r})}{k'} \qquad (5)$$

The projected potential is proportional to the negative of the image intensity, i.e. black features in HREM positives (low intensity) correspond to atoms (high potential). The corresponding image is called the structure image.

Accurate atomic coordinates can be determined from the HREM images, with the help of crystallographic image processing. The experimental procedures for structure solution of inorganic crystals by HREM and crystallographic image processing are summarized in Figure 2.

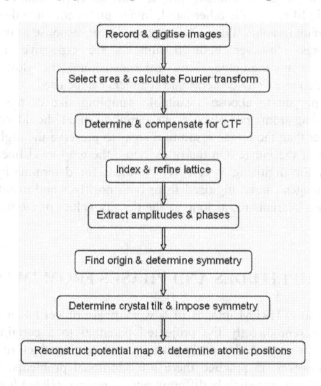

Figure 2 Flow diagram of structure determination by crystallographic image processing.

6. RECORDING AND QUANTIFICATION

Taking good HREM images is a critical step of any structure determination. The thinnest parts of the crystals should be used, to avoid strong multiple scattering. Only then are we close to the kinematical condition, where the relation between the amplitudes and phases extracted from HREM images and the structure factor amplitudes and phases is simple.

Whenever possible, an amorphous area at the edge of the crystal should be included in the HREM images when images are recorded. This will help in the determination of the contrast transfer function. A set of images with different defocus values should be recorded, although it is often possible to solve structures from just a single image. The reasons will be described in section 9.

HREM images can be recorded on different media, such as photographic films, video rate CCD-cameras, slow-scan CCD cameras and image plates. For on-line digitisation, slow-scan CCD cameras provide good linear response and large dynamical range, but cover a smaller area than photographic films. On the other hand, image plates combine the large view area of the photographic films with the good linear response of the slow scan CCD cameras. However, both instruments are expensive. For off-line digitisation of photographic films, microdensitometers, slow-scan CCD cameras, video rate CCD-cameras and scanners can be used.

It is important to choose a suitable sampling size of the image, i.e. number of Ångströms per pixel. Each sampling pixel should be about 2-3 times smaller than the image resolution so as to preserve the high resolution information of the image. On the other hand, the grey-level linearity of the instruments for digitising is not very critical for determination of atom positions. Images can be digitised from both positives and negatives, using any digitising instruments, as long as the density values of the image are not saturated.

7.   AMPLITUDES AND PHASES FROM IMAGES

Theoretically, HREM images of a weak-phase-object taken at Scherzer defocus represent directly the projected potential to a certain resolution (which may or may not be sufficient to reveal all the structure features of interest). However, in practice there are additional problems. Features in different unit cells are slightly different and symmetry-related features in the same unit cell are not exactly identical as they should be, as seen in Fig. 2. Lattice averaging over all the unit cells can be applied to produce an average structure. A further improvement can be reached by crystallographic image processing, imposing the crystallographic symmetry of the projection. These two steps are performed in reciprocal space. The different steps involved in solving an unknown crystal structure from HREM images and the refinement against SAED data will be outlined in the rest of the chapter, using several inorganic structures as examples; Ti_2S[26], $K_7Nb_{15}W_{13}O_{80}$[16],

$K_2O \cdot 7Nb_2O_5{}^{27}$ and $Ti_{11}Se_4{}^{12}$. The crystallographic image processing was carried out with the computer program CRISP[28], which has been designed especially for electron crystallography.

Figure 3a shows an HREM image of Ti_2S crystal taken along the short c axis. This image is first digitised and the thinnest area is selected from the image. The Fourier transform (FT) from this thinnest area is calculated. The image density is a set of real numbers, while the Fourier transform of the image is a set of complex numbers which can be expressed as an amplitude part and a phase part. The amplitude part of the FT is shown in Fig. 3b.

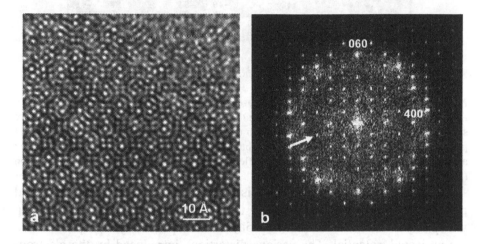

Figure 3 (a) HREM image of Ti_2S along the c axis taken on a Philips CM30/ST microscope at 300kV. (b) The Fourier transform of the image in (a). The dark ring in the FT, corresponding to the first crossover of the contrast transfer function, is indicated by an arrow.

The crystal structure information is periodic in the image and thus is concentrated at discrete diffraction spots in the FT. The amplitude and phase part of the FT around one such diffraction spot is shown in Fig. 4. The lattice in the FT is refined using all spots and the exact position of each reflection is predicted from the refined lattice. Integrated amplitudes and phases for all reflections are extracted from the numerical data around the expected center of the diffraction spots[28].

Digital Map of FFT

Ampl	40	41	42	43	44	45	46	47	48
-67	24	37	19	25	68	55	25	3	16
-66	30	74	41	77	73	38	27	44	32
-65	54	41	94	50	29	50	100	19	27
-64	88	100	77	191	69	139	40	80	50
-63	45	136	58	152	771	186	177	129	49
-62	52	84	164	326	1588	666	262	102	38
-61	87	54	100	60	930	382	243	72	36
-60	64	53	120	161	122	122	106	113	66
-59	34	64	152	32	157	47	99	43	48
-58	58	44	46	54	104	64	68	19	3
-57	31	86	69	61	60	94	80	31	44
Phas	40	41	42	43	44	45	46	47	48
-67	45	-155	98	165	123	136	140	-96	111
-66	-95	76	16	-108	-93	-126	93	2	141
-65	79	-30	-148	135	100	147	125	-3	53
-64	-24	-179	147	-42	-49	-64	22	82	-81
-63	-101	101	-56	175	91	93	-117	38	-180
-62	-71	-1	-108	58	32	16	-161	-17	80
-61	-149	-139	-170	18	-22	-43	135	-21	75
-60	127	137	-40	97	72	93	35	-54	54
-59	22	100	-113	-98	27	55	-155	-126	21
-58	-77	-27	47	154	-145	-178	77	-2	54
-57	-162	-151	-52	-22	-142	-127	-61	142	129

Figure 4 Extraction of amplitudes and phases from the Fourier transform of the image. Amplitudes and phases around the reflection (3 5 0) at pixel position (44 -62) in the FT are shown in digits. The amplitude for reflection (3 5 0) is extracted by first integration of 3x3 pixels around position (44 -62) and then subtraction of the averaged background estimated around the diffraction spot. The phase for reflection (3 5 0) is the phase value at the position (44 -62) i.e. 32°.

If an inverse Fourier transform is calculated using the amplitudes and phases extracted from the FT for all the reflections, a lattice averaged map with *p1* symmetry is obtained (Fig. 5a). This map is not yet proportional to the projected potential. The various distortions introduced by the electron-optical lenses, crystal tilt etc. must first be corrected for.

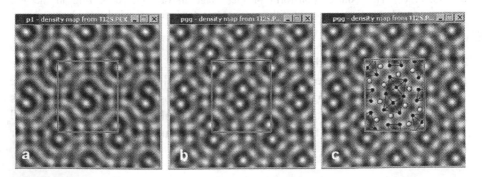

Figure 5 (a) The lattice averaged map of Ti_2S with *p1* symmetry, obtained by inverse Fourier transformation of the amplitudes and phases of all reflections extracted from the FT of the image. (b) The projected potential map reconstructed from the image after compensating for the contrast transfer function and imposing the crystal symmetry *pgg*. (c) The structure model is deduced from the reconstructed potential map (b) and superimposed on (b). The 24 strongest peaks (of which 6 are unique) in the unit cell are assigned to Ti atoms (in solid circles), which form octahedral clusters. The 12 weaker peaks (two unique) are S atoms (marked by open circles).

8. CORRECTING DEFOCUS AND ASTIGMATISM

As mentioned in section 6, the structure factors $F(\mathbf{u})$ are proportional to the Fourier components $I_{im}(\mathbf{u})$ of the HREM image and the projected potential is proportional to the negative of the image intensity, if the image is taken Scherzer defocus where the contrast transfer function $T(\mathbf{u}) \approx -1$. In general, the Fourier components $I_{im}(\mathbf{u})$ are proportional to the structure factors $F(\mathbf{u})$ multiplied by the contrast transfer function (CTF). The contrast transfer function $T(\mathbf{u}) = D(\mathbf{u})\sin\chi(\mathbf{u})$ is not a linear function. It contains two parts[29]: an envelope part $D(\mathbf{u})$ which dampens the amplitudes of the high resolution components:

$$D(\mathbf{u}) = \exp[-\tfrac{1}{2}\pi^2\Delta^2\lambda^2\mathbf{u}^4]\exp[-\pi^2\alpha^2\mathbf{u}^2(\varepsilon + C_s\lambda^2\mathbf{u}^2)^2] \qquad (6)$$

and an oscillating part $\sin\chi(\mathbf{u})$ which determines the contrast of the image, where

$$\chi(\mathbf{u}) = \pi\varepsilon\lambda\mathbf{u}^2 + \frac{\pi C_s\lambda^3\mathbf{u}^4}{2} \qquad (7)$$

ε is the defocus value, C_s the spherical aberration constant, Δ the focus spread and α the electron beam convergence.

The objective lens transfers different structure factors $F(\mathbf{u})$ into the HREM image in different ways, depending on the value of the contrast transfer function $D(\mathbf{u})\sin\chi(\mathbf{u})$. Phases of the Fourier components $I_{im}(\mathbf{u})$ of the image are related to the phases of the structure factors in the following way: those Fourier components in the range where $\sin\chi(\mathbf{u}) > 0$ will have the same phases as the phases of structure factors, giving rise to the same contrast in the image as the projected potential; those Fourier components in the range where $\sin\chi(\mathbf{u}) < 0$ suffer a phase change of 180°, giving a reversed contrast in the image. As a result, an HREM image is usually formed by the combination of Fourier components with both correct and inverted phases with respect to the structure factors.

Amplitudes of the structure factors are sampled by $|D(\mathbf{u})\sin\chi(\mathbf{u})|$ when they are transferred to the image. The most significant effect of the lens to the amplitudes is caused by the $|\sin\chi(\mathbf{u})|$ part, which oscillates with $\mathbf{u}$. Reflections in the resolution regions where $|\sin\chi(\mathbf{u})| \approx 1$ are maximally transferred by the lens, while those at resolutions where $\sin\chi(\mathbf{u}) \approx 0$ are not transferred at all. This can be seen in the Fourier transform of HREM images from amorphous materials (Fig. 6), where the highest amplitudes (brightest areas) correspond to $|\sin\chi(\mathbf{u})| \approx 1$, while the lowest amplitudes (darkest areas) correspond to $|\sin\chi(\mathbf{u})| \approx 0$. If there is no astigmatism in the objective lens, a set of alternating bright and dark rings may be found in the FT of the image (Fig. 6b). If there is astigmatism, these rings become a set of ellipses (Fig. 6c) or in more severe cases hyperbolas (Fig. 6d).

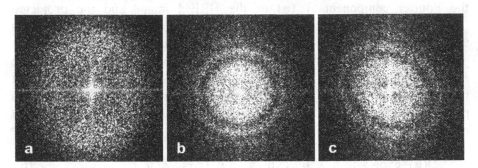

Figure 6 Fourier transforms of HREM images from amorphous carbon films taken (a) near Scherzer defocus, (b) at non-Scherzer without astigmatism and (c) with astigmatism.

In general, an image is a complicated mixture of structure factors which have been sampled by the contrast transfer function; some giving correct contrast and some giving reversed contrast. In summary, the contrast transfer function $T(\mathbf{u})$ is strongly affected by the defocus value and astigmatism, resulting in drastic contrast changes in HREM images.

The defocus value can be determined experimentally from HREM images, using different methods[30, 31, 32, 33]. Here we will use a method similar to that used by Erickson and Klug[30] and Krivanek[31] to determine the defocus and astigmatism from the amorphous region of the image. This will be demonstrated first on the HREM image of Ti_2S (Fig. 3) which was taken with very little astigmatism and then on an image of $K_7Nb_{15}W_{13}O_{80}$[16] (Fig. 8) which is more astigmatic.

In the Fourier transform (FT) of an image containing both crystalline and amorphous regions, the sharp diffraction spots come from the periodic features, while the diffuse background in the FT comes from the amorphous region, as seen in Figs. 3b and 8b. The effects of the CTF are visible in the diffuse background of the FT, seen mostly as dark rings which correspond to where $\sin\chi(\mathbf{u}) \approx 0$.

The u values at the dark rings can be read out from the Fourier transform of the images where

$$\chi(u) = \pi\varepsilon\lambda\,u^2 + \frac{\pi C_s \lambda^3\,u^4}{2} = n\,\pi \qquad (8)$$

and $n = 0, \pm1, \pm2, \ldots$ are integers. In general, the first crossover corresponds to $\chi(\mathbf{u}) = 0$ (n = 0) if the defocus is near zero, $\chi(\mathbf{u}) = -\pi$ (n = -1) for underfocus and $\chi(\mathbf{u}) = \pi$ (n =1) for overfocus. If both λ and C_s are known, the defocus value ε can be determined from the position of the first crossover by:

$$\varepsilon = \frac{n}{\lambda u^2} - \frac{C_s \lambda^2}{2} u^2 \qquad\qquad n = 0, \pm 1 \qquad (9)$$

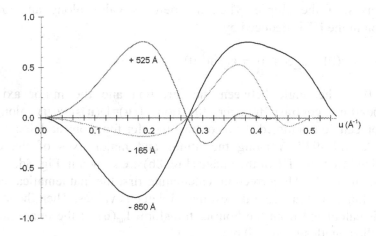

Figure 7 Contrast transfer functions T(u) at defocus values $\varepsilon = $ -850 Å, -165Å and -525Å. The optical parameters are from a Philips CM30/ST microscope: U = 300 kV, C_s=1.15 mm, Δ = 70 Å and α = 1.2 mrad. All the three contrast functions have a common first crossover position at u = 0.272 Å^{-1}. The defocus value -850 Å was determined to be the correct defocus for the image of Ti$_2$S shown in Fig. 2.

Different values of n give different solutions for the defocus. For example, the HREM image of Ti$_2$S shown in Fig. 3a was taken on a Philips CM30/ST microscope operated at 300 kV. The electron wavelength λ is 0.197 Å and the spherical aberration constant C_s is 1.15 mm. The first crossover is determined at u = 0.272 Å^{-1} from the Fourier transform of the image (Fig. 3b). Three possible defocus values, -165 Å (n = -1), -850 Å (n = 0) and +525 Å (n = 1), are deduced from Eq. 9, all giving a first crossover at u = 0.272 Å^{-1}. The corresponding contrast transfer functions at these three defocus values are shown in Fig. 7. The value -850 Å is chosen to be the correct defocus since the calculated CTF at this defocus gives the best fit to the intensity distribution of both the diffraction spots and the background noise in the FT of the image (Fig. 3b). The CTF at the defocus -165 Å would result in much too low amplitudes at low resolution while that at the defocus 525 Å would give much too low amplitudes in the high resolution range, which do not agree with the FT of the image (Fig. 3b). The decision of which of the three possible defocus values is correct can also be based on the positions of the second and third zero crossovers, if visible in the FT [16].

For an image such as that of $K_7Nb_{15}W_{13}O_{80}$ along the c axis shown in Fig. 8a, the Fourier transform (Fig. 8b) shows a dark elliptical ring together with the diffraction spots. This implies that the defocus values are different along different directions in the Fourier transform. First the defocus values ε_u and ε_v along the minor and major axes of the ellipse (or hyperbola) are determined from the positions of the first crossovers along the minor and major axes of the ellipse. Then the defocus value along any arbitrary direction in the FT is deduced by

$$\varepsilon(\theta) = \varepsilon_u \cos^2(\theta) + \varepsilon_v \sin^2(\theta) \tag{10}$$

where θ is the angle between the direction and the minor axis. The corresponding contrast transfer function $T(\mathbf{u})=D(\mathbf{u})\sin\chi(\mathbf{u})$ along this direction can be calculated. Two contrast transfer functions at defocus values -1321 Å and -947 Å along the minor and major axes of the ellipse, determined from the FT of the image (Fig. 8b) are shown in Fig. 8d.

Mathematical CTF correction: calculating first the mathematical contrast transfer function $T(\mathbf{u})$ from the estimated defocus values. Then the structure factor is calculated from the Fourier transform $I_{im}(\mathbf{u})$ of the image for all $\mathbf{u}$ except those with $\sin\chi(\mathbf{u}) \approx 0$ by:

$$F(u) = \frac{1}{k'} \cdot \frac{I_{im}(u)}{T(u)} \tag{11}$$

The projected potential of the crystal can be calculated by inverse Fourier transformation:

$$V(\mathbf{r}) = \frac{1}{k'} \sum_{\mathbf{u}} \left\{ \frac{I_{im}(\mathbf{u})}{T(\mathbf{u})} \exp\left[-2\pi i(\mathbf{u} \cdot \mathbf{r})\right] \right\} \tag{12}$$

The potential map obtained after the CTF correction (Fig. 8f) can be readily interpreted in terms of atomic structures while the map before the CTF correction and imposing the symmetry (Fig. 8e)

In most cases it is possible to retrieve the projected potential map from a single image taken under non-optimal conditions[16, 34]. However, the structure factors can be determined more accurately and an even more accurate potential projection can be obtained by combining a series of through-focus images[16]. Information contributed by kinematical scattering can be maximally extracted and the non-linear effects minimized by combining a series of through-focus images. Thus the structure can be determined more accurately and reliably[18, 35, 36].

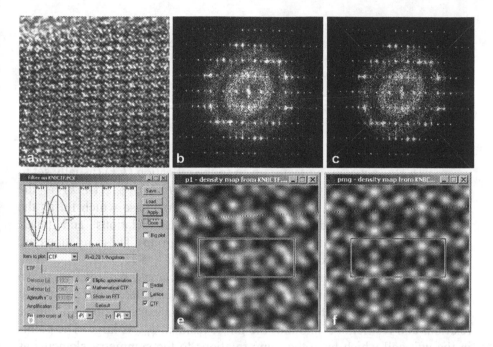

Figure 8   (a) HREM image of $K_7Nb_{15}W_{13}O_{80}$ along the c axis taken at non-Scherzer defocus[16]. (b) The Fourier transform of $I_{im}(u)$ of the image in (a). An elliptic dark ring can be seen in the background noise of the FT, which correspond to the first crossover of the CTF. (c) A set of ellipses are fitted to the dark rings. (After Zou et al. [16]) (d) The defocus values along the minor and major axes are estimated from the innermost ellipse to be -1321 and -947 Å. The two corresponding CTF curves are shown. (e) The lattice averaged map $K_7Nb_{15}W_{13}O_{80}$ obtained from the image in (a). (f) the projected potential map reconstructed from (a) after compensating for the CTF and astigmatism and imposing the crystal symmetry *pmg*.

9.   SYMMETRY DETERMINATION

Symmetry can be determined by different methods. In X-ray crystallography, the symmetry determination is carried out using symmetry-relations of amplitudes combined with systematic absences. In electron diffraction and HREM images, due to multiple scattering, symmetry forbidden reflections are often not absent. Since the systematic absences are often unreliable in electron crystallography experiments, amplitude relations alone are often not sufficient for differentiating between different symmetries. However, the phases, experimentally observed in HREM images, have much better quality and can be used for symmetry determination.

The quality of the measured phases can be characterised by the averaged phase error (phase residual ϕ_{Res}) of symmetry-related reflections:

$$\phi_{Res} = \frac{\sum_{hk}\left[w(h\;k)\left|\phi_{obs}(h\;k) - \phi_{sym}(h\;k)\right|\right]}{\sum_{hk}w(h\;k)} \tag{13}$$

where w($h\;k$) is a weighting factor given to the reflection ($h\;k$) (usually set to be equal to the amplitude of the reflection ($h\;k$)), ϕ_{obs} ($h\;k$) is the experimentally observed phase and ϕ_{sym} ($h\;k$) is the phase which fulfils the symmetry relations and restrictions. The phase relations and phase restrictions are different in each of the 17 plane groups These relations are tabulated and listed for example in Table 3.1 in Zou[18](1995)

Unlike amplitudes, phases are not absolute values, but relative to an origin. When the Fourier transform of an image is calculated, the origin is at an arbitrary position in the unit cell. Phases do not have to obey the phase relations and restrictions, and thus the phase residual ϕ_{Res} is large. The points in the unit cell which have the same relations to the symmetry elements as the origin specified in the International Table for Crystallography is located as described below (for example in centrosymmetric plane groups, the origin should coincide with a center of symmetry). The origin is shifted 360° by 360° in small steps over the entire unit cell and at each step the phase residual ϕ_{Res} is calculated. When all positions are tested within the unit cell, the position ($x_0\;y_0$) which gives the lowest phase residual ϕ^0_{Res} is considered to be the correct origin. Finally all phases are recalculated relative to this origin. This procedure is known as origin refinement.

The symmetrized phase ϕ_{sym} ($h\;k$) is estimated from the experimental phases ϕ_{obs}($h\;k$) as follows:

• If a reflection ($h\;k$) is not related to other reflections by the symmetry (except by Friedel's law):

$$\phi_{sym}(h\;k) = \phi_{obs}(h\;k) \tag{14}$$

• If a reflection ($h\;k$) is symmetry-related to other reflections, the phases for this group of reflections are judged together. $\phi_{sym}(h\;k)$ is determined by vector summation of all these reflections:

$$
\phi_{sym}(h\,k) = \tan^{-1}\left[\frac{\displaystyle\sum_j w^j s^j \sin\left(\phi^j_{obs}(h\,k)\right)}{\displaystyle\sum_j w^j s^j \cos\left(\phi^j_{obs}(h\,k)\right)}\right] + \begin{cases} 0^\circ & \left[\text{if } \displaystyle\sum_j w^j s^j \cos\left(\phi^j_{obs}(h\,k)\right) > 0\right] \\[2mm] 180^\circ & \left[\text{if } \displaystyle\sum_j w^j s^j \cos\left(\phi^j_{obs}(h\,k)\right) < 0\right] \end{cases}
\tag{15}
$$

where the summation is for all symmetry-related reflections in the group (including the $(h\ k)$ reflection). w^j is a weighting factor which can be set either to 1 or for example to the amplitude $\left|F^j_{obs}(h\ k)\right|$ of the corresponding reflection. $s^j = 1$ if the phases $\phi_{sym}(h\ k)$ and $\phi^j_{obs}(h\ k)$ should be equal and $s^j = -1$ if they should differ by 180°.

• For centrosymmetric projections, $\phi_{sym}(h\ k)$ is finally set to 0° if $-90° < \phi_{sym}(h\ k) < 90°$, otherwise it is set to 180°.

Each symmetry has a unique set of phase relations and phase restrictions. Thus, the phase residuals calculated for an image will be different for different symmetries. Once the phase residuals for each of the 17 plane group symmetries have been calculated, the projected symmetry (plane group) of the crystal can be deduced by comparing these phase residuals. Usually the symmetry with the lowest phase residual is the correct symmetry. If phase residuals for several plane groups are similar, the highest symmetry is normally chosen.

The procedure of symmetry determination is demonstrated (Fig. 9) for the [001] projection of Ti_2S, by analysing phases extracted from the HREM image shown in Fig. 3a. Since the lattice of Ti_2S is primitive and the cell dimensions a and b are not equal, the possible plane groups are *p1*, *p2*, *pm*, *pg*, *pmm*, *pmg* and *pgg*. Among those 7 possible plane groups, *p2*, *pg* and *pgg* give relatively low phase residuals (12.5°, 7.7° and 11.3°, respectively). The symmetry of the projection is most probably *pgg*, according to the criteria mentioned above. Notice that the two lower symmetries *p2* and *pg* are subgroups of *pgg*.

In the plane group *pgg* phase restrictions and phase relations for all reflections (once the origin has been shifted to a point with the same relation to the symmetry element in the unit cell as specified in the Int. Table for Crystallography) are: all phases have to be 0° or 180° and phases of all

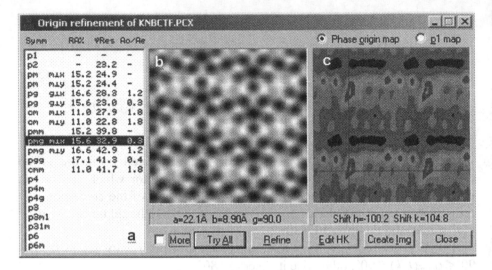

Figure 9   Symmetry determination and origin refinement of $K_7Nb_{15}W_{13}O_{80}$. a) The crystallographic R-value on symmetry-related reflections R_{sym} (here called RA%) is similar for all these plane groups, because they all have the same mm-symmetry relations of amplitudes. In contrast, the phase residual ϕ_{Res} is different in different plane groups and thus can be used to determine the symmetry. b) the map after imposing the symmetry *pmg*. c) Phase residual map showing how ϕ_{Res} varies when the origin is shifted throughout one unit cell. The lowest value of ϕ_{Res} is found at position (-100.2°/360°, 104.8°/360°), so that the position is chosen as the phase origin.

symmetry-related pairs *(h k)* and *(-h k)* are related by $\phi(h\ k) = \phi(-h\ k) + (h + k)\cdot180°$[18]. Furthermore, all symmetry-related pairs *(h k)* and *(-h k)* should have the same amplitude. After the symmetry *pgg* has been imposed to the amplitudes and phases (see Table 1), the inverse Fourier transform gives a density map (Fig. 5b and 1*l*) which is quite similar to the projected potential map shown in Fig. 1k.

10.   INTERPRETATION OF PROJECTED POTENTIAL MAPS

The projected potential map obtained from HREM images after image processing must now be interpreted in terms of chemical structure. At this stage it is of great value to be familiar with the chemical system under investigation. Only in the most fortunate cases there is a one-to one correspondence between the peaks in the map and atoms in the structure. In the many cases where the structure consists of two or more atomic species with very different scattering factors, the lighter atoms are often not seen at all. The resolution is also an important factor here; the interatomic distances in metal oxides are often about 2 Å for metal-oxygen and 4 Å for metal-metal. Thus, at 2.5 Å resolution we may expect to see peaks corresponding to MeO_n polyhedra, but we should not expect to see resolved oxygen atoms. This is the situation for the metal oxides presented in this chapter. For Ti_2S the situation is better; Ti and S are about equally large and sufficiently well separated so that the 1.9 Å resolution of the images in Fig. 5b, is sufficient to resolve all atoms. Atomic positions can be determined directly from the peaks (white spots) in this density map.

In most cases the chemical composition, unit cell dimensions and symmetry are known. We can then estimate the number of formula units in the unit cell from the fact that each atom (except hydrogen) occupies about 15 - 20 Å^3. If the HREM image is taken along a short unit cell axis (< 5 Å), the whole structure may be resolved in that single projection.

Unfortunately, HREM images are black and white only, so there is no direct evidence of which peaks correspond to which atom species. In principle we might expect the heights of the peaks to be proportional to the scattering power of the atoms, but this is not always the case, due to the relatively poor quality of the amplitudes in HREM images. Here again chemical knowledge is indispensable. In the case of Ti_2S, it is known that the Ti atoms often arrange in octahedral clusters, while the S atoms prefer to be inside trigonal bipyramids. With this chemical background, the peaks in the potential map (Fig. 5b) can easily be assigned to Ti or S, as shown in Fig. 5c.

11. CRYSTAL THICKNESS AND CRYSTAL TILT

Crystal tilt is one of the main reasons why HREM images often cannot be directly interpreted in terms of projected crystal structure. The alignment of the crystal in the microscope is usually judged from the SAED pattern, which comes from an area much larger than the area selected for image processing. Even if the SAED pattern is well aligned, the thin area of the crystal selected for image processing may still be slightly tilted if the crystal is bent. An HREM image from a slightly misaligned crystal is similar to the image from a thinner and well aligned crystal. This indicates that the weak-phase-object approximation will be valid for an even thicker crystal if the crystal is a slight crystal tilt [37,38]. This also causes that the crystal thickness estimated by image matching using image simulation are often smaller than the true value[37].

The main effect of crystal tilt is to smear out the structural information in the direction perpendicular to the tilt axis. Atoms from different unit cells no longer project exactly on top of each other. The projection of an atom column becomes a line rather than a point. This smearing is equivalent to loosing the fine details in the direction perpendicular to the tilt axis. As a consequence, the symmetry of the crystal is often lost in the images (Fig. 10b); the amplitudes of reflections away from the tilt axis are attenuated (Fig. 10d). The effect of crystal tilt depends on the crystal thickness; the thicker the crystal, the more rapidly $I_{im}(\mathbf{u})$ is attenuated. The overall effect of crystal tilt on the image is given by the product of the crystal thickness t and the tilt angle γ, $t \cdot \sin\gamma$[18]. Even for the smallest tilts and thinnest crystals, the effect on amplitudes is significant. Pairs of symmetry-related reflections no longer have the same amplitudes if one of the reflections is close to the tilt axis and the other further away.

Furthermore, the effect of crystal tilt on a specific reflection depends on the distance of that reflection to the tilt axis. If the reflection lies on the tilt axis, it will not be affected by crystal tilt. The further away the reflection is from the tilt axis, the more attenuated is $I_{im}(\mathbf{u})$. The thickness of the crystal can be determined directly from the image, if images from at least two different crystal tilts are recorded, as described by Hovmöller and Zou[27].

The effects of crystal tilt on phases is quite different. The phases are practically unaffected for small tilts and thin crystals[18]. However, as long as the product $t \cdot \sin\gamma$ is small, the phases are unchanged. Both phase relations and phase restrictions are still valid. Thus, it is possible to determine the (projected) crystal symmetry also from an image of a tilted crystal, using the phases.

For most thin crystals, the distortion of the image due to crystal tilt can be compensated by imposing the crystal symmetry on the amplitudes and phases extracted from the image.

The projected potential can be reconstructed. This reconstruction method is demonstrated in Fig. 10 on HREM images of $K_2O \cdot 7Nb_2O_5$. This method is especially powerful for crystals with high symmetries.

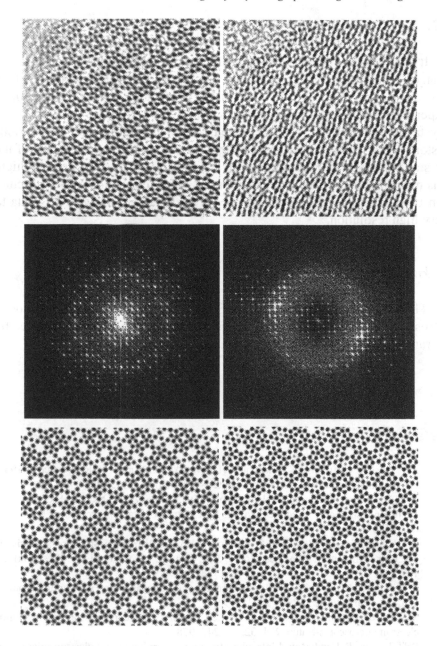

Figure 12 HREM images of K2O·7Nb2O5 along the c-axis from (a) a well-aligned crystal and (b) the same crystal tilted 5°. Atom columns which are separated in (a) are smeared out into lines perpendicular to the tilt axis. (c) and (d) The corresponding Fourier transforms of images (a) and (b). The tilt axis is indicated by a line in (d) and (e). Reflections further away from the tilt axis are attenuated. (e) and (f) Projected potential maps reconstructed by imposing the projection symmetry of the crystal, p4g, on the amplitudes and phases extracted from (c) and (d), respectively. The white dots in the maps are Nb atoms. The positions of the Nb atoms determined from both maps are very similar, within 0.02 Å.

12.   CONCLUSIONS

It has been shown that crystal structures can be solved from HREM images. The experimental conditions defocus and sample thickness, which are only guessed in image simulation procedures, can be determined experimentally, as can also astigmatism and crystal tilt. Also the projected crystal symmetry can be determined. After correcting for the various distortions, a reconstructed projected potential map can be calculated. If the crystal is thin and the resolution of the electron microscope sufficiently high, this map will have peaks at the positions of the heaviest atoms. The structure can be confirmed and further improved by refinement, using SAED data to very high resolution.

Acknowledgements

This project was supported by the Swedish Research Council. XD Zou is a Research Fellow of the Royal Swedish Academy of Sciences supported by a grant from the Knut and Alice Wallenberg Foundation.

References

1.   ZG Pinsker. Diffraktsiya elektronov (Izd-vo Akad. Nauk SSSR., Moscow 1949) [English transl,: Electron diffraction (Butterworths, London 1952)].
2.   BK Vainshtein. Strukturnaya elektronographiya, Izd-vo Akad. Nauk SSSR., Moscow, 1956 [English transl,: Structure Analysis by Electron Diffraction. Pergamon Press Ltd., Oxford, 1964].
3.   BK Vainshtein, BB Zvyagin, AS Avilov. Electron diffraction structure analysis, in Electron Diffraction Techniques, Vol. 1, J.M. Cowley, ed., Oxford Univ. Press, Oxford, 1992, pp. 216-312.
4.   S Iijima. High resolution electron microscopy of crystal lattice of titanium-niobium oxide. J Appl Phys 42: 5891 - 5893, 1971.
5.   MA O'Keefe, P Buseck, S Ijima. Computed crystal structure images for high resolution electron microscopy. Nature 274: 322 – 324, 1978.
6.   PNT Unwin, R Henderson. Molecular structure determination by electron microscopy of unstained crystalline specimens. J Mol Biol 94: 425-440, 1975.
7.   W Dong, T Baird, JR Fryer, CJ Gilmore, DD McNicol, G Bricogne, DJ Smith, MA O'Keefe, S Hovmöller. Electron microscope at 1 Å resolution by entropy maximization and likelihood ranking. Nature 355: 605-609, 1992.
8.   S Hovmöller, A Sjögren, G Farrants, M Sundberg, B-O Marinder. Accurate atomic positions from electron microscopy. Nature 311: 238 – 241, 1984.

9. DN Wang, S Hovmöller, L Kihlborg, M Sundberg. Structure determination and correction for distortions in HREM by crystallographic image processing. Ultramicroscopy 25: 303-316, 1988.

10. JJ Hu, FH Li, HF Fan. Crystal structure determination of $K_2O\cdot7Nb_2O_5$ by combining high-resolution electron microscopy and electron diffraction. Ultramicroscopy 41: 387-397, 1992.

11. H-R Wenk, KH Downing, M Hu, MA O'Keeefe. 3D structure determination from electron-microscope images: electron crystallography of staurolite. Acta Cryst. A48: 700 -716, 1992.

12. T.E. Weirich, R. Ramlau, A. Simon, S. Hovmöller, X.D. Zou. A crystal structure determined to 0.02 Å accuracy by electron crystallography, Nature 382: 144-146, 1996.

13. J.M. Cowley. Diffraction Physics. 2nd edition, North-Holland, Amsterdam, 1984.

14. D.B. Williams, C.B. Carter. in Transmission Electron Microscopy. Vol. II, Plenum Press, New York, 1996, pp.203.

15. XD Zou. On the phase problem in electron microscopy: the relationship between structure factors, exit waves, and HREM images. Microsc. Res. Tech. 46: 202-219, 1999.

16. XD Zou, M Sundberg, M Larine, S Hovmöller. Structure projection retrieval by image processing of HREM images taken under non-optimum defocus conditions. Ultramicroscopy 62: 103-121, 1996.

17. D Van Dyck, M Op de Beeck. A simple intuitive theory for electron diffraction. Ultramicroscopy 64: 99 - 107, 1996.

18. XD Zou. Electron crystallography of inorganic structures - theory and practice. Chem. Comm. 5, Stockholm, Ph.D. thesis, 1995.

19. AL Patterson. A direct method for the determination of the components of interatomic distances in crystals. Z. Kristallogr. 90: 517 - 542, 1935.

20. JP Owens. BR Conrad, HF Franzen. The crystal structure of Ti_2S, Acta Cryst. 23: 77-82, 1967.

21. DL Dorset. Structural electron crystallography, Plenum Press, 1995.

22. XD Zou. Crystal Structure determination by crystallographic image processing. in "Electron Crystallography", eds. DL Dorset, S Hovmöller, XD Zou, Nato ASI Series C, Kluwer Academic Publishers, Dordrecht, 1997, pp163-181.

23. RVincent, PA Midgley. Double conical beam-rocking system for measurement of integrated electron diffraction intensities. Ultramicroscopy 53: 271 - 282, 1994.

24. XD Zou, Y Sukharev, S Hovmöller. ELD - a computer program system for extracting intensities from electron diffraction patterns. Ultramicroscopy 49: 147-158, 1993.

25. XD Zou, Y Sukharev, S Hovmöller., Quantitative measurement of intensities from electron diffraction patterns for structure determination - new features in the program system ELD. Ultramicroscopy 52: 436 - 444, 1993.

26. TE Weirich. Metallreiche systeme mit kondensierten clustern. Osnabrück University, Germany, Ph.D. thesis, 1996.

27. S Hovmöller, XD Zou. Measurement of crystal thickness and crystal tilt from HREM images and a way to correct for their effects. Microsco Res Tech 46: 147-159, 1999.

28. S Hovmöller. CRISP: crystallographic image processing on a personal computer. Ultramicroscopy 41: 121-135, 1992.

29. MA O'Keefe. "Resolution" in high-resolution electron microscopy. Ultramicroscopy 47: 282-297, 1992.

30. HP Erickson, A Klug. Measurement and compensation of defocusing and aberrations by Fourier processing of electron micrographs, Phil. Trans. Roy. Soc. Lond. B261: 105-118, 1971.

31. OL Krivanek. A method for determining the coefficient of spherical aberration from a single electron micrograph. Optik 45: 96-101, 1976.

32. FS Han, HF Fan, FH Li. Image processing in high-resolution electron microscopy using the direct method. II. Image deconvolution, Acta Cryst. A42: 353-356, 1986.

33. JJ Hu, FH Li. Maximum entropy image deconvolution in high resolution electron microscopy. Ultramicroscopy 35: 339-350, 1991.

34. A Klug. Image analysis and reconstruction in the electron microscopy of biological macromolecules. Chimica Scripta 14: 245 -256, 1978-79.

35. W Coene, G Janssen, M Op de Beeck, D Van Dyck. Phase retrieval through focus variation for ultra-resolution in field-emission transmission electron microscopy. Phys. Rev. Letts. 69: 3743-3746, 1992.

36. WO Saxton. What is the focus variation method? Is it new? Is it direct? Ultramicroscopy 55: 171-181, 1994.

37. MA O'Keefe, V Radmilovic. Specimen thickness is wrong in simulated HREM images. Proceedings of the 13th ICEM, Paris), Vol. 2B: 361-262, 1994.

38. XD Zou, EA Ferrow, S Hovmöller. Correcting for crystal tilt in HREM images of minerals: the case of orthopyroxene. Physics and Chemistry of Minerals 22: 517-523, 1995.

3D RECONSTRUCTION OF INORGANIC CRYSTALS:

Theory and application

Xiaodong Zou and Sven Hovmöller
Structural Chemistry, Arrehnius Laboratory, Stockholm University, SE-106 91 Stockholm, Sweden

Abstract: All chemical structures are 3-dimensional; inorganic, organic and macromolecular. Each electron micrograph is a 2D projection of a 3D structure. When the crystal structure is more complex, with all unit cell dimensions larger than about 6 Å, there is generally no single projection in which all atoms are well resolved. Then it is necessary to combine several projections into one 3D map. Here the principles of constructing a 3D structure model from several projections will be presented.

Key words: 3D reconstruction,

1.   INTRODUCTION

All chemical structures are 3-dimensional; inorganic, organic and macromolecular. Each electron micrograph is a 2D projection of a 3D structure. Most inorganic structures which have been solved by crystallographic image processing have had one short (< 4 Å) unit cell axis (Hovmöller *et al.*, 1984; Wang *et al.*, 1988; Weirich *et al.*, 1996; Zandbergen *et al.*, 1997), in which case it is possible to solve the 3D structure from a single projection, since there are no overlapping atoms in the projection down such a short axis. When the crystal structure is more complex, with all unit cell dimensions larger than about 5 Å, there is generally no single

T.E. Weirich et al. (eds.), Electron Crystallography, 301–320.

projection in which all atoms are well resolved. Then it is necessary to combine several projections into one 3D map.

One way to solve such structures is to collect electron diffraction patterns from different zone axes of the crystal to get an essentially complete 3D electron diffraction data set. Direct methods or the Patterson method can then be applied to phase the data, similar to what is done in X-ray diffraction (Gjønnes *et al.*, 1998; Gemmi *et al.*, 2000; Wagner *et al.*, 1999).

As a structure becomes more complex and the number of unique atoms increases, phases derived by direct methods become less reliable, especially when the electron diffraction data deviate from the kinematical approximation because of dynamic effects. HREM combined with crystallographic image processing provides a unique method for determining such structures. HREM images from a number of projections along different zone axes may be combined into a 3D potential map.

2.   DIFFERENT TYPES OF 3D OBJECTS

There are many different types of 3D objects, and the methods of producing a 3D model of them from EM data are different. The objects may be 3D crystals, i.e. having identical unit cells repeating in all three directions, they may be 2D crystals, i.e. repeating in two dimensions but only one unit cell thick in the third dimension, they may consist of identical units arranged in some other regular way such as a helix or a sphere or they can be completely lacking a crystallographic repeat. An interface between two 3D crystals can be considered as a 2D crystal, but embedded in 3D crystals on each side rather than embedded in vacuum or amorphous water as is the case with membrane proteins. The principal idea of 3D reconstruction is the same in all cases, but the practical procedures differ for each type of 3D structures.

2.1   Serial sections and tomography for non-crystalline objects

If there is no crystalline repeat in the structure, the methods of reconstructing their 3D structures are serial sections if the object is large (thicker than 100 nm in the thinnest dimension) or tomography if the object is smaller. These techniques are widely used in biology, and may have important applications also in materials science.

2.2   Fibres and spherical particles

The birth of the science we now know as electron crystallography can be traced back to the work by Aaron Klug and co-workers at the MRC laboratory of Molecular Biology in Cambridge in the 1960-ies. Soon after the double helix of DNA and the first protein structures were solved in that laboratory, Klug started to study viruses by electron microscopy. Initially they used negatively stained specimens, with a resolution of around 3 nm, and deduced the icosahedral building principle of spherical viruses from inspecting EM images of single virus particles by eye. Later optical diffraction and computerised image processing was developed. The first 3D reconstruction by electron microscopy was published in a classical paper by DeRosier and Klug (1968). They digitised an electron micrograph of a helical object, using a microdensitometer, and calculated its Fourier transform (FT). Amplitudes and phases of the FT were extracted. A single image of a helix is in fact a full tilt series, since different parts of the helix are seen from different directions. The mathematics for disentangling the different views and reconstructing them into a 3D structure was developed and applied here for the first time. They also discuss various ways of determining the crystallographic structure factor phases which are needed for solving crystal structures and there mention for the first time that these phases can be extracted directly from the Fourier transform of an EM image. 3D reconstruction of 3D crystals.

In order to see a structure in 3D we need more than one projection. For this purpose, humans and most animals are equipped with two eyes, allowing a stereo view of the world. Although this is a remarkable and efficient tool in our everyday life, it does not really give a 3D view of the objects we look at. What we see is just the surfaces of the objects. Furthermore, the resolution is not as good in the direction of view as it is in the left-right and up-down directions.

In chemistry, we are often interested in the bulk of the material, and for this purpose we must view the structures with a probe that penetrates through the object. Transmission electron microscopy is ideal for this. The 3D structure of a transparent object is much more complex than the surface, which can be considered as a 2D object, although it often is not at all flat. In the case of a crystal, the object may be hundreds of atoms thick, resulting in a massive overlap of atoms in any direction we chose to look at the crystal from. It is easily realized that even three orthogonal views are not sufficient for resolving all overlapping reflections, unless the structure is very simple. The larger the unit cell is, the more projections are needed in order to obtain a structure with all atoms resolved.

2.3   3D reconstruction of 3D crystals

In order to see a structure in 3D we need more than one projection. For this purpose, humans and most animals are equipped with two eyes, allowing a stereo view of the world. Although this is a remarkable and efficient tool in our everyday life, it does not really give a 3D view of the objects we look at. What we see is just the surfaces of the objects. Furthermore, the resolution is not as good in the direction of view as it is in the left-right and up-down directions.

In chemistry, we are often interested in the bulk of the material, and for this purpose we must view the structures with a probe that penetrates through the object. Transmission electron microscopy is ideal for this. The 3D structure of a transparent object is much more complex than the surface, which can be considered as a 2D object, although it often is not at all flat. In the case of a crystal, the object may be hundreds of atoms thick, resulting in a massive overlap of atoms in any direction we chose to look at the crystal from. It is easily realized that even three orthogonal views are not sufficient for resolving all overlapping reflections, unless the structure is very simple. The larger the unit cell is, the more projections are needed in order to obtain a structure with all atoms resolved.

3.   THEORY

The 3D reconstruction of an object is performed more conveniently in reciprocal (Fourier) space. The 2D Fourier transform of a projection of an object is *identical* to a plane of 3D Fourier transform of the original object normal to the projection direction (electron beam). The origin of each 2D Fourier transform of a projection is identical to the origin of the 3D Fourier transform of an object, provided that the projections are aligned so that they have the same (common) phase origin. This is known as the *Fourier slice theorem* or the *central projection theorem*.

3D reconstruction can be performed by restoring the 3D Fourier space of the object from a series of 2D Fourier transforms of the projections. Then the 3D object can be reconstructed by inverse Fourier transformation of the 3D Fourier space. For crystalline objects, the Fourier transforms are discrete spots, i.e. reflections. In electron microscopy, the Fourier transform of the projection of the 3D electrostatic potential distribution inside a crystal, or crystal structure factors, can be obtained from HREM images of thin crystals. So one can obtain the 3D electrostatic potential distribution $\varphi(\mathbf{r})$ inside a crystal from a series of projections by

$$\varphi(xyz) = \frac{\lambda}{\sigma\,\Omega} \sum_{h,k,l} F(hkl) \exp\left[-2\pi i(hx + ky + lz)\right] \tag{1}$$

where F(hkl) is the structure factors of the crystal, λ electron wave length, σ interaction constant and Ω unit cell volume.

The number of projections that are needed for a 3D structure determination is easier to answer in reciprocal space than in real space. It depends on the symmetry of the crystal and how the strong reflections are distributed. At least all strong reflections within the asymmetric unit should be included, since each strong reflection gives significant contribution to the 3D potential map. Missing strong reflections may result in a wrong 3D map. On the other hand, weak reflections contribute less to the 3D map and for solving structures, they can be ignored. The resolution that is needed to resolve the desired atoms depends on the inter-atomic distances within the crystal, should be at least better than the inter-atomic distances, for example 1.5 Å for Si-O and 2.0 for inter metallic compounds.

4.   PRACTICE

Here the principles of constructing a 3D structure model from several HREM images of projections of inorganic crystals will be presented. Some of the principles may also be applied to non-periodic objects. A complex quasicrystal approximant v-AlCrFe is used as an example (Zou et al., 2003). Procedures for *ab initio* structure determination by 3D reconstruction are described in detail. The software CRISP, ELD. Triple and 3D-Map are used for 3D reconstruction. The 3D reconstruction method was demonstrated on the silicate mineral (Wenk et al. 1992). It was also applied to solve the 3D structures of a series mesoporous materials (Keneda *et al.* 2002).

4.1   Data recording

To obtain good HREM images and electron diffraction patterns is one to the most important step in 3D reconstruction. The data should be as kinematical as possible, so HREM images and electron diffraction (ED) patterns should only be taken from very thin crystals.

As mentioned above, the number of projections, or zone axes, which need to be collected depends on the symmetry of the crystal and distribution of strong reflections. In order to determine which zone axes should be chosen for collecting HREM images, a nearly complete 3D electron diffraction data set needs to be collected.

Three tilt series of ED patterns, tilted along the (h 0 0), (0 0 1) and (2h –h 0) axes, were collected on a JEOL 2000FX electron microscope (Fig. 1). Due to the limited specimen tilt angles of the microscope (±45°), several crystals with different orientations are used for collecting a complete ED patterns. From these tilt series, 13 zone axes that contained strong and/or many reflections were chosen for 3D reconstruction: [001], [010], [011], [012], [013], [014], [021], [023], [120], [121], [122], [241] and [5 18 0].

HREM images and ED patterns of the v-AlCrFe crystals were collected on JEOL 3010 (point resolution 1.7 Å) and JEOL 4000 EX (point resolution 1.6 Å) electron microscopes at 300kV and 400kV, respectively. Since the tilt range of these microscopes is very limited (±15°), it is only possible to collect data from one or a few of all the zone axes that are needed for 3D reconstruction from each crystallite.

For each of the 13 zone axes, a through-focus series with three to five HREM images were recorded on film at 500,000 magnification. In order to minimize dynamic effects, only the thinnest edges of the crystals were selected for further processing. The corresponding ED patterns were also recorded, as exposure series using 1, 2, 4, 8, 16, 32, 64, 128 and 180 seconds. The number of ED patterns needed in an exposure series depends on the intensity difference between the strongest and the weakest reflections.

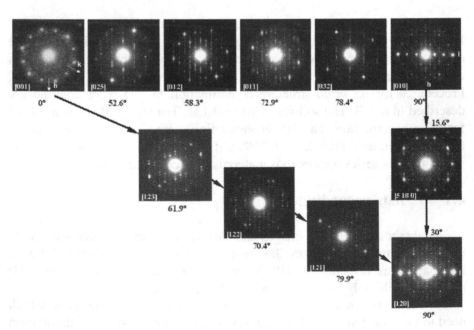

Figure 1 Some of the ED patterns used for sampling the reciprocal space of the v-AlCrFe phase. The corresponding zone axes and tilt angles from the [001] and [010] zone axes are given.

4.2   Digitisation of HREM images and ED patterns

Negatives of HREM images were digitised at 0.6 Å per pixel on a light box by an 8-bit CCD camera (MTI CCD 72) and a SHARK frame grabber using the program CRISP (Hovmöller, 1992). Preliminary examinations with Fourier transformations were conducted to assess the general quality, *i.e.* crystallinity, resolution, defocus, astigmatism and so on.

Since the intensity difference between strong and weak spots in an ED pattern is very large, negatives of ED patterns were digitised with a 12-bit slow-scan CCD camera Kite, (Calidris, 2000a), which has a dynamic range of more than 2.5 optical density units. The data was first corrected for the response of each of the 1280 by 1024 individual pixels of the CCD camera, and then converted from transmission to optical density values in the following way. For each exposure series, a black image (I_b) with the lens covered by the lens cap was taken in order to measure the dark current of each pixel in the CCD camera. A white image (I_w) with the direct light from the light box was taken to measure response for transmitted light. The dark image was subtracted both from each digitised ED pattern (I_{ED}) and from the white image, using the Image Calculator function in CRISP. Finally, the ED pattern is converted from transmission to optical density units I_{OD} by logarithmation

$$I_{OD}(xy) = \log \frac{I_w(xy) - I_b(xy)}{I_{ED}(xy) - I_b(xy)} \qquad (2)$$

$I_{OD}(xy)$ is scaled to fit to 0-255 grey scale (8 bit).

Intensities of all reflections for each thus corrected negative were extracted by the program ELD (Zou *et al.*, 1993). To make sure that the intensities were within the usable range of the CCD camera, intensities of the strongest reflections were only extracted from the negatives with the shortest exposure times, while intensities of the weakest reflections were obtained only from those with longer exposure times. Data sets from negatives with different exposure times were merged, using the program Triple (Calidris, 2000b). The scale factors were calculated by comparing the intensities of the common reflections (typically over 100 per pair of negatives). All R_{merge} values for pairs of films, differing in exposure time, were less than 8% for all orientations.

4.3   2D image processing

Before performing 3D reconstruction, HREM images from each zone axis were studied by crystallographic image processing using CRISP. First, a Fourier transform (FT) was calculated for each image and the quality of the image was judged from the resolution, thickness and alignment of the crystal. The defocus and astigmatism values were estimated from the contrast transfer function that could be seen from the FT of the amorphous region of the crystal. Two images of the highest quality for each zone axis were selected for further processing. Here the [001] zone axis is used as an example to demonstrate the steps of 2D image processing, to obtain amplitudes and phases of the structure factors for each zone axis.

Two HREM images of the [001] zone axis taken at different defocus values; image 1 (Fig. 2a) and image 2 (Fig. 2b) were used for image processing. For each image, a Fourier transform was calculated from the thinnest part of the crystal (Fig. 2c for image 1 and Fig. 2d for image 2).

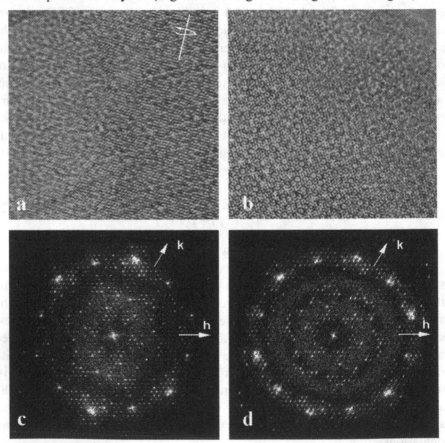

Figure 2 (a, b) HREM images of the [001] zone axis taken at two different defocus values and their corresponding Fourier transforms ((c) from (a) and (d) from (b)). The 12 very strong reflections are all at about 2.1 Å resolution. One dark ring in (c) and two dark rings in (d) are seen. These correspond to the zero cross-overs of the CTF. Both images are astigmatic, since the dark rings are elliptical. The crystal in (a) is more misaligned than that in (b) so that the image in (a) is more smeared out in the direction perpendicular to the tilt axis (indicated in (a)).

4.3.1   Determination of the defocus and astigmatism

Since the area selected was from the thinnest part of the crystal, it could be considered as a weak phase object. For a weak phase object, the Fourier transform of the image I(u) is related to the Fourier transform of the projected potential $\Phi(u)$ by

$$I(u) = D(u)\sin \chi(u)\, \Phi(u) \tag{3}$$

where $D(u)\sin \chi(u)$ is the contrast transfer function (CTF), $D(u)$ is the damping function of the CTF and

$$\chi(u) = \pi\varepsilon\lambda u^2 + \frac{\pi C_s \lambda^3 u^4}{2} \tag{4}$$

ε the defocus value, C_s the spherical aberration constant and λ wavelength.

The amplitudes of the Fourier transform I(u) are zero where $\sin \chi(u) = 0$, i.e. no information at these positions in reciprocal space are transferred to the image. These positions are zero-crossover(s) of the CTF. The defocus value(s) can be estimated from the position(s) u of the CTF crossover, and C_s and λ (Zou *et al.*, 1996).

In the Fourier transform of image 1 (Fig. 2c), one dark ring is seen at the d-value 3.1 Å, marking the zero crossover of the CTF. The position of the dark ring was very clear when a large amorphous region was included. The defocus value was estimated to be about -700 Å, with an astigmatism of ± 57Å. The Fourier transform of image 1 (Fig. 2c) did not show perfect 6-fold symmetry, due to a slight crystal misalignment. The CTF for image 2 was determined in a similar way, from the two dark rings at d-values about 4 Å and 2.5 Å. The defocus value was estimated to be –932 Å, with an astigmatism of ± 65Å.

4.3.2   Compensation for the defocus and astigmatism

Once the CTF is determined, the Fourier transform of the projected potential $\Phi(u)$ can be retrieved by

$$\Phi(u) = \frac{I(u)}{D(u)\sin \chi(u)} \tag{5}$$

This method applies for both crystalline and non-crystalline samples. At those regions in the Fourier transform where the CTF is positive, $\Phi(u)$ and I(u) have the same phases, while at those regions where the CTF is negative, the phases of $\Phi(u)$ are shifted by 180° by the CTF compared with the phase of I(u). We compensated for the effect of the CTF by adding 180° to the phases at those regions where the CTF is negative. Since the amplitudes of $\Phi(u)$ are obtained by dividing the amplitudes of $I(u)$ by the CTF, this correction is very sensitive to the accuracy of the CTF determination and to which extent the weak phase object approximation is valid, especially when the CTF is close to the zero crossovers. We found that the final results (projected potential maps) often became worse when also the amplitudes were compensated for the effects of the CTF. Thus we only compensated the phases for the effects of the CTF. The amplitudes for all the reflections were taken from the corresponding electron diffraction patterns, which are not affected by the CTF.

4.3.3   Lattice refinement and extracting amplitudes and phases

After the phases of the Fourier transform were compensated for the effects of the CTF, the 2D reciprocal lattice was determined and refined using all the diffraction spots in the Fourier transform. The amplitude of each reflection was extracted by integrating a 3x3 pixel area around the reflection (with the background subtracted) and the phase was taken from the refined position of the reflection in the Fourier transform, as described by Hovmöller (1992).

4.3.4   Symmetry determination and origin refinement

The 2D symmetry for each projection can be determined by analyzing the phase values. For different symmetries, the phase relations between symmetry-related reflections are different. When the projection is centrosymmetric, the phases are restricted to 0° or 180°. Since phases are not absolute values but relative to an origin, the phase restriction and phase relations are only valid when the origin is on the specific symmetry element of the respective 2D symmetry, as given in the International Table for Crystallography, Volume A. For each 2D symmetry, CRISP searches for the best origin, at which the experimental phases deviate least from the ideal phase relations and phase restrictions. The average phase error (φRes) is calculated. The 2D symmetry of the projection can be determined by comparing the phase errors of different symmetries.

In the case of the [001] zone axis, the projection symmetry is *p6*. In *p6* the origin should coincide with the 6-fold axis. All phases should be 0° or 180° and the symmetry related reflections should have the same phases:

$$\phi(h\ k\ 0) = \phi(\ k\ -h-k\ 0) = \phi(\ h+k\ -k\ 0) \tag{6}$$

These phase restrictions and phase relations were imposed on the experimental phases by setting first the symmetry-related reflections equal and then to 0° or 180°, depending on which value they were closer to. The amplitude-weighted phase residuals for the reflections within different resolution ranges indicated that the phases are reliable up to a resolution of about 1.95Å. All reflections outside 1.95Åresolution were weak and they were excluded in the final processing.

Although the amplitudes from the two images in Fig. 1 differ quite a lot, the phases agree extremely well; among the 110 strongest of the 220 reflections from both images, 84 of the 88 common reflections have the same phases. The symmetry was also imposed on the amplitudes, by averaging the amplitudes of the symmetry related reflections. This step will eliminate most of the effects of crystal tilt.

4.3.5   Projected potential map

Different maps were calculated by inverse Fourier transformation of the amplitudes and phases of all reflections at different stages of image processing. Lattice averaged maps were obtained from the amplitudes and phases directly as extracted from the images, before the CTF correction and imposing the symmetry.

The lattice averaged maps from image 1 and image 2 are very different (Fig. 3a and b). One of the reasons is crystal tilt. When a crystal is tilted, black and white dots in the image/lattice averaged map are smeared out in the direction perpendicular to the tilt axis. Consequently, amplitudes especially of those reflections further away from the tilt axis in the Fourier transform are dampened, (Zou, 1995). From both the images/lattice averaged maps and the Fourier transforms, it is obvious that image 1 suffers more crystal tilt than image 2 (Fig. 1 and Fig. 3a and b). However, phases are not affected by a slight crystal tilt.

The effects of crystal tilt are largely compensated for by imposing the symmetry on the amplitudes and phases (Hovmöller and Zou, 1999). After imposing the *p6* symmetry, both maps obtained from image 1 and 2 show well-resolved dots (Fig. 3c and d). However, the map from image 1 (Fig. 3c) shows white dots while the map from image 2 shows black dots   (Fig. 3d). This is because the images were taken at different defocus conditions. When the CTF is also compensated for, by adding 180° to the phases of those reflections which are at the negative CTF, the maps from image 1 (Fig. 3e) and image 2 (Fig. 3f) become very similar, and these are the final projected potential maps. Since white contrast corresponds to high projected potential in the crystal, white dots in the map correspond to atom rows in the projections. The white dots in Fig. 3e are better separated than those in Fig. 3f, indicating that the resolution is somewhat higher in the map from image 1 than that from image 2. The great similarity of the projected potential maps reconstructed from HREM images of two different crystals taken at quite different conditions demonstrates the power of image processing. The amplitudes and especially the phases obtained from HREM images by crystallographic image processing are reliable.

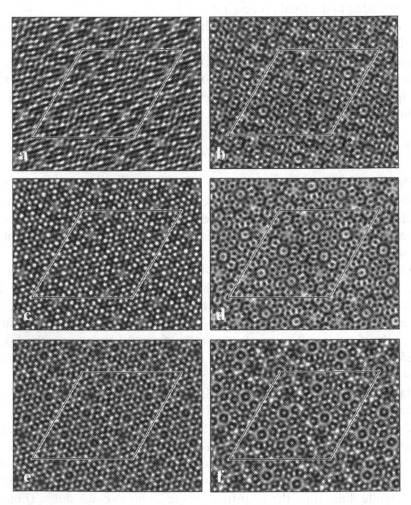

Figure 3 Projected potential maps of the v-phase at various stages of image processing. (a and b) Only lattice averaging of the original images. (c and d) After imposing the p6 symmetry. (e and f) After correction of CTF and imposing the p6 symmetry.

Images taken along the other 12 zone axes were processed in a similar way and amplitudes and phases of the structure factors were extracted. The results of image processing are summarised in Table 1 and Fig. 4. Three zone axes, [010], [120] and [5 18 0] have *pmg* symmetry while the other 9 zone axes have only *p2* symmetry. The number of unique reflections from each zone axis varied between 19 and 224. The amplitude weighted phase residuals varied between 17.9° and 30.6° for different zone axes (Table 1).

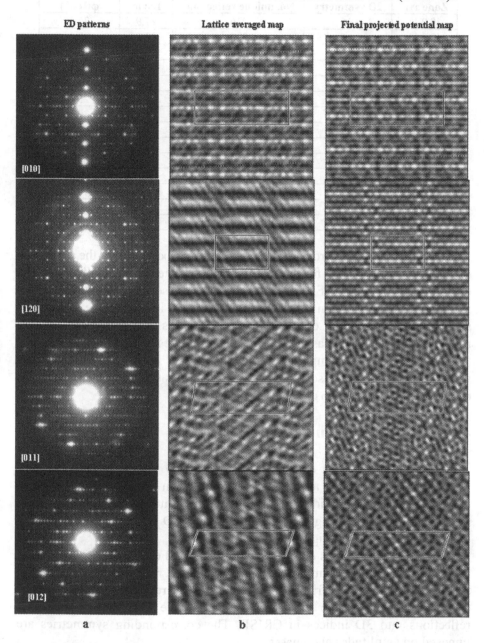

Figure 4 Image processing of four different zone axes [010], [120], [011] and [012]. Column a: ED patterns of the corresponding zone axes. Column b: lattice averaged maps of the original images. Column c: the final projected potential maps using amplitudes from ED patterns and phases from images after CTF correction and imposing the symmetry.

Table 1 Summary of the image processing of HREM images from all the 13 zone axes.

Zone axis	2D symmetry	No. unique reflections	Rsym	φRes(°)
[0 0 1]	*p6*	216	27.9	25.3
[0 0 1]	*p6*	220	22.6	27.1
[0 1 0]	*pmg*	100	65.8	24.0
[1 2 0]	*pmg*	51	25.2	27.1
[5 18 0]	*pmg*	19	43.4	25.8
[0 1 1]	*p2*	224	-	25.5
[0 1 2]	*p2*	132	-	30.6
[0 1 3]	*p2*	201	-	21.5
[0 1 4]	*p2*	143	-	17.9
[0 2 1]	*p2*	129	-	28.5
[0 2 3]	*p2*	120	-	24.3
[1 2 1]	*p2*	137	-	19.8
[1 2 2]	*p2*	204	-	18.1
[2 4 1]	*p2*	52	-	20.8

Also the amplitudes from HREM images are modulated by the CTF. It is difficult to compensate for these effects completely, especially near the cross-overs of the CTF. Furthermore, for those projections with *p2* symmetry, crystal tilt can not be compensated by imposing the symmetry, since reflections have only their Friedel pairs as symmetry related reflections. For these reasons, we used amplitudes from the ED patterns (Fig. 4) instead of the amplitudes from the HREM images. The amplitudes were calculated as the square root of the intensities extracted from the corresponding ED patterns of an exposure series.

4.4   3D reconstruction

Due to the large unit cell of ν- AlCrFe, many atoms overlap in every projection. Images have to be combined to obtain a 3D potential map in order to resolve the individual atoms. Two important steps are needed for 3D reconstruction: converting the 2D indices into 3D indices and putting all images into a common origin.

Firstly, the 3D indices of two reflections in the Fourier transform of each image are determined from the 3D reciprocal space reconstructed from three tilt series of electron diffraction (Fig. 1). Then all reflections in the Fourier transform of the image are converted from the indices of these two reflections into 3D indices in CRISP. The corresponding symmetries are imposed on amplitudes and phases.

Since phases are related to the origin, we have to find a common origin for all the projections and shift the phases accordingly to the common origin. We choose the origin given for $P6_3/m$ by the International Tables for

Crystallography, Vol. A (1983). The origin is on the 6_3 axis, halfway between the mirror planes. There are two such origins in each unit cell, at $z = 0$ and $z = 0.5$.

CRISP locates the origin of the projected unit cell at the origin of the corresponding 2D plane group defined by the International Tables for Crystallography, Vol. A. In most cases the origin coincides with the highest symmetry element. For the projection along the [001] zone axis, the origin is unique, at the 6-fold axis (6_3 axis in 3D). In projections along all other zone axes (having *pmg* and *p*2 symmetry), there are 4 mathematically equivalent origins per unit cell; at $(x\ y) = (0\ 0)$, $(½\ 0)$, $(0\ ½)$ and $(½\ ½)$, respectively. The detailed procedures are the following:

The origin in 3D is fixed from the [001] and the [010] projections. The origin in the *ab* plane was fixed to the 6_3 axis, which is at the 6-fold axis of the [001] projections. The origin along the *c*-axis was set to the same origin determined from the [010] projection. This leaves only two possibilities for the origin of the [010] zone axis, at $(0\ 0)$ or $(½\ 0)$. The final origin is determined by considering the agreement of the phases of the common reflection row h00 ($h = 2n+1$) with those already fixed in the [001] projections (the h00 reflections with even h indices are seminvariants, *i.e.* their phases are fixed independent of the choice of origin, so they cannot be used for fixing the origin).

The origin for each of the other [0 v w] zone axes was determined in a similar way, by comparing the phases of those reflections which are common with the [001] and [010] zone axes.

The origin for each of the [u, 2u, w] projections was determined by comparing the phases of those reflections which are in common with the [001] and [0vw] zone axes. For example, (-2h h 0) reflections exist both in the [1 2 1] and [001] zones. We could shift the origin until most of the common reflections had the same phases.

When the origin has been found for a projection, the phases of all reflections in this projection are shifted from the 2D origin at (00), with phase $\phi_{old}(hk)$, on to the new common 3D origin at $(x\ y)$. This results in a phase shift. The new phases ($\phi_{new}(hk)$) are calculated from using:

$$\phi_{new}(hk) = \phi_{old}(hk) + 360°(hx + ky) \tag{7}$$

The ED intensities obtained from each zone axis were scaled together by the program Triple (Calidris, 2000b) using all common reflections. The merging procedure plays an important role in the final result, since intensities of the common reflections from different zone axes are different due to dynamic effects. Unfortunately, the [001] zone which contains more reflections than any other zones, is the one suffering mostly from dynamical effects. Thus we waited to the end before merging these data. We first merged reflections of all the [0 v w] zone axes, using the common row of reflections (h00). The R_{merge} values were all in the range from $11 - 23$ %. Then reflections from all the [u 2u w] zone axes were merged, using the common row (2h -h 0), in the order [120], [121], [122] and [241]. The R_{merge} values (%) were 17.3, 16.6 and 19.4, respectively. Finally these two data sets were merged with the [001] and [5 18 0] zone axis, in the order [001], [0 v w], [u 2u w] and [5 18 0]. The R_{merge} values (%) were 28.1, 26.6 and 24.4, respectively.

The phases derived from images were combined with the corresponding amplitudes from ED patterns and the data set of 3D structure factors was obtained. Since the phase residual increased abruptly beyond the 1.95Å resolution, and weak reflections are readily distorted by dynamic scattering, noise and optical distortions and they do not contribute very much to the final map, only the 304 strongest out of 641 unique reflections to 1.95 Å resolution were used for further processing. The sum of the intensities of these strongest reflections accounted for 78% of the total intensities. A 3D potential map was calculated from the amplitudes and phases of the 304 strongest reflections by 3D inverse Fourier transformation, using the E-map software (Oleynikov et al., 2003). The sampling of the 3D map was chosen to be 0.1 Å/pixel along the three main axes, sufficient to keep the 0.2 Å accuracy of atomic positions obtained from HREM with a resolution of 2 Å.

4.4.1   Structure model of the v-phase obtained by 3D reconstruction

The 3D potential map was examined section by section perpendicular to the c-axis. There are totally 6 layers stacked along the c axis in each unit cell. Only two of these 6 layers are unique, one flat layer occurring twice (at $z = 0.25$ and 0.75) and one puckered layer occurring four times (at $z \approx 0.10$, 0.40, 0.60 and 0.90). Sections corresponding to the flat (F) and puckered (P) layers are shown in Figs. 6a and b, respectively. The flat layers coincide with mirror planes. The stacking sequence is $PFP^m(PFP^m)'$, where P^m relates to P via a mirror reflection on the flat layer, and the $(PFP^m)'$ block is related to the PFP^m block by a 6_3 operation along the c axis.

The potential maps were remarkably clear (Figs. 5a and b) with most of the peaks perfectly round and well resolved. 63 and 68 peaks were peak found from the density maps in the flat (Fig. 5a) and puckered layers (Fig. 5b), respectively. Out of the 131 peaks, 7 were removed due to too short interatomic distances. It was possible to distinquish Al and transition metal atoms according to the peak height and chemical knowledge. There are in total 129 unique atoms in the unit cell, 62 atoms in the flat layer and 67 atoms in the puckered layer. The structure model and the potential map are superimposed in Fig. 5.

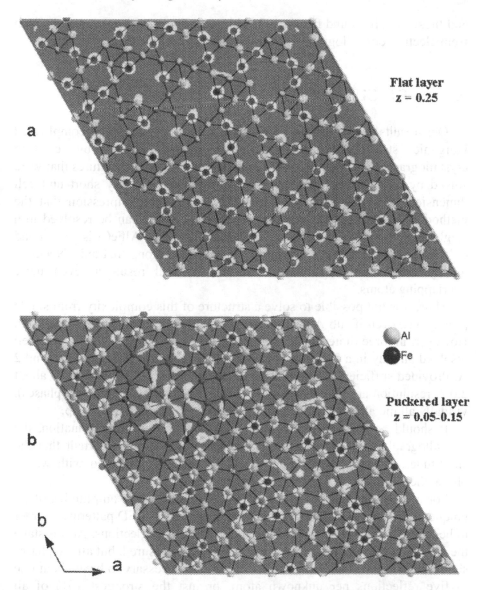

Figure 5 Sections of the 3D potential map obtained by electron crystallography; (a) the flat layer (F) at z=0.25; (b) the puckered (P) at z = 0.10 ± 0.05. In general the strong peaks correspond to Fe/Cr atoms while the weak peaks are Al. 131 unique peaks were found, out of which 7 were removed due to too short inter-atomic distances. 124 unique atoms were assigned from these potential maps, out of which 115 atomic positions agree very closely (differ by less than 0.85 Å) to those obtained by X-ray crystallography (Mo et al., 2000).

Compared with the X-ray model, 110 atoms differ by less than 0.31 Å (mean error 0.14 Å), 90 of those atoms differ by less than 0.20 Å (mean error 0.12 Å). 5 atoms differ between 0.31 - 0.85 Å (mean error 0.59Å). There are 14 atoms which differ from the X-ray model by more than 1.2 Å, and those are all around the three 3-fold axes. The structure model obtained from electron crystallography is more chemically reasonable.

5.   DISCUSSION

Our results show that it is possible to solve extremely complicated inorganic structures, with over 100 unique atoms, by electron crystallography. The fact that nearly all inorganic crystal structures that were solved by electron crystallography until now had one very short unit cell dimension (typically about 4Å) may have given the impression that the method is limited to such structures, where all atoms can be resolved in a single projection. The shortest unit cell dimension of v-AlFeCr is 12.5 Å and 6 different layers of atoms overlap when projected along that axis. Needless to say, any other projection (axial or not) will result in even more overlapping atoms.

Thus, it is not possible to solve a structure of this complexity from single projections, even if sub-Ångström resolution electron microscopes are used. However, in three dimensions all atoms in intermetallic compounds are well resolved already in a 2 Å map, since the inter-atomic distances are around 2 Å. Provided sufficiently many of the most important reflections out to about 2.0 Å resolution are included in the 3D reconstruction (with correct phases), virtually all metal atoms will be seen already in the first density map.

It should be stressed that at this step of a structure determination, the crystallographic structure factor phases are much more important that the amplitudes. If just a single very strong reflection is included with wrong phase, the density map is typically uninterpretable.

For a more general structure, the message is that the amplitudes of all unique reflections must be estimated, for example from ED patterns, in order to be sure that the 3D map includes all the strongest reflections. At this stage the amplitudes do not need to be very accurately measured, but all the phases should be correct. For most structures it will be necessary to have about one to five reflections per unknown atom, or just the strongest 10% of all reflections. This is similar to the data used for generating triple-relations in direct methods in X-ray crystallography; also there only some 10% of all reflections are used, namely those with the highest E-values.

A complete structure determination contains the two distinct steps: solving and refining the structure. The refinement can only be started after the structure has been solved. By solving a structure we mean that most of the most strongly scattering atoms are found to within an accuracy of 0.2 to 0.3 Å. All methods for solving crystal structures from X-ray diffraction data in most cases give just a fraction of the complete structure. Patterson methods give only the heaviest metals; direct methods don't give hydrogen atoms for organic structures etc. This is no problem, since the remaining atoms are readily seen in difference Fourier maps, once these initial atoms are positioned. In the case of v-AlFeCr, over 96% of all atoms were clearly resolved in the 3D map calculated from only a limited amount of all reflection data to 2.0 Å resolution. It is quite remarkable that this map shows both the lighter and heavier atoms and that even the atomic type Al or Fe/Cr in most cases is evident from the peak heights.

These uncertain atoms remain to be verified by a careful structure refinement. For a structure refinement, as many reflections as possible should be included. The phases are not needed at the refinement stage, but if possible complete 3D data out to 1 Å resolution should be used. Strong and weak reflections are equally important. Such data can be obtained by electron diffraction, which is not affected by the contrast transfer function of the electron microscope, but suffers from dynamical scattering. The higher the accuracy of the amplitudes, the more accurate will the atomic positions become.

6. CONCLUSION

3D reconstruction from HREM images offers a new possibility for *ab initio* structure determination of inorganic compounds existing only with very small crystal sizes. 3D reconstruction can be applied to structures too complex to be solved by direct methods even using single crystal X-ray diffraction, using a moderately high voltage electron microscope. There is no limit in terms of the number of unique atoms in a structure that can be solved by 3D reconstruction.

Acknowledgements

The work on the 3D reconstruction of the v-AlFeCr structure was done together with Zhimin Mo, Xingzhong Li and Kehsin Kuo. Peter Olynikov is thanked for writing the software 3D-Map.

References

Calidris. (2000a) Manual for the Kite CCD camera.
Calidris. (2000b) Triple manual.
DeRosier, D.J. and Klug, A., *Nature (London)* **217** (1968) 130

Gemmi, M. Righi, L. Calestani, G. Migliori, A. Speghini, A., Santarosa M. and Bettinelli M. (2000). *Ultramicroscopy*, **84**, 133-142.

Gjønnes, J., Hansen, V., Berg, B. S., Runde, P., Cheng, Y. F., Gjønnes, K., Dorset D. L. & Gilmore, C. J. (1998). *Acta Cryst.* A**54**, 306 - 319.

Hovmöller, S., Sjögren, A., Farrants, G., Sundberg, M. & Marinder, B.-O. (1984). *Nature* (London), **311**, 238-241.

Hovmöller, S. (1992). *Ultramicroscopy*, **41**, 121-135.

Hovmöller, S. & Zou, X.D. (1999) Microscopy Research and Technique 46, 147-159.

International Tables for Crystallography, Vol. A. (1983) T.Hahn, Ed. D.Reidel, Dordrecht, The Netherlands.

Kaneda M., Tsubakiyarna, T., Carlsson, A., Sakamoto, Y., Ohsuna, T., Terasaki, O., Joo, S.H. and Ryoo, R., *J. Phys. Chem. B* **106** (2002) 1256-1266.

Mo, Z.M., Zhou, H.Y. & Kuo, K.H. (2000). *Acta Cryst.* B**56**, 392-401.

Olynikov, O., Hovmöller, S. & X.D. Zou (2003), 3D-MAP, a program for generating 3D Fourier map and determining peak positions.

Wagner, P., Terasaki, O., Ritsch, S., Nery, J.G., Zones, S.I., Davis, M.E. & Hiraga, K. (1999). *J. Phys. Chem.* B**103**, 8245-8250.

Wang, D.N., Hovmöller, S., Kihlborg, L. & Sundberg, M. (1988). *Ultramicroscopy*, **25**, 303-316.

Wenk, H.-R., Downing, K.H., Hu, M. S. & O'Keefe, M.A. (1992). *Acta Cryst.* A**48**, 700-716.

Weirich, T.E., Ramlau, R., Simon, A., Hovmöller, S. & Zou, X.D. (1996). *Nature* (London), **382**, 144-146.

Zandbergen, H.W., Andersen, S.J. & Jansen, J. (1997). *Science*, **277**, 1221-1225.

Zou, X.D., Sukharev, Yu & Hovmöller, S. (1993). *Ultramicroscopy*, **49**, 147-158.

Zou, X.D., *Chemical Communication* No. **5**, Stockholm University (1995).

Zou, X.D., Sundberg, M., Larine, M. & Hovmöller, S. (1996). Ultramicroscopy, **62**, 103-121.

Zou, X.D., Z.M. Mo, S. Hovmöller, X.Z. Li & Kuo, K.H. *Acta Cryst.*A**59** (2003) 526.

SOLVING AND REFINING CRYSTAL STRUCTURES FROM ELECTRON DIFFRACTION DATA

Christopher J. Gilmore[1] and Andrew Stewart[2]

[1]*Department of Chemistry, University of Glasgow, Glasgow G12 8QQ, Scotland, UK and* [2]*Cornell High Energy Synchrotron Source, Cornell University, Ithaca,NY 14853,USA. Email:chris@chem.gla.ac.uk*

Key words: electron diffraction, structure determination, direct phasing, structure refinement

1.   INTRODUCTION

Electron crystallography uses the electron microscope to generate diffraction patterns or high resolution images which can be used to solve crystal structures from nano-sized crystallites. The diffraction data collected are subject to systematic errors in intensity arising from n-beam dynamical scattering, secondary scattering, sample bending and radiation damage, but nonetheless the technique yields structural information when all other diffraction methods fail. Solving such structures can be routine especially from inorganic materials and intermetallic compounds, but, in general, the process can be difficult as can the validation of the proposed model. Direct methods, maximum entropy, Patterson techniques, model building and the use of image data can all be employed, and structures of high complexity can be solved and refined without the use of multislice methods in favourable cases and these are described here.

The process of solving and refining crystal structures from electron diffraction data proceeds in a series of steps:
1. Intensity data collection. Image data can also be useful in providing some phase information.
2. The determination of the unit cell parameters.

T.E. Weirich et al. (eds.), Electron Crystallography, 321–335.
© 2006 *Springer. Printed in the Netherlands.*

3. Space or plane group determination. There can be ambiguities here especially if data are limited in their sampling of reciprocal space. In this case all the possibilities need to be explored with respect to solution and refinement procedures.
4. Structure solution.
5. Refinement both to improve the accuracy of the model and also to validate it.

This chapter is concerned with the latter two steps. It is not possible to be fully comprehensive in a short chapter of this nature, but we provide suitable references for both the theory and for examples.

2.   SOLVING CRYSTAL STRUCTURES

Not unlike X-ray crystallography, the following methods are used to solve crystal structures for small molecules from electron diffraction data:
1. Routine direct methods computer packages.
2. Non-routine direct methods.
3. Maximum entropy. This will be discussed in a separate chapter.
4. Patterson methods.
5. Model building.
6. Using image data.

All of them assume that the data are at least pseudo-kinematic, and use techniques that are robust against systematic and random errors. The list is not exhaustive and new techniques are still being developed, but it covers more than 95% of published structures. We will examine each technique in turn, but postpone the routine application of solution methods at this stage since it teaches us very little about how structure solution methods actually work.

2.1   Direct Methods

2.1.1   Normalisation

We have collected a set of observed structure factors $\{|F_h|_{obs}\}$ for which we have only amplitude information. Denote the required phase angle for reflection h by ϕ_h. Associated with each angle is a normalised magnitude $|E_h|$:

$$\left|E_{\mathbf{h}}\right|^2 = \frac{k\left|F_{\mathbf{h}}\right|_{obs}}{\varepsilon_{\mathbf{h}} \sum\limits_{j=1}^{N} f_j^2 e^{2B\sin^2\theta/\lambda^2}}$$

where k is a scale factor that puts the observed intensities $\left|F_{\mathbf{h}}\right|^2$ on an absolute scale, $\varepsilon_{\mathbf{h}}$ is the statistical weight for reflection **h** derived from point group considerations, f_j is the electron scattering factor for atom j. There are N atoms in the unit cell with an overall, isotropic temperature factor B applied to the thermal motion of each atom. B and k are calculated using Wilson's method [1]. This technique is routine for X-ray data but is usually problematic for electrons: because of data limitations B usually calculates as a negative number, which is, of course, physically impossible. Often a small positive value of B is imposed on the normalisation. Ofetn the approximation $\left|F_{\mathbf{h}}\right| \propto I_{\mathbf{h}}$ is used.

2.1.2   Relationships between phases

Triplets are the key phase relationship in direct methods and they take the form

$$\theta = \phi_{\mathbf{h}} + \phi_{\mathbf{k}} + \phi_{-\mathbf{h}-\mathbf{k}} \approx 0$$

It is clear that the indices of the three reflections involved sum to zero. Associated with each triplet is a concentration parameter $\kappa_{\mathbf{h},\mathbf{k}}$

$$\kappa_{\mathbf{h},\mathbf{k}} = \frac{2\left|E_{\mathbf{h}} E_{\mathbf{k}} E_{-\mathbf{h}-\mathbf{k}}\right|}{\sqrt{N}}$$

The larger the value of $\kappa_{\mathbf{h},\mathbf{k}}$ the more reliable is the estimate of θ *i.e.* the more likely, in a statistical sense, that the phase sum is zero. This is shown in Figure 1.

It can be seen that the mode of the distribution is always zero and that as $\kappa_{\mathbf{h},\mathbf{k}}$ decreases the information content of the triplet relation also decreases until at $\kappa_{\mathbf{h},\mathbf{k}}=1$ very little useful information can be obtained.

Quartets are an extension of this idea and involve four phases instead of three:

$$\Phi_4 = \phi_{\mathbf{h}} + \phi_{\mathbf{k}} + \phi_{\mathbf{l}} + \phi_{-\mathbf{h}-\mathbf{k}-\mathbf{l}}$$

Defining the principal terms

$$\left|E_{\mathbf{h}}\right|, \left|E_{\mathbf{k}}\right|, \left|E_{\mathbf{l}}\right|, \left|E_{-\mathbf{h}-\mathbf{k}-\mathbf{l}}\right|$$

and the cross-terms

$$\left|E_{\mathbf{h}+\mathbf{k}}\right|, \left|E_{\mathbf{k}+\mathbf{l}}\right|, \left|E_{\mathbf{l}+\mathbf{h}}\right|$$

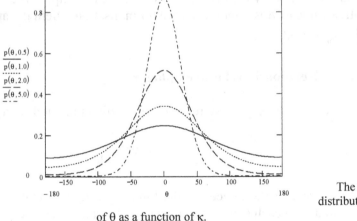

Figure 1: The probability distribution of θ as a function of κ.

The quartets for which all three cross terms are small (say < 0.5) and the principal terms are large (say >2.0) are called 'negative' quartets and $\Phi_4 \approx \pi$, whereas if both the principal and cross-terms are large, then the quartets are deemed positive with $\Phi_4 \approx 0$ [2]. Care is needed with negative quartets when using electron diffraction data because small intensities are often grossly over-estimated and therefore can give erroneous quartet indications.

2.1.3   The tangent formula

The tangent formula [3] is a key formula in direct methods that lets us refine phase values and determine new ones. Consider the situation in which we have a series of triplets with a common reflection $\phi_{\mathbf{h}}$. They can be written

$$\phi_h = \phi_{h2} - \phi_{h-h2}$$

$$\phi_h = \phi_{h3} - \phi_{h-h3}$$

$$\phi_h = \phi_{h4} - \phi_{h-h4}$$

etc.

Consider also a situation in which all the phases on the RHS of all the equations above are known, at least approximately, then the tangent formula gives us an estimate for ϕ_h which in its simplest form is

$$\tan \phi_h = \frac{\sum\limits_k |E_k E_{h-k}| \sin(\phi_k + \phi_{h-k})}{\sum\limits_k |E_k E_{h-k}| \cos(\phi_k + \phi_{h-k})}$$

It can be extended to include quartets of all types and various weighting schemes can be used which help impose stability on a formula which can be prone to problems with ED data.

2.1.4   Figures of merit

In general direct methods are multisolutional: they give rise to multiple phase sets and we need to select those which are most likely to give useful structural information. Figures of merit serve this purpose and are used to rank phase sets. There are numerous such indicators including, for example

The 'absolute figure of merit' ABSFOM, defined by the following equations:

$$ABSFOM = \sum_h (\alpha_h - \alpha_{Rh}) \Big/ \sum_h (\alpha_{Eh} - \alpha_{Rh})$$

where:

$$\alpha_{Eh}^2 = \sum_k \kappa_{hk}^2 + \sum_k \sum_{k'} \kappa_{hk} \kappa_{hk'} \frac{I_1(\kappa_{hk}) I_1(\kappa_{hk'})}{I_1(\kappa_{hk}) I_1(\kappa_{hk'})}$$

$$\alpha_{Rh} = \left(\sum_h A_{hk}^2 \right)^{1/2}$$

The negative quartet figure of merit [4]

$$NQEST = \frac{\displaystyle\sum_{negative} B\cos\left(\phi_h + \phi_k + \phi_l + \phi_{-h-k-l-m}\right)}{\displaystyle\sum_{negative} B}$$

where the summation spans all those quartets assumed negative. An optimal phase set should have a minimum value of NQEST and should be negative. The optimum value is -1.0.

RESID is an R-factor defined either on α or on E-magnitude. For the former case:

$$R_\alpha = 100 \times \sum_h \left|\alpha_h - \alpha_{Eh}\right| \bigg/ \sum_h \alpha_{Eh}$$

The relationship to the conventional crystallographic R factor is obvious. A value for R_α around 20-25% is a good indicator of phase consistency.

The Ψ_0 figure of merit is defined as:

$$\Psi_0 = \frac{\displaystyle\sum_l\left(\left|\sum_k E_k E_{l-k}\right|\right)}{\displaystyle\sum_l\left(\sum_k |E_k E_{l-k}|^2\right)^{1/2}}$$

The reflections involved in these triplets have $|E_k|$ and $|E_{l-k}|$ large but $|E_l|$ *small*. This differs from the other figures of merit in that it establishes a consistency between the amplitudes and phases of the strong reflections with a small set of weak reflections. Correct phase sets usually give a value of Ψ_0 close to unity.

Usually several figures of merit are calculated for a given phase set and these are combined to given an overall figure called a CFOM, the solution(s) with the best CFOM are selected and used to compute a Fourier map.

2.1.5   E-maps

So far our discussions have involved reciprocal space quantities; the transform into real space is carried out using E magnitudes *via* E-maps

$$\rho(\mathbf{x}) \approx \frac{1}{V}\sum_{\mathbf{h}} |E_h| e^{i\phi_h} e^{-2\pi i \mathbf{h}.\mathbf{x}}$$

The use of E magnitudes and the limits we shall impose on the reflections entering the summation mean that the electron density is only approximate (At the very least there are serious series termination errors.) but hopefully is sufficient to reveal structural features so that model building can begin.

For a complete discussion of direct methods see Giacovazzo's monograph [5].

2.1.6 Simplifying the problem

The problem of direct phasing can be simplified by the following heuristic rules:
1. Only the top 8-$10N_a$ need to bc phased where N_a is no of atoms in asymmetric unit.
2. In centrosymmetric space groups all the phases are centric with phases restricted to 0, π. Non centrosymmetric space groups often have centrosymmetric projections giving rise to centric reflections which have restricted phase choices *e.g.* $0, \pi; \pm \pi/2$.
3. We can tolerate relatively large (*ca.* 40^0) random errors, but smaller systematic errors.

2.1.7 Symbolic addition

It should be apparent that if we know some initial phase information, we can extend phases with the tangent formula or the simple triplet relationship. One source of this initial information is the use of origin defining reflections which represent the reciprocal space equivalent of fixing the origin of the unit cell in three-dimensional space. In symbolic addition [6] origin definition is coupled with the assignment of algebraic symbols to a small number (typically 4-8) phases that have large associated E magnitudes and which interact strongly through triplets and quartets. These are used to determine 20-100 new phases as functions of these symbols. The accuracy is not high, but as long as systematic errors are avoided, mean phase errors of around 40^0 can still yield interpretable potential maps. The symbols are converted into numerical values using relationships between them made manifest by the symbolic addition procedure or by giving unassigned symbols permuted values in the range $0 - 2\pi$. In X-ray crystallography symbolic addition is not much used currently for solving small molecules, although in the1960s it was the technique of choice; it has been superseded by methods that are much easier to automate. It does, however, have the virtues of stability and robustness, and this is that is exploited by Dorset.

There are numerous, fully worked examples in Dorset's book [7] including C_{60} [8], waxes [9], perfluorotetracosane [10], and poly(3,3-bis(chloromethyl)oxacyclobutane [11]. In the latter case, for example, the data set comprised 29 hk0 intensities - a situation from which no routine direct methods program could extract meaningful phases. Twenty seven triplets were used to define19 phases in terms of a single symbol, *a*. Since there was no reliable indication of the value of *a* except that it is constrained by space group symmetry to have a phase of either 0 or π, two Fourier maps were computed examined using both values, and only one of these maps was chemically interpretable, and used to give the structure.

2.1.8   The Sayre equation

The Sayre equation [12] is algebraic rather than probabilistic in origin and is derived from the expression relating the electron density and its square:

$$F_{\mathbf{h}} = \frac{\theta}{V} \sum_{\mathbf{k}} F_{\mathbf{k}} F_{\mathbf{h}-\mathbf{k}}$$

where θ is a constant, and V the unit cell volume. Note that the structure factors are complex - the phase information is included. The Sayre equation can be successfully employed with poor data sets. (see, for example, Yi-Wei *et al.* [13]). It is usually used in this context to improve phases derived from symbolic addition, and extend them to reflections that were not accessible to symbolic addition. As an example in EC see the structure determination of poly(3,3)-bis(chloromethyl)oxacyclobutane) [11] where a partial set of 19 phases was extended to 29 with only three phases incorrectly indicated.

2.1.9   Using the tools to solve crystal structures

There are numerous procedures for solving structures *via* direct methods. A typical, though somewhat simplified, sequence is as follows:
1. The data are normalised using Wilson's method to give E-magnitudes. These are sorted in descending order.
2. Triplets are generated for the top *$10N_a$* reflections. Quartets (usually just the negative ones) are optionally generated as well. This is not always possible in the ED case.
3. The top *$8-10N_a$* reflections are given random phases.
4. The phases are refined to convergence using the tangent formula in one of its many variants.
5. Figures of merit are calculated for this phase set and combined together to give an overall figure of merit CFOM.
6. Steps 3-5 are repeated 24-1000 times depending on the difficulty and complexity of the structure.
7. The phase sets are sorted on CFOM.

8. An E-map is computed for the best set and the peaks picked. We then use our knowledge of molecular dimensions and conformations to extract a trial structure. This is the first point at which chemical knowledge is used actively *i.e.* the direct methods procedure is model free until this point.
9. The structure is completed and refined in the usual way.

If no identifiable fragment can be found then the next ranked phase set from step 7 is used and steps 8 and 9 repeated. This can be done for the top 10 or more phase sets.

2.1.10 What is needed for this method to work?

The procedure is usually routine if the following criteria are met:
1. Atomicity: We need intensity data to a resolution of 1.1-1.2Å
2. Completeness: The data must be complete to this resolution.
3. Accuracy: Accurate data are required.
4. Complexity: The number of non-hydrogen atoms in asymmetric unit should be <1000.

Clearly the first three of these criteria do not apply to most electron diffraction data sets where a resolution of 2Å is common, low angle data may be missing and where accuracy is limited by poor crystalline specimens.

2.1.11 Routine direct methods computer packages

To solve a crystal structure by direct methods, 'difficult' data are those which are incomplete in the sampling of reciprocal space, have non-atomic (*i.e.* < 1.3Å resolution) and are noisy with large (systematic) errors in the data measurements. As we have seen, this definition spans many electron diffraction data sets, but there are some of sufficient quality that they can be solved routinely using conventional direct methods packages. Often these are of inorganic materials or intermetallic compounds that are relatively resistant to radiation damage.

Some recent examples include the metal rich compound $Ti_{45}Se_{16}$ [14]. The unit cell dimension are a = 36.534, b = 3.453, c = 16.984Å and β=91.73°. The space group is C2/m. The automatic direct methods program SIR97 [15] was modified to use electron scattering factors, and used in fully automatic mode gave a potential map in which all 31 atoms comprising the asymmetric unit were revealed, although, not unusually, the peak heights were not indicative of atom type. The structure refined to an R factor based on F of 0.33 which is a good figure for such data. To reduce the effects of dynamical scattering the approximation $|F_h| \propto I_h$ was made in structure solution.

In a similar vein Weirich *et al.* [16] have solved the structure of the β-phase of Ti_2Se using an identical procedure. SIR97 revealed all 9 atoms of the asymmetric unit and refined to R=0.24. Bougerol-Chaillout *et al.* [17] have solved Nd_2CuO in a similar routine way.

With the right sample, it is even possible to solve structures from electron powder data. Weirich *et al.* [18] have collected such data on anastase (TiO_2) to get a data set comprising 20 unique reflections which also solved with SIR97 by using the data as single crystal reflections.

However, most samples cannot give data of the required quality, and under these circumstances less routine applications of direct methods are needed.

2.1.12   Patterson methods

The range of values of the atomic scattering factors for electrons is less than that of X-rays. The Patterson method works optimally when

$$\sum f_{heavy}^2 \Big/ \sum f_{light}^2 \approx 1.0$$

and this is difficult to achieve in practice with electron data. However, under favourable circumstances, there is still sufficient variation to allow the identification of heavy atom vectors, and because the method uses all the data, it can be sufficiently robust with respect to errors that one can expect vector methods to be a viable alternative to direct methods provided that the data are at least *quasi*-kinematic, and this is indeed the case.

As an example see the Patterson method applied to the high T_c superconductor $(CuC)Ba_2Ca_2Cu_3O_x$ by Gautier *et al.*, [19]. A Patterson map based on 50 unique intensities arising from averaging and merging of 5 zones from which outliers had been excluded gave vectors which yielded the Ba, Ca and two of the three Cu cations. For further examples see Bougerol-Chaillout (2001).

An interesting case study in which a high quality data set of an intermetallic compound, Al_mFe, is submitted to Patterson, maximum entropy and direct methods has been published by Gjønnes *et al.* [20].The data set comprised 598 unique intensities collected by a novel precession method. The space group was $I\bar{4}2m$ with a = 8.84, c=21.6 Å. In this case all the methods behaved in a comparable way with no one method exhibiting inherent superiority.

2.1.13   Model building

Model building remains a useful technique for situations where the data are not amenable to solution in any other way, and for which existing related crystal structures can be used as a starting point. This usually happens because of a combination of structural complexity and poor data quality. For recent examples of this in the structure solution of polymethylene chains see Dorset [21] and [22]. It is interesting to note that model building methods for which there is no prior information are usually unsuccessful because the data are too insensitive to the atomic coordinates. This means that the recent advances in structure solution from powder diffraction data (David *et al.* [23]) in which a model is translated and rotated in a unit cell and in which the torsional degrees of freedom are also sampled by rotating around bonds which are torsionally free will be difficult to apply to structure solution with electron data.

2.1.14   Using image data

The most frequent use of an electron microscope is to generate high magnification images. With suitable instrumentation images can be obtained from nano-crystalline areas at 2.0Å resolution or even less. At such resolution, and with suitable samples (usually inorganic materials) the atomic positions in projection may be determined and used as a starting model for phasing the corresponding higher resolution electron diffraction data or, more reliably, the Fourier transform of the image can be used to extract phase information which can then generate a trial crystal structure. The images are only interpretable however if the samples are very thin and the defocus of the instrument (the Scherzer defocus) is optimised. Some early work in this field used either image data with traditional direct methods [24] or with the Sayre equation [13] although these examples used simulated data which can be misleading.

An elegant example of the use of image data can be found in the work of Weirich *et al.* [25] in studying $Ti_{11}Se_4$. A total of 113 two-dimensional diffraction intensities to a resolution of 1.75Å were obtained from the FT of selected images suitably averaged, and these gave a starting model for the 23 atoms in the asymmetric unit which were then refined against 408 unique electron diffraction intensities at 0.75Å resolution. The third atomic coordinate was obtained from packing and geometrical arguments, and since there were very few degrees of freedom in this direction.

For other examples involving materials see, for example, the determination of the crystal structure of the high T_c superconductor $YBa_2Cu_3O_7$ by Huang *et al.* [26], and $K_2O.Nb_2O_5$ [27].

2.2   Structure refinement

By far the best way to refine structures using electron diffraction data is to use multislice calculations. These will be discussed in the next chapter. However, some useful information can be obtained by regular crystallographic least squares with the assumption of kinematic data.

The process of structure refinement aims to modify model parameters to obtain both the optimal agreement between calculated and observed intensities and improve model accuracy. To do so we minimise

$$R_1 = \sum_{\mathbf{h}} w_{\mathbf{h}} \left(|F_{\mathbf{h}}|_{obs} - |F_{\mathbf{h}}|_{calc} \right)^2$$

where $w_{\mathbf{h}}$ is the weight of reflection $\mathbf{h}$ usually defined as the inverse variance

$$w_{\mathbf{h}} = 1.0 / \sigma_{\mathbf{h}}^2$$

The lack of accurate variances for electron diffraction data can be a problem. Equation 16 is called 'refinement on F'. In recent years an alternative has become more widely used

$$S' = \sum_{\mathbf{h}} w_{\mathbf{h}}' \left(|F_h|_{obs}^2 - |F_h|_{calc}^2 \right)^2$$

which is 'refinement on F^2. The success of a structure refinement is usually judged by the R-factor

$$R1 = \frac{\sum_{\mathbf{h}} \left(|F_{\mathbf{h}}|_{obs} - |F_{\mathbf{h}}|_{calc} \right)}{\sum_{\mathbf{h}} |F_{\mathbf{h}}|_{obs}}$$

or, more usually now the weighted R factor, which is usually twice the value of R1

$$wR2 = \left\{ \frac{\sum_{\mathbf{h}} w_{\mathbf{h}}' \left(|F_{\mathbf{h}}|_{obs}^2 - |F_{\mathbf{h}}|_{calc}^2 \right)^2}{\sum_{\mathbf{h}} w_{\mathbf{h}}' |F_{\mathbf{h}}|_{obs}^2} \right\}^{\frac{1}{2}}$$

The R-factor, R1, is a useful indicator for deciding whether a particular model needs adjustment. If the factor is above R=0.83 for centrosymmetric space group and R=0.59 of non-centrosymmetric space group the atoms are essentially random in relation to the diffraction pattern. R-factors of R=0.45 indicate the model has some merit. R=0.35 is taken to mean that the final model can be deduced from the current point. R=0.25 means that the model is within 0.1Å of the correct atomic positions. In X-ray experiments R-factors of <0.05 are expected for a successful structure refinement, giving accurate bond lengths and geometry. In the case of electron diffraction experiments R-factors of <0.20 are exceptional with the values being more commonly 0.25 and above.

We also need the concept of constraints and restraints. Constraints include fixing parameters to be the same or related, fixing a rigid body - a phenyl group for example. Restraints are softer, and include such examples as limiting the values of bond lengths or defining the range of a temperature factor. Both are needed in refining many electron diffraction data sets.

SHELXL [28] is the most commonly used structure refinement program. We have devised a strategy for addressing the refinement:

1. Use electron scattering factors.
2. Refine without any restraints or constraints.
3. Fix all temperature factors. In general temperature factor behaviour is unpredictable in refinement.
4. Damp the diagonal of the least squares matrix. This stops the calculation from locating false minima by moving parameters too far in a given cycle of refinement.
5. Introduce restraints on bond lengths where required.
6. Introduce bond angle restraints where required.
7. Introduce bond and bond angle constraints as required.

The use of constraints and restraints arise from unrealistic bond lengths and angles. It is a reasonable use of statistical analysis to include prior information in a refinement procedure, and geometry is one such restraint/constraint. Poor geometries are often a consequence of missing data down one axis, invariably the result of the missing cone or lack of an epitaxially grown crystal. This lack of data has a profound effect on the refinement process

2.2.1 Examples of refinement

All obtainable three-dimensional data sets were used to test this methodology. They were the following: Aluminium Iron alloy [20], Brucite $(Mg(OH)_2)$, CNBA, DMABC, DMACB, Copper Perchlorophthalocyanine, Poly(1-butene) form III, Polyethylene, Silicon surface, Poly (1,4- trans-cyclohexanediyl dimethylene succinate) (T-cds)

Compound	# of data	# of atoms	R1 (F>4σF)	# of param.	# of restr.
Al$_m$Fe	743	13	0.46	32	0
Brucite, Mg(OH)	59	3	0.11	4	0
CNBA, 10-cyano-9,9'-Bianthryl	144	30	0.29	123	57
Basic Copper Chloride	119	12	0.27	15	0
DMABC C$_{20}$ N$_2$ O$_1$	133	14	0.26	42	18
DMACB C$_{15}$ N$_2$	118	17	0.26	53	24
Mannan I	58	11	0.20	34	9
Copper Perchlorophthalocyanine	197	16	0.20	45	0
Polybut-1-ene Form III	146	8	0.29	27	28
Polycaprolactone	47	10	0.20	33	2
Polyethylene	51	3	0.14	6	0
Poly(1,4,trans-cyclohexanediyl dimethylene succinate)	85	8	0.16	33	0

References

1. Wilson, A.J.C. (1949). The probability distribution of X-ray intensities. *Acta. Cryst.* **2**, 318-321.
2. Schenk, H. (1973). The use of phase relationships between quartets of reflexions *Acta Cryst.* **A29**, 77-82.
3. Karle, J. & Hauptman, H.A. (1956). A theory for the four types of non-centrosymmetric space groups 1P222, 2P22, 3P$_1$2, 3P$_{22}$. *Acta Cryst.* **9**, 635-651.
4. DeTitta, G.T., Langs, D.A., Edmonds, J.W. & Duax, W.L. (1975). Use of negative quartet cosine invariants as a phasing figure of merit: NQEST. *Acta Cryst.* **A31**, 472-479.
5. Giacovazzo, C. (1998). *Direct phasing in crystallography. Fundamentals and applications* Oxford: Oxford University Press and IUCr.
6. Karle, J. and Karle, I.L. (1966). The symbolic addition procedure for phase determination for centrosymmetric and noncentrosymmetric crystals *Acta Cryst.* **21**, 849-859.
7. Dorset, D.L. (1995). *Structural Electron Crystallography.* New York: Plenum Press.
8. Dorset, D.L. and McCourt (1994). Disorder and the molecular packing of C$_{60}$ buckminsterfullerene: a direct electron crystallographic analysis' *Acta Cryst.* **A50**, 344-351.
9. Dorset, D.L. (1987). Role of symmetry in the formation of n-paraffin solutions. *Macromolecules* **20**, 2782-2788.
10. Zhang, W.P. and Dorset, D.L. (1990). Epitaxial growth and crystal structure analysis of perfluorotetracosane *Macromolecules*, **23**, 4322-4326.
11. Dorset, D.L. (1992.) Electron crystallography of linear polymers: direct determination for zonal data sets *Macromolecules* **25**, 4425-4430.
12. Sayre, D. (1952). The squaring method: a new method for phase determination. *Acta Cryst.* **5**, 60-65.

13. Yi-Wei, L., Hai-Fu, F. and Chao-De, Z. (1988) Image processing in high-resolution electron microscopy using the direct method. III. Structure-factor extrapolation *Acta Cryst.* **A44**, 61-63.

14. Weirich, T.E. (2001). Electron crystallography without limits? Crystal structure of $Ti_{45}Se_{16}$ redetermined by electron diffraction structure analysis. *Acta Cryst.* **A57**, 183-191.

15. Altomare, A., Burla, M.C., Camalli, M., Cascarano, G., Giacovazzo, C., Guagliardi, A., Moliterni, A.G., Polidori, G. and Spagna, R. (1999) *SIR97*: a new tool for crystal structure determination and refinement *J.Appl. Cryst.* **32**, 115-118.

16. Weirich, T.E., Zou, X. , Ramlau, R., Simon, A, Cascarano, G.L., Giacovazzo, C. and Hovmöller, S. (2000). Structures of nanometre-sized crystals determined from selected-area electron diffraction data *Acta Cryst.* **A56**, 29-35.

17. Bougerol-Chaillot, C. (2001) Structure determination of oxide compounds by electron crystallography *Micron*, **32**, 473-479.

18. Weirich, T.E., Winterer, M., Seifried, S. and Mayer, J. (2002) Structure of nanocrystalline anatase solved and refined from electron powder data. *Acta Cryst.*, **A58**, 308-315.

19. Gautier, E., Tranqui, D., and Chaillout, C. (1997) First steps in the structure determination of an oxycarbonate superconductor from electron diffraction intensities. In *Electron crystallography* 375-378, edited by D.L. Dorset, S. Hovmöller and X. Zou NATO ASI series, **347**.

20. Gjønnes, J., Hansen, V., Runde, P., Cheng, Y.F., Gjønnes, K., Gilmore, C.J. and Dorset, D.L. (1998). Structure model for the phase Al_mFe derived from three-dimensional electron diffraction intensity data collected by a precession technique. Comparison with convergent beam diffraction *Acta Cryst.* **A54**, 306-319.

21. Dorset, D.L. (2000). Electron crystallography of the polymethylene chain. 4. Defect distribution in lamellar interfaces of paraffin solid solutions Z. Krist. **215**, 190-198.

22. Dorset, D.L. (2001). Electron crystallography of the polymethylene chain. 5. Three dimensional structure of a Fischer-Tropsch wax Z. Krist. **216**, 234-239.

23. David, W.I.F., Shankland, K., McCusker, L.B., and Baerlocher, Ch. (2002.) Structure determination from powder diffraction data. Oxford: Oxford University Press and IUCr.

24. Hai-Fu, F., Zi-Yang, Z., Chao-De, Z. and Fang-Hua, L. (1985). Image processing in high-resolution electron microscopy using the direct method. I.Phase extension *Acta Cryst.* **A41**, 163-165.

25. Weirich, T.E, Ramlau, R., Simon, A., Hovmöller, S. and Zou, X. (1996). A crystal structure determined with 0.02Å accuracy by electron microscopy. *Nature*, **382**, 144-146.

26. Huang, D.X., Liu, W., Guy, X., Xiong, J.W., Fan, H.F. and Li, F.H. (1996). A method of electron diffraction intensity correction in combination with high resolution electron microscopy. *Acta Cryst.* **A52**, 152-157.

27. Hu, J.J., Li, F.H. and Fan, H.F. (1992). Crystal structure determination of $K_2O.7Nb_2O_5$ by combining high-resolution electron crystallography and electron diffraction. *Ultramicroscopy*, **41**, 387-397.

28. Sheldrick, G. M. (1997). SHELXL-97. A computer program for the Refinement of Crystal Structures. University of Göttigen, Germany.

THE MAXIMUM ENTROPY METHOD OF SOLVING CRYSTAL STRUCTURES FROM ELECTRON DIFFRACTION DATA

C. J. Gilmore

Department of Chemistry, University of Glasgow, Glasgow G12 8QQ, Scotland, UK. .Email: chris@chem.gla.ac.uk.

Key words: electron diffraction, structure determination, maximum entropy method

1.   INTRODUCTION

This chapter will illustrate the process of solving crystal structures using the maximum entropy (ME) method. In the first section the theory is described; this is followed by a practical example of the method in action, and there is then a brief review of other applications.

1.1   Bayesian statistics

Maximum entropy (ME) is a tool of Bayesian Statistics, and thus is built around Bayes' Theorem. Since, as diffractionists, we are interested in maps and particularly in obtaining an optimum map from measured data, we can state this theorem in the following way

$$p(map|data) \propto p(map)\, p(data|map)$$

posterior   prior   likelihood

where p(.) means 'probability of'. In Bayesian statistics we define a map with our current knowledge to give the *prior*; we then consult the data with *likelihood* and the use of Bayes' theorem gives us an updated *posterior map*. The whole process can be repeated in a cyclical way. There is, however, a

T.E. Weirich et al. (eds.), Electron Crystallography, 337–353.

problem: bias in our prior belief about the map; we need to minimise this, and the ME formalism can help.

1.2   The maximum entropy principle

Consider a situation for which we know a set of N probabilities $P=\{p_1,p_2 ......p_N.\}$. We can represent this by a probability distribution, and this has an entropy, S given by:

$$S = -\sum_{i=1}^{N} p_i \log p_i$$

This equation was first defined and used in a non-thermodynamic context by Shannon [1] when working with problems of the capacity of communication channels, and the transmission of signals down noisy lines. Suppose we have m constraints, expressed as expectation values, $\langle A_r \rangle$, which represent our knowledge derived from experiment. These constraints can be written:

$$\sum_{j=1}^{N} A_{ri} p_i = \langle A_r \rangle \qquad r=1,2..........m$$

and from this we wish to derive the set of probabilities {P}. Often the situation under study is a mathematically indeterminate one in which $N >> m$. The ME approach consists of maximising the entropy subject to the constraints to determine {P}.

1.3   The Brandeis Dice problem

This classical problem may help to illuminate this formalism Consider a die. When it is thrown there are six possible results i where $1 \leq i \leq 6$. If it is thrown N times, an "honest" dice will give all $p_i = 1/6$, and an average throw of 3.5 when thrown repeatedly. Suppose, however, we have a dishonest or biased one for which the mean is 4.5 *i.e.*

$$\sum_{i=1}^{6} i p_i = 4.5$$

Given this information *and nothing else* what probability, $p_{i,i=1,6}$ should we assign to the i spots on the next throw?

Clearly the problem is indeterminate, and it is possible to devise an infinity of solutions. (For example, the die only gives $i=5$ or $i=4$ with equal probabilities). The ME principle, however, generates a unique solution which, in the jargon of the subject, is *maximally unbiased*, has a *minimum information content,* and is declared *maximally non-committal with respect to missing information:*

There are two constraints: the value of the mean and the normalisation condition:

$$\sum_{i=1}^{6} p_i = 1$$

Assuming uniform prior probabilities, we maximise S subject to these constraints. This is a standard variation problem solved by the use of Lagrangian multipliers. A numerical solution using standard variation methods gives $\{p_1....p_6\}=\{0.05435, 0.07877, 0.11416, 0.16545, 0.23977, 0.34749\}$, with an entropy of 1.61358 natural units.

Now what does this result mean? To quote Jaynes [2] " *..our result as it stands is only a means of describing a state of knowledge about the outcome of a single trial. It represents a state of knowledge in which one has only (1) the enumeration of the six possibilities, and (2) the mean value constraint...and no other information. The distribution is maximally noncommittal with respect to all other matters; it is as uniform ... as it can get without violating the given constraint.* "

1.4 Crystallography: maps

A map has entropy. Take a map and partition it into pixels. Let the *i-th* pixel be p_i then we can normalise each pixel to get

$$p_i' = p_i / \sum p_i$$

and the map has an entropy

$$S = -\sum_i p_i' \log p_i'$$

The maximum entropy principle: To produce an image or map which is maximally non-committal or minimally biased with respect to missing data, maximize the entropy of the map subject to the constraint that this map must reproduce the data which generated it within experimental error.

2.   APPLICATION TO ELECTRON CRYSTALLOGRAPHY

2.1   Some necessary theory and notation

Let us start with a set of unitary structure factors $|U_h|^{obs}$ which are derived from the intensity data by standard normalisation procedures. (For N point atoms of unit weight in space group P1 $|U_h|^{obs} = 1/\sqrt{N}|E_h|^{obs}$) We also have a set of phase angles φ_h most of which are either unknown (the *ab initio* case) or only approximately determined. For *ab initio* structure solution, because of rules of origin and enantiomorph definition, some phases can usually, but not always (it depends on the space group) be assigned subject to certain well known rules. In electron diffraction, the Fourier transform of the corresponding image may give useful phase information.

Those reflections which are phased comprise the basis set {H}; the disjoint set, the non-basis set, of unphased amplitudes is {K}; the phase problem is one of phasing $|U_{h \in K}|^{obs}$ from $U_{h \in H}$. In electron diffraction the unmeasured set {U} may also be important. This is shown diagramatically in Figure 1.

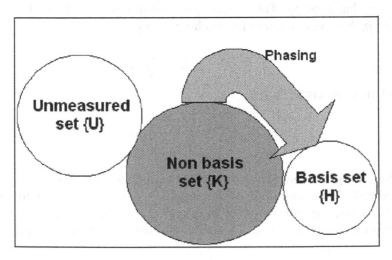

Figure 1. Partitioning the intensity data.

A set of reflections is chosen to form the basis set. These reflections are used as constraints in an entropy maximisation to generate a maximum entropy map $q^{ME}(\underline{x})$. Because it is a maximum entropy map $q^{ME}(\underline{x})$ satisfies the following conditions:

1. It is optimally unbiased.
2. Its Fourier transform reproduces the constraints to within experimental error.

The Fourier transform of $q^{ME}(\underline{x})$ generates estimates of amplitudes and phases for reflections in {K}. This process is called *extrapolation*.

This is shown diagramatically in Figure 2. At this stage, for a small basis set, the extrapolation is rather weak, and the structure is certainly not visible in any map one cares to generate. So what can be done? The most obvious answer is to examine the most strongly extrapolated reflections, and to use the phase information from them coupled with the observed U magnitudes, and to pass these into the basis set {H}. This usually leads to disaster: the ME solution gets trapped in a local optimum point in phase space, whole subsets of reflections become wrongly phased, and the structure remains unsolved. This is a manifestation of the *branching problem*: phases are being selected without exploring the relevant U-space in sufficient detail, so that what appears to be an unambiguous choice of new phase is, in reality, no such thing. So what can be done? Extra unphased reflections must be added to the starting set with permuted phases giving rise to a multisolution environment just as in conventional direct methods, and these reflections should be those which are very weakly extrapolated, so that their inclusion in a new basis set offers maximum surprise to the calculations. How then do you select the correct permutation set? Can you choose those which give maps of maximum entropy? The answer to this is *'no'*, but likelihood functions can help.

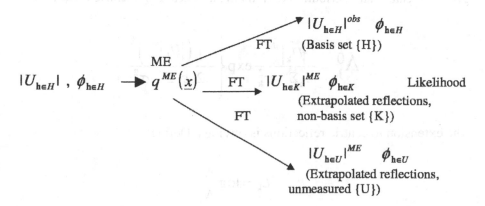

Figure 2. Phase extrapolation

2.2   Likelihood

For each acentric extrapolated, non-basis set reflection $\mathbf{k}$ the likelihood measure, in its diagonal approximation, can be written [3,4,5]

$$
\Lambda_{\mathbf{k}} = \frac{|U_{\mathbf{k}}|^{obs}}{\varepsilon_{\mathbf{k}}\Sigma + \sigma_{\mathbf{k}}^2} \exp\left\{-\frac{1}{2}\frac{\left(|U_{\mathbf{k}}|^{obs}\right)^2 + |U_{\mathbf{k}}^{ME}|^2}{\varepsilon_{\mathbf{k}}\Sigma + \sigma_{\mathbf{k}}^2}\right\} I_0\left(\frac{|U_{\mathbf{k}}|^{obs}|U_{\mathbf{k}}^{ME}|}{\varepsilon_{\mathbf{k}}\Sigma + \sigma_{\mathbf{k}}^2}\right)
$$

where $\varepsilon_{\mathbf{k}}$ is the statistical weight of reflection k, $\sigma_{\mathbf{k}}^2$ the variance of $|U_{\mathbf{k}}|^{obs}$ and Σ a refinable measure of unit cell contents $\Sigma \approx 1/(2N)$ for N point atoms in the unit cell. This distribution is a Rice distribution comprising a Gaussian (the exponential term) with an offset represented by the Bessel function term (I_0). Note also that this expression is a measure of agreement between $|U_{\mathbf{k}}|^{obs}$ and $|U_{\mathbf{k}}^{ME}|$, indeed it has a maximum where $|U_{\mathbf{k}}|^{obs} = |U_{\mathbf{k}}^{ME}|$. For the centric case the Bessel function is replaced by a cosh term:

$$
\Lambda_{\mathbf{k}} = \frac{2|U_{\mathbf{k}}|^{obs}}{\pi\left(2\varepsilon_{\mathbf{k}}\Sigma + \sigma_{\mathbf{k}}^2\right)} \exp\left\{-\frac{1}{2}\frac{\left(|U_{\mathbf{k}}|^{obs}\right)^2 + |U_{\mathbf{k}}^{ME}|^2}{2\varepsilon_{\mathbf{k}}\Sigma + \sigma_{\mathbf{k}}^2}\right\} \cosh\left(\frac{|U_{\mathbf{k}}|^{obs}|U_{\mathbf{k}}^{ME}|}{2\varepsilon_{\mathbf{k}}\Sigma + \sigma_{\mathbf{k}}^2}\right)
$$

In the spirit of traditional likelihood analysis, we define a corresponding null hypothesis for the situation of null extrapolation, $|U_{\mathbf{k}}^{ME}| = 0$, which gives the Gaussian distribution of Wilson statistics. For acentric reflections:

$$
\Lambda_{\mathbf{k}}^0 = \frac{|U_{\mathbf{k}}|^{obs}}{\varepsilon_{\mathbf{k}}\Sigma + \sigma_{\mathbf{k}}^2} \exp\left\{-\frac{1}{2}\frac{\left(|U_{\mathbf{k}}|^{obs}\right)^2}{\varepsilon_{\mathbf{k}}\Sigma + \sigma_{\mathbf{k}}^2}\right\}
$$

The extension to centric reflections is obvious. Define:

$$
L_{\mathbf{k}} = \log\frac{\Lambda_{\mathbf{k}}}{\Lambda_{\mathbf{k}}^0}
$$

Then the global log-likelihood gain (LLG) is:

$$LLG = \sum_k L_k$$

The LLG will be largest when the phase assumptions for the basis set lead to predictions of deviations from Wilson statistics for the unphased reflections, and in this context it is used as a powerful figure of merit.

However, rather than just choose those phase sets with high associated LLG, which is a somewhat subjective process, the Student t-test is used [6]. The LLGs are analysed for phase indications using the t-test. The simplest example involves the detection of the main effect associated with the sign of a single centric phase. The LLG average, μ^+, and its associated variance V^+ is computed for those sets in which the sign of this permuted phase under test is +. The calculation is then repeated for those sets in which the same sign is - to give the corresponding μ^-, and variance V^-. The t-statistic is then:

$$t = \frac{|\mu^+ - \mu^-|}{\sqrt{V^+ + V^-}}$$

The use of the t-test enables a sign choice to be derived with an associated significance level. This calculation is repeated for all the single phase indications, and is then extended to combinations of two and three phases. An extension to acentric phases is straightforward by employing two signs to define the phase quadrant *(i.e.* $\pm\pi/4$, $\pm3\pi/4$) both in permutation and in the subsequent analysis. In general, only relationships with associated significance levels <2% are used, but this is sometimes relaxed. This calculation is repeated for all the single phase indications, and is then extended to combinations of two and three phases. Each of the m phase relationships, i, so generated is given an associated weight w_i

$$w_i = \left(1 - \frac{I_1(s_i)}{I_0(s_i)}\right)$$

where I_1 and I_0 are the appropriate Bessel functions and s_i is the significance level of the *i-th* relationship from the t-test. This weighting function reflects the need for a scheme in which the absolute values of the significance levels are not given undue emphasis since they are themselves subject to errors arising from the nature of the likelihood function used and the lack of error estimates for the LLGs themselves. Each node n is now given a score, s_n

$$s_n = LLG_n \sum_{j=1}^{m} w_j$$

where the summation spans only those phase relationships where there is agreement between the basis set phases and the t-test derived phase relationships. The scores are sorted and only the top 8-16 nodes are kept; the rest are discarded. New reflections are then permuted and a corresponding set of ME solutions is generated. In this way we build a *phasing tree* in which each phase choice is represented as a *node*, and has a score, or figure of merit, based on its log likelihood gain (LLG). The root node of the tree is defined by the origin defining reflections. The first set of phase permutations defines the second level. Those which do not pass the analysis of likelihood are discarded, then further phase permutations are used to generate the third level, and this continues until a recognisable structure or structural fragment appears. Figure 3 shows a simple outline of a three-level phasing tree.

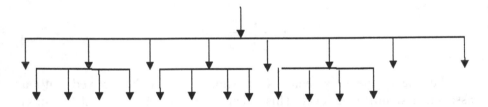

Figure 3. A Three level, 21 node phasing tree.

2.3   Centroid maps

Now $q^{ME}(\underline{x})$ is a probability distribution, and not a map in the traditional sense (although its peaks do correspond to atom positions), and thus it needs conversion to a more conventional one. The trial electron density maps generated in this approach are called *centroid maps*, and can be visualised as Sim filtered maximum entropy maps. For k acentric, the Fourier coefficients are:

$$\left|U_k\right|^{obs}\left[I_1(X_k)/I_0(X_k)\right]\exp\left(i\varphi_k^{ME}\right)$$

where:

$$X_k = (2N/\varepsilon_k)\left|U_k\right|^{obs}\left|U_k^{ME}\right|$$

For k centric, these coefficients become:

$$|U_\mathbf{k}|^{obs} \tanh(X_\mathbf{k}) \exp(i\varphi_\mathbf{k}^{ME})$$

with:

$$X_\mathbf{k} = (N/\varepsilon_\mathbf{k})|U_\mathbf{k}|^{obs} |U_\mathbf{k}^{ME}|$$

The practical differences between centroid and maximum entropy maps can be seen with reference to membrane data in [7].

2.4   Fitting

One final practical point needs to be made here concerning the tightness with which one fits the constraints in the entropy maximisation which can be very important in a phasing environment. If the fit between $|U_\mathbf{h}|^{obs}$ and $|U_\mathbf{h}^{ME}|$ is very slack the ME extrapolation is weaker than it can be, and the phasing power and the discriminating power of likelihood is correspondingly reduced. If however there is overfitting, spurious details (often looking like small stones) appear in $q^{ME}(x)$ which give false phase ndications. The latter situation can often be detected by the use of likelihood: if the LLG is monitored through the iterative cycles of entropy maximisation, a maximum is often reached and this can be used as a place to stop. Alternatively, the χ^2 statistic can be used with a default choice of unity as a place to stop.

$$\chi^2 = \frac{1}{n} \sum_{h \in H} \frac{1}{\sigma_h^2}(|U_\mathbf{h}|^{obs} - |U_\mathbf{h}^{ME}|)^2$$

where n is the total number of degrees of freedom (2 for each acentric reflection in the basis set, and one for each centric). In practice, for small basis sets either method works well, but for large basis sets maximum LLG is preferred.

The MICE computer program is a practical implementation of the ME formalism [8].

3.   AN EXAMPLE

The structure in question is the membrane protein Omp F porin [9] from a set of two-dimensional electron diffraction data at *ca.* 6Å resolution. It follows the method outlined by Gilmore, Nicholson & Dorset [10]

3.1   Data Normalisation

We start with the process of normalisation in which the raw intensities are converted to unitary structure factors via the equation:

$$\left(\left|U_{\mathbf{h}}\right|^{obs}\right)^2 = k\sigma_2\left(\left|F_{\mathbf{h}}\right|^{obs}\right)^2 \bigg/ \sigma_1 \exp\left(-2B\sin^2\theta/\lambda^2\right)\varepsilon_{\mathbf{h}}\sum_{j=1}^{N}f_j^2$$

where B is an overall, isotropic temperature factor, k a scale factor (both obtained by a Wilson plot), f_j is the electron scattering factor for atom j, the summation spans the N atoms in the unit cell, θ is the Bragg angle, ε is the staistical weight and:

$$\sigma_n = \sum_{j=1}^{N} z(eff)_j^n$$

where $z(eff)_j$ is the effective atomic number of atom j. Each U magnitude has an associated phase angle $\varphi_{\mathbf{h}}$ which is to be determined. Table 1 shows the unit cell and symmetry information needed for normalisation using MITHRIL [11], and Table 2 lists the top U magnitudes along with the experimentally observed phase angles. We are not going to use the latter directly, but they will act as a useful check on how well the *ab initio* phasing is progressing. Notice also that we employ the standard deviations of the structure factors in the ME calculations. These are problematic, but we have found that

$$\sigma\left(\left|F_{\mathbf{h}}\right|^{obs}\right) = 0.1\left|F_{\mathbf{h}}\right|^{obs}$$

works effectively.

```
CELL  72 72 5 90 90 120  !Cell parameters
LATT  A P          !Acentric, primitive cell
ELECTRON                 ! Electron data
SYMM -Y,X-Y,Z      !Symmetry operations
SYMM -X+Y,-X,Z
SYMM Y,X,Z
SYMM X-Y,-Y,Z
SYMM -X,-X+Y,Z
CONT  1500 C       ! 1500 C atoms in cell
DATA 1                   ! Data format
(3I3,F7.2,F5.2,I5)
```

Table 1. The Normalisation Data File for MITHRIL

The Wilson plot is shown in Figure 4. At first sight all seems well except that the overall temperature factor is $B=327\pm57\text{Å}^2$. This is clearly unlikely, but is a common feature when carrying out normalisation with such a sparse data set. To overcome the problem, a temperature factor of $B=0.0\text{Å}^2$ is imposed on the data.

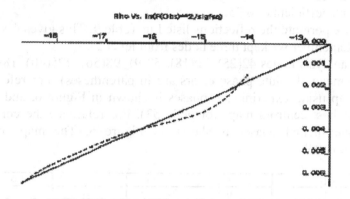

Figure 4. The Wilson plot for Omp F Porin

3.2   Phasing

Normally one would begin by defining the origin, but the plane group here is p31m and all the reflections are structure seminvariants so no such reflections are available. To overcome this the strongest 5 centric reflections: (1) 5 0 0 (4) 4 0 0 (8) 3 0 0 (10) 2 0 0 (14) 7 0 0 are given permuted phases. This gives 2^5 nodes on a rootless phasing tree. Each of these phase combinations is subjected to constrained entropy maximisation and a log-likelihood gain (LLG) computed for each calculated for each one.

The phasing tree is shown in Table 3. It can be seen that there is initially little discernible variation in LLG, but the phase analysis using the t-test correctly obtains the correct pattern of phases as in shown in tables 4 and 5. Analysis gives no significant indications for main effects, but, for the double and triple sign analysis of likelihood, there are significant and correct indications which can be used to derive single phase angles correctly.

Note that traditional direct methods are wrong 2 out of 3 times whereas likelihood is always correct. The indications for reflection (1) are very strong so we define the phase of (1) as 180^0. This becomes node 1 of the phasing tree and we now permute the reflections listed in Table 6. This gives $4^2\text{x}2=32$ nodes *i.e.* nodes 2-33. The resulting LLGs after entropy maximisation are given in Table 7.

Analysis of the LLGs gives 4 nodes which pass the t-test. They are listed in table 8. One has a negative LLG and so is discarded.

Nodes 29 (13), 31(49), and 32(27) are kept (the mean absolute phase errors are in parentheses).

Now let us examine the centroid map for node 29. The centroid map is shown in figure 5 for 4 unit cells. For reference the true map using image derived phases is shown in Figure 6. The agreement is excellent with a map correlation coefficient of 0.95.

Now we permute the reflections listed in Table 9. This gives 4^3x2x3=184 nodes. (Remember we kept three nodes from level 2).

After analysis nodes 42(25), 50(18), 53(9), 98(56), 125(40), 189(26) are kept (the mean absolute phase errors are in parentheses). For reference the correct map using experimental phases is shown in Figure 6, and Figure 7 shows the best centroid map (for node 53). For reference the correct map using experimental phases is shown in Figure 6. The map correlation coefficient is 0.94.

No	h	k	|U(obs)|	d	Constraint
1	5	0	0.10330	12.47	180/360
2	4	-2	0.06775	18.00	None
3	6	-3	0.05809	12.00	None
4	4	0	0.04797	15.59	180/360
5	5	-2	0.04424	14.30	None
6	8	-4	0.04341	9.00	None
7	5	-1	0.04154	13.61	None
8	3	0	0.04115	20.78	180/360
9	6	-2	0.03350	11.78	None
10	2	0	0.03275	31.18	180/360
11	7	-3	0.03104	10.25	None
12	7	-2	0.02619	9.93	None
13	3	-1	0.02552	23.57	None
14	7	0	0.02456	8.91	180/360
15	9	-3	0.02383	7.86	None
16	2	-1	0.02363	36.00	None
17	8	-3	0.02313	8.91	None
18	11	-4	0.02217	6.47	None
19	8	-2	0.02135	8.65	None
20	6	-1	0.02075	11.20	None
21	8	-1	0.01857	8.26	None
22	7	-1	0.01812	9.51	None
23	10	-5	0.01730	7.20	None
24	9	-4	0.01650	7.98	None
25	10	-2	0.01465	6.81	None
26	10	-3	0.01438	7.02	None
27	9	-2	0.01411	7.62	None
28	4	-1	0.01262	17.29	None
29	11	-5	0.01185	6.54	None

Table 2. The top 29 U-magnitudes, phase restrictions and resolution (Å) for Omp F porin.

Node	Entropy(x10^{-2})	LLG	NS + L	Error
1	-0.031	0.01	-0.4604	52
2	-0.023	0.01	-0.3433	126
3	-0.029	0.02	-0.4303	17
4	-0.023	0.00	-0.3468	92
5	-0.025	0.00	-0.3703	81
6	-0.023	-0.01	-0.3466	156
7	-0.028	0.00	-0.4212	47
8	-0.023	0.00	-0.3459	121
9	-0.028	0.00	-0.4231	75
10	-0.024	0.00	-0.3527	150
11	-0.024	0.02	-0.3768	41
12	-0.024	0.00	-0.3573	115
13	-0.030	0.01	-0.4437	105
14	-0.023	0.01	-0.3433	180
15	-0.029	0.01	-0.4410	70
16	-0.023	0.01	-0.3504	145
17	-0.030	0.02	-0.4481	34
18	-0.024	0.01	-0.3530	109
19	-0.029	0.02	-0.4311	0
20	-0.023	0.01	-0.3446	74
21	-0.025	0.00	-0.3698	64
22	-0.023	0.00	-0.3516	138
23	-0.025	0.00	-0.3678	29
24	-0.023	0.00	-0.3514	104
25	-0.023	0.00	-0.4344	58
26	-0.024	0.00	-0.3563	132
27	-0.028	-0.01	-0.4236	23
28	-0.028	0.00	-0.3546	98
29	-0.032	0.01	-0.4821	87
30	-0.024	0.01	-0.3493	162
31	-0.029	0.01	-0.4363	53
32	-0.023	0.00	-0.3437	127

Table 3. The phasing tree for permuting 4 centric reflections without an origin

No	No	No	PosAv	NegAv	Signif	Sign
1	10		0.006	0.002	0.153	+
8	10		0.010	-0.002	0.000	+
10	14		0.002	0.006	0.173	-
1	8	10	0.002	0.006	0.126	-

Table 4. The phase relationships from the t-test.

No	h	k	Deduced phase	Correct	Σ_1	Prob
1	5	0	180	Yes	0	0.776
4	4	0	0/180			
8	3	0	180	Yes	0	0.527
10	2	0	180	Yes	180	0.505
14	7	0	0/180			

Table 5. The derived phases and a comparison with the Σ_1 formula.

No	h	k	|U(obs)|
2	4	-2	0.068
3	6	-3	0.058
4	4	0	0.048

Table 6. Phases permuted for the second level of the phasing tree.

Node	Entropy	LLG	NS+L	Error
1	-0.025	0.00	-0.3690	0
2	-0.038	0.01	-0.5676	85
3	-0.038	-0.01	-0.5715	99
4	-0.038	0.01	-0.5654	77
5	-0.037	0.02	-0.5529	63
6	-0.038	-0.01	-0.5695	85
7	-0.038	0.01	-0.5721	99
8	-0.038	0.02	-0.5753	77
9	-0.038	0.01	-0.5635	63
10	-0.038	0.01	-0.5637	66
11	-0.038	0.02	-0.5753	80
12	-0.038	0.01	-0.5721	58
13	-0.038	-0.01	-0.5695	44
14	-0.037	0.02	-0.5527	66
15	-0.038	0.01	-0.5658	80
16	-0.038	-0.01	-0.5715	58
17	-0.038	0.01	-0.5676	44
18	-0.038	0.01	-0.5714	53
19	-0.038	0.03	-0.5707	68
20	-0.038	0.01	-0.5770	46
21	-0.038	-0.02	-0.5755	31
22	-0.038	0.03	-0.5666	53
23	-0.038	0.01	-0.5790	68
24	-0.039	-0.02	-0.5817	46
25	-0.038	0.01	-0.5716	31
26	-0.038	0.01	-0.5716	35
27	-0.039	-0.02	-0.5818	49
28	-0.038	0.01	-0.5791	27
29	-0.038	0.03	-0.5666	13
30	-0.038	-0.02	-0.5755	35
31	-0.038	0.01	-0.5770	49
32	-0.038	0.03	-0.5707	27
33	-0.038	0.01	-0.5714	13

Table 7. The second level phasing tree.

Node	LLG	Entropy	Score	No.of violations
29	0.032	-0.038	1.000	0
30	-0.017	-0.038	1.000	0
31	0.007	-0.038	1.000	0
32	0.033	-0.038	1.000	0

Table 8. Analysis of the second level nodes.

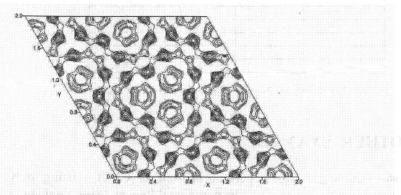

Figure 5. The centroid map for node 29.

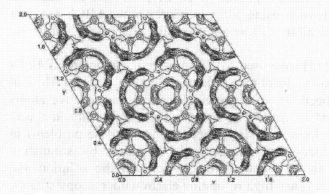

Figure 6. The true map for Omp F porin using image derived phases.

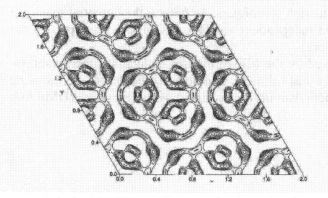

Figure 7. The centroid map for node 53 with a mean phase error of 90 and a map correlation coefficient of 0.94.

No	h	k	|U(obs)|
7	5	-1	0.042
8	3	0	0.041
9	6	-2	0.034
10	2	0	0.033

Table 9. The reflections permuted in level 3 of the phasing tree.

4.   OTHER EXAMPLES

The first example is that of perchlorocoronene, $C_{24}Cl_{12}$, using only projection data down the c axis [12]. Experimental images were obtained at 3.2Å together with a diffraction pattern extending to 1.0Å. The Fourier transform of the image yielded 4 phases which were used for the basis set {H}. Routine phase extension using ME procedures in the MICE program produced a map in which all the diffraction intensities are phased, and whose resolution exceeds 1Å.

Voigt-Martin *et al.* [13] have used MICE to solve the structure of 4-(4'-(N,N-dimethyl)aminobenzylidene)-pyrazolidine-3,5-dione at 1.4Å in projection using 42 reflections. The potential maps do not resolve atoms with these data and models have to be fitted to the map density in a way reminiscent of macromolecular crystallography. This can pose problems in structure validation which were overcome in this case by simulation calculations. There is an excellent agreement between the solution and independent model building and high resolution electron microscopy studies.

In a similar way, Voigt-Martin *et al.* [14] have solved the structure of [9,9'-bianthryl]-10-carbonitrile in three dimensions using 150 unique diffraction intensities, and independently verified the result with model building and image simulation techniques. As before, the potential maps are difficult to interpret, and independent validation is an important part of the structure solving procedure.

Other structures solved by the ME formalism include 4-dimethylamino-3-cyanobiphenyl [15] 2,6-bis(4-dimethylamino-benzylidene)-cyclohexanone [16] and 2,6-bis(4-dimethylamino-benzylidene)-cyclohexanone (DMBAC) [17].

References

1. Shannon, C.E. & Weaver, W. (1949). *The Mathematical Theory of Communication.* University of Illinois Press, Urbana, USA.
2. Jaynes, E.T. (1986). Where do we stand on maximum entropy? in J.H.Justice (ed.), *Maximum Entropy and Bayesian Methods in AppliedStatistics,* Cambridge University Press, Cambridge, pp 26-58.

3. Bricogne, G. (1984) Maximum entropy and the foundations of direct methods, *Acta Cryst.* A40, 410-445.

4. Bricogne, G. and Gilmore, C.J. (1990) A multisolution method of phase determination by combined maximisation of entropy and likelihood. I. Theory, algorithms and strategy, *Acta Cryst.* A46, 284-297.

5. Gilmore, C.J., Bricogne, G. and Bannister, C. (1990) A multisolution method of phase determination by combined maximisation of entropy and likelihood. II Application to small molecules, *Acta Cryst.* A46, 297-308.

6. Shankland, K., Gilmore, C.J., Bricogne G. and Hashizume, H. (1993) A multisolution method of phase determination by combined maximisation of entropy and likelihood. V Automatic Likelihood analysis *via* the student t-test, with an application to the powder structure of magnesium boron nitride, Mg_3BN_3. *Acta Crystallogr.* A49, 493-501.

7. Gilmore, C.J., Shankland, K. and Fryer, J.R. (1993), Phase extension in electron crystallography using the maximum entropy method and its application to two-dimensional purple membrane data from *Halobacterium halobium. Ultramicroscopy*, 49, 147-178.

8. Gilmore, C.J. & Bricogne, G. (1997) The MICE Computer Program in *Methods in Enzymology vol. 277* C.W.C. Carter Jnr.(ed.) Academic Press, New York 65-78.

9. Sass, H. J., Büldt, G., Beckmann, E., Zemlin, F., Van Heel, M., Zeitler, E., Rosenbusch, J. P., Dorset, D. L., and Massalski, A. (1989). Densely packed beta-structure at the protein-lipid interface of porin is revealed by high-resolution cryo-electron microscopy. *J. Mol. Biol.* 209, 171-1

10. Gilmore, C.J., Nicholson, W.V., and Dorset, D.L., (1996) Direct methods in protein electron crystallography: the ab initio structure determination of two membrane protein structures in projection using maximum entropy and likelihood, *Acta Cryst.* A52, 937-946.

11. Gilmore, C.J. (1984) MITHRIL - an integrated direct-methods computer program *J.Appl.Cryst.* 17, 42-46.

12. Dong, W., Baird, T., Fryer, J.R., Gilmore, C.J., MacNicol, D.D., Bricogne, G., Smith, D.J., O'Keefe, M.A. and Hovmöller S. (1992) Electron microscopy at 1Å resolution by entropy maximisation and likelihood ranking. *Nature*, 355, 605-609.

13. Voigt-Martin, I.G.,Han, D.H., Gilmore,C.J., Shankland, K. and Bricogne, G (1994) The Use of Maximum Entropy and Likelihood Ranking to Determine the Crystal Structure of 4-(4'-(N,N-dimethyl)aminobenzylidene)-pyrazolidine-3,5-dione at 1.4Å Resolution from Electron Diffraction and High Resolution Electron Microscopy Image Data, *Ultramicroscopy*, 56, 271-288.

14. I.G.Voigt Martin, D.H.Yan, C.J.Gilmore and G.Bricogne (1995) Structure Determination by Electron Crystallography Using both Maximum Entropy and Simulation Approaches, *Acta Cryst*, A51, 849-868.

15. Voigt-Martin, I.G., Zhang, Z.H., Kolb, U. and Gilmore, C.J. (1997) The use of maximum entropy statistics combined with simulation methods to determine the structure of 4-dimethylamino-3-cyanobiphenyl *Ultramicroscopy*, 68, 43-59.

16. Voigt-Martin, I.G., Gao Li, I.G., Kolb, U., Kothe, H., Yaminsky, A.V., Tenkovtsev, A.V. and Gilmore, C.J. (1999) Structure determination to calculate nonlinear optical coefficients in a class of organic molecule *Phys. Rev. (B)*, 59, 6722-6735.

17. Voigt-Martin, I.G., Kothe, H., Yaminsky, A.V., Tenkovtsev, A.V., Zandbergen, H., Jansen, J. and Gilmore, C.J. (2000) Comparison of electron diffraction data from non-linear optically active organic DMABC Crystals obtained at 100kV and 300kV *Ultramicroscopy*, 83, 33-59.

STRUCTURE REFINEMENT BY TAKING DYNAMICAL DIFFRACTION INTO ACCOUNT

Multi-Slice Least-Squares Refinement

J. Jansen

Nationaal Centrum voor HREM, Kavli Institute of Nanoscience, Technische Universiteit Delft, Rotterdamseweg 137, 2628 AL Delft, Nederland.

Key words: electron diffraction, dynamical diffraction, structure refinement

1. INTRODUCTION

Structure determination from X-ray and neutron diffraction data is a standard procedure. Starting with a rough model, the accurate structure is determined using a least-squares structure refinement, which is based on kinematic diffraction and in which the differences between calculated and experimental intensities are minimized. X-ray and neutron diffraction are not applicable to all crystals. To determine crystal structures of thin layers on a substrate or small precipitates in a matrix (see figure 1) only electron diffraction (ED) can lead you to the crystal structure.

For X-ray and neutron diffraction the resulting patterns can be approximated by the kinematic diffraction theory. In this case the intensities of the reflections increase linearly with thickness. Dynamic scattering, which occurs in electron diffraction (ED), will change the intensities of all reflections with respect to each other as a function of the specimen thickness. An example of this change for $Ce_5Cu_{19}P_{12}$, the example structure of this contribution, is shown in Figure 2. Therefore the kinematic refinement software can only be used for electrons in the regime where the dynamic scattering is small, which is in the case of $Ce_5Cu_{19}P_{12}$ for specimen thicknesses smaller than about 7 nm.

T.E. Weirich et al. (eds.), Electron Crystallography, 355–371.
© 2006 *Springer. Printed in the Netherlands.*

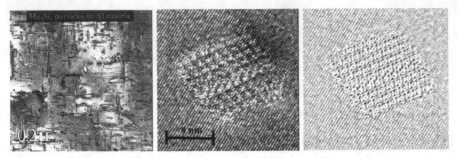

Figure 1. Mg$_5$Si$_6$ particles in am aluminum matrix. **a.** Overview. The needle-shape of the precipitates is obvious. **b.** Single HREM image. The viewing direction is along the needle-axis. **c.** HREM image enhanced by through focus exit-wave reconstruction.

We have developed a software package MSLS [1], in which multi-slice calculation software is combined with least squares refinement software used in X-ray crystallography. With multi-slice calculations which are standardly used for image calculations of HREM images, dynamic diffraction is taken into account explicitly.

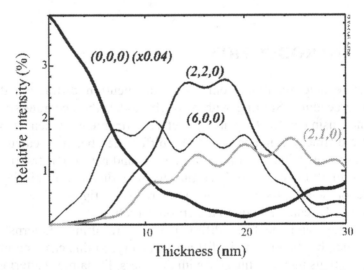

Figure 2. Intensities of several reflections including the central beam of Ce$_5$Cu$_{19}$P$_{12}$ as a function of thickness. Note that the intensities do not show a linear increase and that the central beam drops from 100% at zero thickness to about 5% at a thickness of 20nm.

Although the dynamic diffraction has the disadvantage that the intensities of the reflections depend on the thickness, which make the conventional kinematic diffraction software only valid for very thin crystals, is has on the other hand some major advantages, some of which will be discussed in the next chapters.

2.   EXPERIMENTAL

We performed the electron microscopy with a Philips CM30UT electron microscope with a field emission gun operated at 300 kV. The field emission gun has a major advantage that with very small spot sizes the convergence angle is still small, such that the illumination is similar to a plane wave, resulting in sharp diffraction spots. If due to the convergence angle the spots are disc-like, one can use the diffraction lens of the electron microscope for focusing the discs to sufficiently small dots, but this leads to distortions as will be discussed below. High resolution images and electron diffraction patterns were recorded with a 1024x1024 pixel Photometrix CCD camera having a dynamic range of 12 bits or with image plates to increase the dynamic range. Exposure times were 0.5 to 1.0 second per HREM image. Electron diffraction was performed with spot sizes between 6 and 15 nm. Exposure times for the electron diffraction ranged from 0.4 to 2 seconds. To reduce the electron microscope induced contamination and amorphisation, the specimens were cooled using liquid nitrogen.

A small spot size for electron diffraction is used for three reasons: i) to have a relatively small variation of thickness since most crystals are wedge shaped, ii) to reduce the amount of unwanted information like that of the matrix around a small precipitate and iii) to have a little variation in the crystal orientation. The latter reason is quite important which one can appreciate by moving the electron beam in nanodiffraction mode over the specimen: although the crystal is well aligned according to the selected area diffraction, fluctuations in orientation over 1 to 2° in all directions occur, even for areas which are very close to each other (10-50 nm). Such orientation variations should be considered as normal rather than an exception.

A typical diffraction pattern is shown in Figure 3. The misorientation of the crystal (0.45°) is evident from the asymmetry of the diffraction spots around the central spot. A misorientation has a strong advantage that it results an increase in the high order reflections on the other side, such that even reflections with d-values well below 0.05 nm can be significant, as can be seen in Figure 3. Note that HREM imaging normally gets you a point to point resolution of 0.14 to 0.19nm, which can be enhanced by image reconstruction methods like the focus variation methods[2] to the information limit which is in the order of 0.11nm.

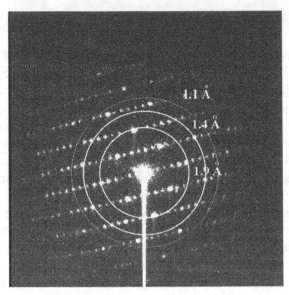

Figure 3. Typical diffraction pattern. The 0.14 and 0.19nm circles indicate the point to point resolution of a normal HREM image. The 0.11 circle is a guess of the information limit in an HREM image

3.   DATA PROCESSING

3.1   The indexing procedure

An essential part of the data reduction is the indexing. In this procedure all positions on the recorded image where intensity from reflected beams can be expected have to be is determined irrespectively whether the reflection is strong or weak. In the two-dimensional reciprocal space every reflection can be indicated by a vector *H* which has two integer elements, h and k, the indices. All position in reciprocal space can be described as:

$$P = O + A\,H \qquad\qquad (1)$$

where *O* is the origin and the matrix *A* consists of the two basis vectors of the reciprocal space. The matrix *A* describes the orientation of the crystal in respect of the recorded image. Since one uses lenses to image the diffraction pattern to the recording plane, image distortions can occur. Luckily these errors do not influence the relative intensity of the diffracted beams, but only their positions. Figure (2) shows an example where the effect is a little overdone. However in most diffraction patterns, even taken with the greatest care small deviations from straight lines for the rows of diffraction spots can be observed. Due to the non-linearities expression (1) is not valid. To overcome this problem one can introduce correction terms to the right hand side of formula (1):

$$P = \sum_{n=0}^{N} \sum_{m=0}^{N-m} a_{nm} \, h^n \, k^m \qquad (2)$$

Where a_{nm} contains the origin and and the reciprocal vectors:

$$a_{00} = O \qquad (3)$$

a_{01} and a_{10} contain the elements of the matrix A. The N in formula (2) indicates the degree of the indexing function. If N equals 1 the linear indexing function of formula (1) is obtained. Since formula (1) is normally a close approximation, one expects that the higher order terms in (2) will be small.

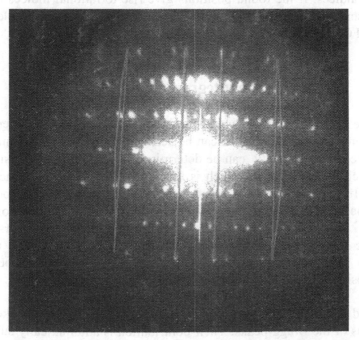

Figure 4. Example of the distortion of the diffraction pattern due to a focusing of the diffraction disks. The vertical lines and curves show that the same problems occurs here.

Table 1. Figure of merit for the fit between the calculated and experimental positions of the reflections using a polynomial of the N-th power

Degree (N)	FOM
1	0.084
2	0.072
3	0.035
4	0.033
5	0.032

To have some measure for the goodness of the resulting indexing a Figure of Merit (FOM) is defined:

$$FOM = \sum_{i=1}^{2} \sum_{H} (H_i - int(H_i))^2 \; / \; \#refl. \qquad (4)$$

This FOM is based on the fact that all reflections can be described by integer indices. If the found positions give rise to rational indices then the FOM starts to rise. A perfect indexing gives a FOM of zero. Table 1 gives typical FOM's for different degree's N.

3.2   Extracting the integrated intensities.

Once the indexing is known all positions in the diffraction pattern where reflections are to be expected can be scanned to obtain indexed intensities. The integrated intensity can be determined by including this position in a box or circle, the size of which is to be determined by the background level. If by enlarging the box or circle no extra intensity above the background is found the optimal size is reached. A larger box would contain the same net intensity with larger calculation errors. Since inelastic scattered electrons should not be included in the resulting intensity the best way is to measure the actual background of a peak close to its position. In practice we take the boundary of the box or circle.

This data reduction algorithm has a big advantage: Many inelastic scattered electrons contribute to the background which is not peaked at the locations of the Bragg-peaks. Their contribution is automatically subtracted. On the other hand form the plasmon-loss electrons a diffraction pattern that gives no significant different result in the refinement process that follows. A more detailed description can be found in literature [3]. An expensive microscope with energy filters is not needed for this analysis.

3.3 The Choice of Intensity calculation algorithm

If one has an atomic model one has to prove it is correct according to the observed data. In the field of crystallography normally a Least-squares program is used to perform this task and to improve the starting model. The choice of the algorithm which calculates the observed data from the atomic model is of utmost importance for the correctness of the method. The easiest way to calculate intensities is by the kinematic diffraction theory. For X-rays this works quite well. However since electrons interact strongly with the crystal, the electrons tend to be diffracted more than once. Therefore the kinematic theory is valid for very thin crystals only. In the example we will show that the kinematic refinements will lead to very poor results. A better way is to incorporate the dynamical scattering theory into the calculation of the reflection intensities. In literature several algorithms to perform this task can be found, which lead to the same reflection intensities. Amongst them are 1) Bloch-waves 2) Multi-slice calculations 3) Column approximation. The algorithm for our program was selected by two criteria : It should be not too much computer time consuming and it should be able to handle all types of crystals and orientations. The latter criterion rules out the column approach, which allows very fast computing times but fails if the atom columns are too close together. The choice between Bloch waves and Multi-slice was decided on computing time in favor of the latter. Therefore our Least-squares program is called MSLS (Multi-Slice Least-Squares).

To allow a comparison between dynamic and kinematic diffraction the MSLS program was also used to simulate a kinematic refinement. In order to approximate a kinematic refinement with the MSLS program, a very small thickness (i.e. 1 nm) or a very low occupancy should be taken. The latter approach was used because in this way the thickness and orientation dependence of the shape of the diffraction spots is properly taken into account. In the calculation of the kinematic R values a 0.1% occupancy of all atom sites and the thicknesses obtained for the dynamic refinement were taken.

3.4 The least-squares algorithm

The least-squares algorithm used is basically the linearized non-linear algorithm which is of common use in crystallographic structure refinement. The main part of the algorithm is a set of linear equation for parameter shifts s :

$$M s = v \qquad (5)$$

where the refinement matrix M is given by:

$$M_{ij} = \sum_m w_m \, \partial I_m / \partial p_j \, \partial I_m / \partial p_i \qquad (6)$$

and the vector by I_m

$$v_{ij} = \sum_m w_m \, (I_m - I_{obs,m}) \, \partial I_m / \partial p_i \qquad (7)$$

$I_{obs,m}$ are the observed reflection intensities and I are the calculated reflection intensities; p is the the parameter to be refined and w are the weights of the reflections. In theory these weights should be 1/ (I), in which (I) is the standard deviation of the intensity. The derivative of the intensity I to the parameters p raises some problems. The multi-slice algorithm is an iterative process for which no analytical function is available. So, the derivatives cannot be calculated analytically. In MSLS these derivatives are numerically calculated using the definition of the derivative:

$$I'(p) = \frac{I(p + \delta) - I(p)}{\delta} \qquad (8)$$

with δ tending to zero. Numerically one can only take a finite value for δ. However, δ cannot be too small, due to the accuracy of a computer. A too small will cause a random difference between $I(p+\delta)$ and $I(p)$ in the order of the numerical accuracy of the computer. In practice this difference will tend to zero and leads to a singular least-squares matrix M. The optimal values of were found by trial and error with simulated data. Just as in a traditional structure refinement program, the parameters in MSLS include atomic positions, Debye-Waller factors, scaling factors etc. As a result of using the multi-slice method, some more parameters may be involved: the crystal thickness, the crystal misalignment and absorption parameters. The misalignment of the crystal is expressed in terms of the center of the Laue circle in the pattern. A crystal tilt results in a corresponding shift of this center. Note that crystal tilt and beam tilt are equivalent for diffraction patterns. Therefore, we do not consider beam tilt in this paper.

The refinement of the thickness requires a special trick. The derivatives, I' (formula (8)), imply that the dependency of the intensity to the thickness is continuous. However, by dividing the crystal in as many slices as in the Multi-slice calculation, the thickness becomes a discrete parameter. If the δ in formula (8) is smaller than the slice size the resulting derivative will be zero because $I(p+\delta)$ and $I(p)$ are equal in many cases due to the calculation method. To overcome this problem, in MSLS a crystal is divided in a certain number of equal slices and a last slice having a thickness of a fraction f of the other slices. It is assumed that this slice contains the same scattering potential as that of the other slices but multiplied by f.

Diffraction patterns may always be multiplied by a constant factor without changing the physics of the system. In MSLS this is a scaling factor which is needed to scale the observed and calculated intensities to the same order of magnitude. This scaling factor is one of the refinable parameters. However, one has to take care with the definition of this factor. It should be defined in such a way that it is not dependent on the change of any other parameter in the refinement procedure, in particularly the crystal thickness and the absorption parameter. Therefore the actual scaling factor, C, was expressed as function of c, the parameter which was used in the refinement process:

$$ C = c \sum_{H \neq 0} I_H^{obs} \Big/ \sum_{H \neq 0} I_H^{calc} \qquad (9) $$

This results in the ideal case in a scaling factor C to be 1.0. In order to get a 3-dimensional crystal structure data from an electron diffraction pattern one zone is not enough in general, though for high symmetric space groups one special zone may be sufficient. MSLS allows for simultaneous refinement of several diffraction patterns, each with its own parameters for scaling, thickness and crystal alignment.

3.5   Correlation between crystal thickness and Debye-Waller factors

A small problem arises when the crystal thickness and temperature factors are refined simultaneously, because these parameters are highly correlated. Raising both the thickness and the temperature factors results in almost the same least-squares sum. This is not an artifact of the calculation method but lies in the behavior of nature. Increasing the Debye-Waller factor of an atom means a less peaked scattering potential, which in turn results in a less sharply peaked interaction with the incident electron wave. It can be shown that a thickness of 5 nm and B=2 Å^2 will give about the same results as a thickness of 10 nm and B=6 Å^2.

4.   EXAMPLES

4.1 $Ce_5Cu_{19}P_{12}$

$Ce_5Cu_{19}P_{12}$ is an intermetallic compounds of which we wanted to know the atomic structure. This compound was available as polycrystalline material, which allows the use of X-ray or neutron powder diffraction methods to determine the structure, provided one has a good starting model. We explored another route of structure determination: by electron diffraction. This has as major advantages that secondary phases are not at all a problem (one can simply select small crystals of the right phase), the crystals can be very small, it does not have the problem of overlapping reflections and the composition can be verified by EDX element analysis. In order to obtain a good starting model for the least squares refinement of the electron diffraction data a special form of high resolution electron microscopy was used: through focus holography.

The through focus holography uses the focus dependence of the image distortion by the electron microscope to reconstruct the exit wave. Using algorithms developed by Van Dyck and Coene [2], all useful information is extracted from a series of high resolution electron microscope images taken at different defocus values with known defocus increments. The result is the exit wave function which contains amplitude as well as phase information up to the information limit of the microscope. The exit wave function - or in short exit wave - is the electron wave at the exit plane of the specimen. Since the electron wave passing through the crystal is continuously changed, the exit wave function is still thickness dependent, but not dependent on the distortion introduced by the electron microscope. As such it provides a major advantage over conventional high resolution electron microscopy (HREM), yielding a more reliable starting model for the electron diffraction refinement.

Electron transparent areas were obtained by crushing. The unit cell of $Ce_5Cu_{19}P_{12}$ was determined by electron diffraction. This was done by scanning reciprocal space by tilting a crystal over about 90° and taking a series electron diffraction patterns for different tilts. The unit cell of $Ce_5Cu_{19}P_{12}$ was determined to be hexagonal with a=1.275nm and c=0.396nm. Because an analogous compound $La_5Cu_{19}P_{12}$ (with a=1.2773(1)nm and c=0.39876(3)nm) [4] adopts the space group P-62m we choose this space group for the refinement. However, the space group could have been determine easily by means of convergent beam electron diffraction.

Through focus series of HREM images were taken with crystals in the [001] orientation. A typical example of the resulting exit waves is shown in Figure 5. One can observe in this exit wave that the thickness of the specimen increases rather strongly over the field of view (about 11 nm). This was also observed in other exit waves and conventional HREM images. This shape of the particles is inherent to the method of sample preparation. Transparent areas with a much smaller thickness variation over for instance 20 nm could have been obtained by ion milling. This method was not chosen in our investigation because $Ce_5Cu_{19}P_{12}$ is rather sensitive to moisture, but with proper precautions ion milling is certainly feasible.

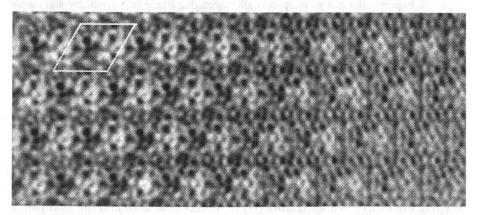

Figure 5. Reconstructed exit wave of $Ce_5Cu_{19}P_{12}$. Note the change in the image from left to right (thinner).

Figure 6 shows a part of the exit wave shown in Figure 5 but after applying three fold symmetry. The white dots in this image represent atom columns or groups of atom columns. One can observe several (almost) round white dots and several elongated ones. Since the c axis is short (0.396 nm) and given the resolution of the exit wave, the elongated ones are assumed to be due to two atoms columns. This gives a number of unique atom positions as indicated in the figure. These estimated atom positions are given in Table 2.

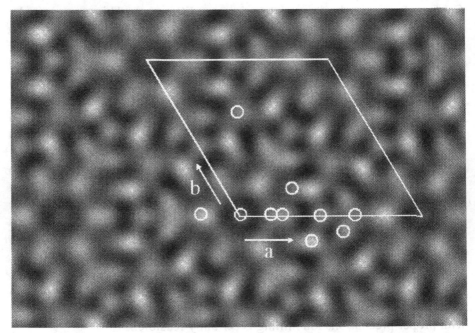

Figure 6. Experimental exit wave (phase) after averaging (three fold symmetry and mirror plane). The unit cell is indicated together with the initial guesses of the atomic positions in the asymmetric unit.

Given the thickness dependence of the contrast of the atom columns in the exit waves as shown in Figure 5, the contrast of the dots in the exit waves cannot be used for a assignment of Ce, Cu or P to the various dots in the exit wave. Therefore the MSLS refinement was started by putting Cu atoms at all sites. First the scale factor, crystal (mis)orientation and thickness for each data set were refined. Nine data sets were taken to have a range of thicknesses. All subsequent refinements were done with these nine data sets simultaneously. Next a refinement was done of the atom positions but keeping the Debye Waller factors (B) of all atoms at zero and the occupancy at 1. This resulted in an overall R value of 10.5 %.

The refined atom positions were used as model for a refinement in which the atom positions were fixed and the occupancies of all atom positions were refined as well as the orientation and thickness of each data set. The resulting overall R value of this refinement was 4.8 %. The resulting occupancies are given in Table 2. One can distinguish three distinct groups of occupancies: about 1.23 (1.20 and 1.26), about 1 (0.94, 0.95, 0.97 and

1.02) and about 0.68 (0.67, 0.67 and 0.69). These positions can be assigned to Ce, Cu and P respectively. One position (Cu4) has a rather deviating value of 0.74; from the composition it can be concluded that this should be Cu. In the following we will see that this position is indeed extraordinary.

The final refinement was done with Ce, Cu and P on the positions as discussed above. All atom positions and B's, including the crystal orientation and thickness of each data set were refined. The final atomic parameters are given in Table 3. The refined parameters of each diffraction set and the individual R values are given in Table 4.

Table 2. Atomic positional parameters estimated from the reconstructed exit wave, those obtained after the refinement with all Cu atoms and subsequently refined occupancies

atom	x from exit wave	y from exit wave	x after refining positions	y after refining positions	x after refining occupancies	y after refining occupancies	n after refining occupancies
Cu1	1/3	2/3	1/3	2/3	1/3	2/3	1.20(2)
Cu2	0.80	0	0.817(3)	0	0.817	0	1.26(1)
Cu3	0.23	0	0.289(4)	0	0.289	0	0.94(1)
Cu4	0.45	0	0.447(2)	0	0.447	0	0.74(1)
Cu5	0.18	0	0.173(3)	0	0.173	0	0.67(1)
Cu6	0.67	0	0.646(4)	0	0.646	0	0.67(1)
Cu7	0.37	0.16	0.377(2)	0.177(3)	0.377	0.177	0.95(1)
Cu8	0.32	0.84	0.317(3)	0.833(3)	0.317	0.833	0.69(1)
Cu9	0.51	0.86	0.517(4)	0.875(3)	0.517	0.875	0.97(1)
Cu10	0	0	0	0	0	0	1/02(2)

All atoms have normal B's except Cu(5), which is denoted as Cu4 in Table 3. The B of Cu(5) is about 7 times higher than that of the other Cu positions. This, however, is in very good agreement with the structure data reported for $La_5Cu_{19}P_{12}$, which structure was determined with X-ray single crystal diffraction. Also for $La_5Cu_{19}P_{12}$ the Cu(5) position has a much higher B (6.5) as can be seen from Table 3, in which both the atomic positions and the B's of $La_5Cu_{19}P_{12}$ are given. The larger B for Cu(5) obtained for $La_5Cu_{19}P_{12}$ can be attributed to the anisotropy of B, since the X-ray refinement shows the strongest movement along the c axis. Since the MSLS refinement of $Ce_5Cu_{19}P_{12}$ is only done on [001] diffraction patterns the z component of B is not influencing the diffraction data and thus not measurable. The atomic positions obtained for $Ce_5Cu_{19}P_{12}$ are very close to those of $La_5Cu_{19}P_{12}$. The differences in the atomic positions can be mainly attributed to the different size of Ce and La.

The z parameters are not determined in the MSLS refinement because only [001]diffraction patterns were used. With the knowledge that the atoms are located at z=0 or z=1/2 and taking the Cu(1) atom as origin, the z coordinates of all other atoms can be simply determined when one considers the plausible inter atomic distances. Although the structure of $Ce_5Cu_{19}P_{12}$ is rather complicated, the inter atomic distances are still within a narrow range, i.e. Ce-Cu 0.308-0.326 nm, Ce-P 0.295-0.305 nm, Cu-Cu 0.279-0.286 nm and Cu-P 0.227-0.249 nm. No short Ce-Ce or P-P distances occur.

Table 3. Final atomic position parameters for $Ce_5Cu_{19}P_{12}$ determined by MSLS and for $La_5Cu_{19}P_{12}$ determined X-ray single crystal diffraction data.

atoms		MSLS refinement Ce5Cu19P12				X-ray single crystal results for La5Cu19P12		
	All Cu	x	y	z	B	x	y	B
Ce(1)	Cu1	1/3	2/3	0	0.06(5)	1/3	2/3	0.65(3)
Ce(2)	Cu2	0.8141(9)	0	1/2	0.06(4)	0.8067(1)	0	0.62(4)
Cu(1)	Cu10	0	0	0	0.44(8)	0	0	0.8(1)
Cu(2)	Cu3	0.2905(9)	0	1/2	0.63(5)	0.2878(2)	0	1.0(1)
Cu(3)	Cu7	0.3812(6)	0.1784(7)	0	0.80(5)	0.3783(2)	0.1725(2)	1.2(1)
Cu(4)	Cu9	0.5203(7)	0.8812(6)	1/2	0.68(4)	0.5174(2)	0.8815(2)	1.2(1)
Cu(5)	Cu4	0.4536(6)	0	0	4.10(11)	0.4508(2)	0	6.5(3)
P(1)	Cu5	0.1782(8)	0	0	0.63(6)	0.1768(2)	0	0.6(1)
P(2)	Cu6	0.6415(5)	0	0	0.99(8)	0.6300(5)	0	0.9(2)
P(3)	Cu8	0.3168(4)	0.8302(4)	1/2	0.95(7)	0.3074(4)	0.8276(4)	0.9(2)

The R values assuming kinematic diffraction are also given in Table 4. For the calculation of these R values the MSLS program was used with occupancies of 0.1%; only the scale factors were refined. Evidently these R values are much higher then the ones taking into account the dynamic scattering.

Table 4. Data on the [001] electron diffraction sets used for the structure reported in table3. Three types of R-values are given: $R(I)= \Sigma(I_{obs}-I_{calc})^2/\Sigma I_{obs}^2$, $R(I)$ using a quasi kinematic calculation of I_{calc} and $R(I)=\Sigma(I_{obs}-I_{calc})^2/\Sigma I_{obs}$ The overall $R(I)$ in the final MSLS refinement is 2.7% for the significant reflections and 3.2% for all reflections.

Thickness (nm)	Number of observed reflections	h Laue circle	k Laue circle	R(I) (%)	R(I) (%)	R(I) kinematic (%)
9.6(2)	305	2.90(13)	-1.43(9)	1.69	2.08	26.9
11.8(2)	348	1.56(5)	-0.76(6)	2.09	2.66	63.3
11.8(2)	264	-0.23(5)	-1.71(7)	4.18	5.07	23.0
12.8(2)	238	-0.18(4)	-0.94(4)	1.09	1.78	33.1
13.0(2)	306	3.20(8)	0.61(6)	6.74	8.34	25.5
13.1(2)	154	1.64(6)	-1.01(5)	0.96	0.99	32.1
15.7(2)	156	0.07(3)	-0.44(3)	1.59	1.81	49.6
16.3(2)	137	-0.20(3)	-0.62(3)	3.31	3.34	40.6
17.4(2)	330	1.84(4)	-0.92(3)	4.04	3.58	63.5

A kinematic refinement starting with the atoms at the positions estimated from the exit wave or at the positions given in Table 2 does not lead to R values below 40% nor reliable atom positions. The reason for this is that we use a range of thicknesses in the refinement. Every element contributes in a different ratio to the diffraction pattern (or image) as can be seen in the simulated image of figure 7.

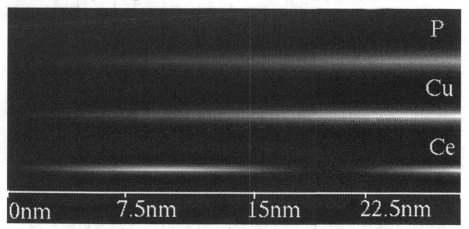

Figure 7. *Simulation of the contributions of Ce, Cu and P to the image as function of thickness.*

4.2 Thickness determination

One could be concerned about the accuracy of the resulting thickness. Is it the real parameter or is it just a parameter to make the R-values nice? The best example to test this are the precipitates shown in figure 1. From HREM images one can measure the needle radius to be around 4nm. MSLS refinements[5,6] yielded for the [001] direction (perpendicular to the needle axis) values ranging from 3.7(3) to 4.9(6) nm which is in the expected order of magnitude. Along the needle axis one measures length in a wide range (6.5-22.5nm), which is also to be expected.

4.3 Advantages of dynamic scattering

One of the advantages of dynamical scattering we have already seen. It is rather easy to determine the positions of Ce, Cu and P without prior knowledge on them. A similar example is a ZnCuAl-alloy of which to crystal structures were proposed, of which the occupancies of the positions were the only differences. MSLS was able to give an answer which of the two was the one in the sample[7].

Another advantage of dynamic scattering is that, for acentric zones, Friedel's law breaks. This allows for an easy way, much more reliable than for X-ray diffraction, to determine the absolute structure configuration[8].

$Ce_5Cu_{19}P_{12}$ is one of the structure for which this works. Of all investigated materials GaN in the [010] orientation shows the effect in the strongest way. A simulation of the correlation between the intensities of the structure and the inverse structure as function of thickness is shown in figure 8. For a well aligned crystal the correlation may drop to only 0.25 for a thickness of around 10nm. A correlation of 0.92 is already low enough to discriminate easily between the 2 configurations.

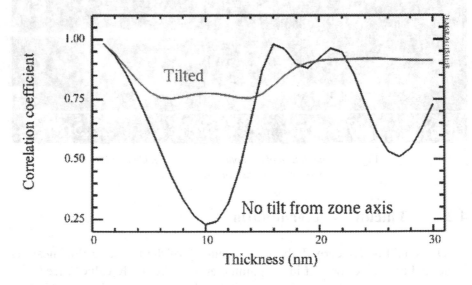

Figure 8. Correlation between corresponding intensities of the 2 possible absolute configurations of GaN in the [010] zone. It is obvious that in in case of perfect zone orientation the differences in intensities are more pronounced.

5.   CONCLUSION

It is possible to determine structures from dynamic electron diffraction data. In most cases low R-values are obtained, which are comparable to those of X-ray single crystal diffraction data. The presence of strong dynamic scattering makes the structure refinement more complex and certainly more time consuming. As such it is a disadvantage. However, it also has a large advantage, because the scattering of a column of atoms depends not only on the scattering potential of the atoms but also on the thickness, resulting in an oscillation in the scattering potential of the column with thickness. Thus for certain thicknesses a column containing only weakly scattering atoms show a higher scattering than a column containing strongly scattering atoms. Consequently by using various thicknesses the weakly scattering atoms can be determined with a higher accuracy. Furthermore it can be used to discriminate between atom types as has been done in the refinement of $Ce_5Cu_{19}P_{12}$, which was started with a model with only Cu atoms on all sites, where by refining the occupancies the locations of the Ce and P atoms were determined.

References

[1] J.Jansen, D.Tang, H.W.Zandbergen, H.Schenk,
    Acta Cryst. **A54**, 91 (1998).

[2] W.M.J.Coene, A.Thust, M.Op De Beeck, D.Van Dyck,
    Ultramicroscopy **64**, 109 (1996)

[3] J.Jansen, H.W.Zandbergen, M.Otten,
    Ultramicroscopy **98**, 165 (2004).

[4] R.J.Cava, T.Siegrist, S.A.Carter, J.J.Krajewski, W.F.Peck Jr.,
    H.W.Zandbergen, *J. Solid State Chem.* **121**, 51 (1996)

[5] H.W.Zandbergen, S.J.Andersen, J.Jansen, *Science* **277**, 1221 (1997).

[6] S.J.Andersen, H.W.Zandbergen, J.Jansen, C.Træholt, U.Tundal,
    O.Reiso, *Acta Materialia* **46**, 3283 (1998).

[7] C.Satto, J.Jansen, C.Lexcellent, D.Schryvers,
    Solid State Comm. **116**, 273 (2000).

[8] J.Jansen, H.W.Zandbergen, *Ultramicrocopy*, **90**, 291 (2002).

TrueImage
A Software Package for Focal-Series Reconstruction in HRTEM

C. Kübel[1] and A. Thust[2]
1) FEI Company, Achtseweg Noord 5, 5651 GG Eindhoven, The Netherlands.
2) Forschungszentrum Jülich GmbH, 52425 Jülich, Germany.

Abstract: The basic principles of focal-series reconstruction in high-resolution TEM are introduced. The paraboloid (PAM) and the maximum-likelihood (MAL) algorithms, which are implemented in TrueImage, are explained. Two application examples are shown to illustrate the benefits of focal-series reconstruction for atomic resolution imaging. Furthermore, a short introduction into linear imaging theory is given as background information.

Key words: Focal-Series Reconstruction, High-Resolution Transmission Electron Microscopy (HRTEM), TrueImage

1.   INTRODUCTION

High-resolution transmission electron microscopy (HRTEM) allows the observation of structural details in a material on an atomic scale. In contrast to "bulk" diffraction techniques like X-ray or neutron diffraction, which result in a diffraction pattern averaged from substantial specimen volumes, HRTEM gives direct and local insight into the structure in the form of a real-space image. With modern high-resolution microscopes, it is possible to resolve distances smaller than one Ångstrom (0.1 nm). A similar resolution can also be obtained by Scanning Tunneling Microscopy (STM). However, whereas STM is mainly sensitive to the surface structure, HRTEM is a projection method, meaning that the HRTEM image reflects the complete specimen volume traversed by the electron beam. HRTEM imaging can provide insight into the atomic structure of materials that cannot be obtained in any other way and has become a very powerful tool for both materials research and in industrial applications.

T.E. Weirich et al. (eds.), Electron Crystallography, 373–392.

Unfortunately, the interpretation of HRTEM images is not as straightforward as of light-optical images. In general, a HRTEM image is not a simple projection of the imaged object, but a complex interference pattern formed by the interaction of the electron beam with the specimen, which is further 'scrambled' by the electron optical system of the microscope. Inevitable lens aberrations are the reason for the later 'scrambling' of the information. The spherical aberration is limiting what came to be called the TEM point resolution, while the chromatic aberration and the coherence of the electron beam define the information-limit of the microscope.

In light optics, the shape of a lens can be controlled and it is possible to correct for spherical aberration by giving the lens an appropriate shape. However, for the electromagnetic lenses in a TEM, the spherical aberration cannot be easily compensated. In electro-magnetic lenses, the spherical aberration is due to the magnetic fields decreasing with increasing distance away from the pole pieces (towards the center of the lens). The only direct solution to decrease the spherical aberration, which is typically in the millimeter range, either requires lenses with enormously strong magnetic fields or a strong reduction of the gap between the pole pieces, which is practically not feasible as the specimen is located there. Besides the development of a spherical aberration corrector (Rose 1990; Haider et al. 1995), the only recourse in HRTEM is to realize that one is limited by spherical aberration and to see how the effects can be effectively countered.

One such approach to counter the effects of spherical aberration is to record a through-focus series, which allows reconstruction of the electron exit-wave function and thereby removal of the spherical aberration. The basic idea of this reconstruction approach, as implemented in the TrueImage software package, will be described after a simplified introduction to HRTEM imaging theory (for more details see Williams and Carter 1996, Reimer 1984, Spence 1988). Furthermore, two application examples are shown to discuss the benefits of focal-series reconstruction and illustrate the information that can be obtained.

2.   HRTEM IMAGING THEORY FOR THIN OBJECTS

Image formation in a transmission electron microscope can be considered as a two-step process. In the first step, the electron beam is interacting with the specimen. This interaction is very strong compared to X-ray or neutron scattering and causes multiple scattering events. In order to understand this process, the classical particle description of the electron is not adequate, and the quantum mechanical wave formalism has to be used. Thus, assuming the specimen is hit by a plane wave of electrons, the electron wave interacts with the specimen leading to a modulation of the amplitude and phase depending on the local optical density of the specimen. This process is governed by the relativistic Schrödinger equation describing the motion of the electron wave, $\Psi(r)$, in the potential, $V(r)$, of the specimen. For periodic crystals with small unit cells, it is possible to obtain an analytical solution to describe the electron wave at the exit plane of the specimen using the Bethe-Bloch formalism (Bethe 1928, Humphreys 1979). For larger, non-periodic objects, this approach is not suitable, but a numerical approximation called the "Multislice Algorithm" (Cowley and Moodie 1957, Goodmann and Moodie 1974) can be used. This approximation converges arbitrarily close to the analytical solution.

For very thin specimens (a few nanometers thick), one can assume that the amplitude of the electron wave is constant and only the phase will be modulated by the potential of the specimen. In this phase object approximation (POA), the electron wave function at the exit plane of the specimen Ψ_e can be described by

(1) $\Psi_e(\vec{r}) \approx \exp[i\sigma V_p(\vec{r})t]$ POA

where V_p is the projected crystal potential, σ the interaction constant and t the traversed thickness of the sample.

For a very thin object that is acting as a weak scatterer, the phase modulation in the exit plane is small, i.e. $\sigma V_p t \ll 1$. In this weak-phase object approximation (WPOA), equation 1 can be further simplified to

(2) $\Psi_e(\vec{r}) \approx 1 + i\sigma V_p(\vec{r})t$ WPOA

In this approximation, the exit wave function can be directly interpreted in terms of the projected potential of the specimen and the imaginary part exhibits sharp maxima at the position of the atomic columns. Unfortunately, most practical specimens do not satisfy the POA or WPOA and, thus, the electron wave function is more difficult to understand. Nevertheless, we are going to work with the WPOA in this paper to simplify the explanation of the imaging process.

In the second step of the image formation process, the exit wave propagates from the bottom of the specimen through the lens and towards the image plane, where the Intensity, I, of the electron wave, Ψ_i, is recorded in the HRTEM image. In this magnification process, the exit wave is affected by aberrations, χ, of the objective lens (spherical aberration, defocus, coma, and astigmatism), which are described by the transfer function, T, of the microscope. The image plane wave function, Ψ_i, can be obtained by convolving the exit wave function, Ψ_e, with the transfer function, T.

(3) $\Psi_i(\vec{r}) = \Psi_e(\vec{r}) \otimes T(\vec{r})$

(4) $I(\vec{r}) = |\Psi_i(\vec{r})|^2 = \Psi_i(\vec{r})\Psi_i^*(\vec{r})$

(5) $I(\vec{r}) \approx 1 + 2\text{Re}\{i\sigma t V_p(\vec{r}) \otimes T(\vec{r})\} + |i\sigma t V_p(\vec{r}) \otimes T(\vec{r})|^2$ WPOA

In linear imaging theory, it is assumed that the object is a weak scatterer, so that the quadratic term in equation 5 is small and can be neglected

(6) $I_L(\vec{r}) \approx 1 + 2\text{Re}\{i\sigma t V_p(\vec{r}) \otimes T(\vec{r})\}$

Since V_p is a real quantity and thus $i\sigma t V_p$ is imaginary, no contrast would be visible without the transfer function of the microscope (Equation 6). The lens aberrations result in a transfer of some imaginary part information at the exit plane into the real part at the image plane, which can be imaged by HRTEM.

To discuss the influence of aberrations in the HRTEM image formation process in more detail, it is convenient to work in Fourier space, where the real-space quantities $I(r)$, $\Psi(r)$, $V_p(r)$, and $T(r)$ are related to their counterparts $I(g)$, $\Psi(g)$, $V_p(g)$ and $T(g)$ by a Fourier Transformation. Distances, d, in direct space correspond to spatial frequencies, g, in Fourier space. With this approach, the electron wave can be expressed as

(7) $\Psi_e(\vec{g}) \approx \delta(\vec{g}) + i\sigma V_p(\vec{g})t$ WPOA

(8) $\Psi_i(\vec{g}) = \Psi_e(\vec{g})T(\vec{g})$

(9) $T(\vec{g}) = \exp[-i\chi(\vec{g})]$

In the WPOA and only considering the linear contributions in coherent imaging, the observed intensity, I_L, can be simplified by combining equations 4 and 7-9 to

(10) $I_L(\vec{g} \neq 0) \approx 2\sigma V_p(\vec{g})t \sin[\chi(\vec{g})]$

Equation 10 can be interpreted as the aberrations of the objective lens multiplying the intensities of the diffracted beams by a phase factor $\sin[\chi(g)]$, which depends on the spatial frequency. Thus, in the WPOA, the observed image is proportional to the projected potential, but is modulated by the phase factor. Without the phase shift, χ, due to the lens aberrations, a weak phase object would not be visible in HRTEM (this is analogous to the interpretation of equation 6).

2.1 Spherical aberration and point resolution

The phase shift induced by the aberration function, χ, can be understood geometrically in terms of the path length difference between a diffracted beam in an ideal lens and in a lens affected by aberrations. This path length difference is a function of the diffraction angle $\theta \approx \lambda g$, which is the reason why it is more convenient to describe the imaging process in Fourier space.

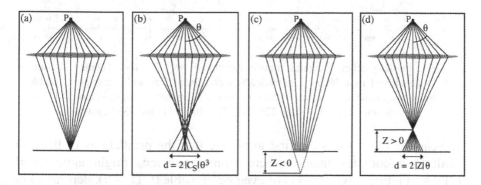

Figure 1. Ray optical description of lens aberrations. (a) Perfect lens imaging a point P in the object plane onto a sharp point in the image plane. (b) Lens affected by spherical aberration. (c) Lens underfocused by a distance Z. (d) Lens overfocused by a distance Z. Spherical aberration and defocusing cause a blurring of the point P in the image plane.

Only considering aberrations due to spherical aberration, C_s, and defocus, z, it can be shown that the aberration function, χ, can be expressed as

$$(11) \quad \chi(g) = 2\pi \left[\frac{C_s \lambda^3 g^4}{4} + \frac{z\lambda g^2}{2} \right]$$

where the vector notation of g has been abandoned (only the absolute value of g is relevant for rotationally symmetric lenses in the framework of linear imaging theory).

Within the weak-phase object approximation, the effect of the aberrations is most conveniently described by the Contrast Transfer Function (CTF), which gives the phase factor as a function of spatial frequency (diffraction angle).

$$(12) \quad \text{CTF}(g) = \sin[\chi(g)] \qquad \qquad \text{Coherent CTF}$$

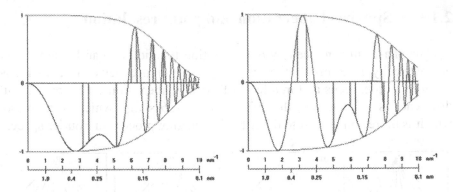

Figure 2. Partially coherent contrast transfer function exemplified for a Tecnai F30 with U-TWIN objective lens at 300kV at extended Scherzer focus (left) and $2*z_s$ (right). The blue vertical lines indicate the effect of the CTF on the diffracted beams of silicon in [110] orientation (the 111, 200, 220, 311, 222, 400, 331, and 333 beams).

Scherzer (1949) was the first to realize that the oscillations of the CTF result in a complex image, where atomic positions might appear dark (CTF = −1), bright (CTF = +1) or even be invisible (CTF = 0), depending on the interatomic distances and the selected defocus. However, there is a defocus setting where the contrast remains constant (or at least of the same sign) over a wide range of spatial frequencies. When the image acquisition is limited to only these spatial frequencies (by an objective aperture), the HRTEM image of a weak-phase object is intuitively interpretable. This so-called Scherzer focus, z_s, is given by

$$(13) \quad z_s = -c\sqrt{C_s\lambda} \qquad \text{with c a constant between 1.0 and 1.2}$$

The first zero transition of the CTF at this defocus is called the point resolution, d_p, of the TEM and can be shown to be

$$(14) \quad d_p = \frac{1}{\sqrt{2c}} C_s^{1/4} \lambda^{3/4}$$

The point resolution of a TEM, which only depends on the spherical aberration, C_s, and the electron wavelength, λ, (which is determined by the accelerating voltage) sets the limit for a straightforward interpretation of a HRTEM image of a thin object. However, this is different from the information limit, which defines the highest frequencies that can be transferred in a microscope.

2.2   Limited coherence and information limit

The imaging theory discussed so far would allow for an infinite resolution since the coherent CTF transfers information up to arbitrarily high spatial frequencies. However, in practice, the information that can be obtained from a TEM is limited due to the partial coherence of the electron source and electronic instabilities. There are two main contributions to the limited coherence. One, the temporal coherence, is due to the effect of the chromatic aberration of the objective lens coupled with the energy spread of the beam and the stability of the objective lens. The other, the spatial coherence, is due to the fact that there is an angular spread of the electron beam resulting in the specimen being illuminated under a semi-convergence angle, α.

The effect this has on a HRTEM image can be understood as follows. When the electron beam is not perfectly monochromatic, the chromatic aberration, C_c, of the objective lens causes the electrons of different wavelength (energy) to be focused at different levels. This results in a focus spread, δz, and the effect can be understood by integrating the CTF for all focus values represented by the focus spread. For low spatial frequencies, this will not result in noticeable differences, but, for high frequencies, where the CTF varies rapidly with defocus, the variations will start to cancel each other out, thereby limiting the information that is transferred in the image.

$$(15) \quad E_c(g) = \exp\left[-\frac{1}{4}\pi^2 \lambda^2 \delta z_{1/e}^{\;2} g^4 \right]$$

The effect of spatial coherence can be understood based on simple geometric considerations. The illumination of an object under different angles results in a distortion and shift of the object in the image plane that depends on the spherical aberration and the defocus. As there is a range of incident angles between $\pm\alpha$, this will result in an image averaging over the (slightly) distorted and shifted object, thereby 'smearing' out the smallest features.

$$(16) \quad E_s(g,z) = \exp\left[-\left(\frac{\pi\alpha}{\lambda}\right)^2 \left(C_s \lambda^3 g^3 + z\lambda g \right)^2 \right]$$

In linear imaging, these two effects can be mathematically described by damping functions, E, applied to the CTF (for details see William and Carter 1996, Spence 1988). Their combined effects are shown in Figure 2 as the envelope function to the CTF. The partial temporal coherence places a limit on the information that can be transferred in a microscope, a value called the information limit. Traditionally, the information limit is defined as the spatial frequency, where the contrast (the temporal envelope function) falls off to $1/e^2$ (~13.5%). Nevertheless, the information obtained in a particular single image may be further limited by the partial spatial coherence, depending on the selected defocus.

For a microscope equipped with a LaB_6 gun, the information limit goes not much beyond the point resolution, but, for a field emission gun (FEG) instrument, the information limit is far below the point resolution. For example, the point resolution of a Tecnai F30 S-TWIN is 2.0 Å, but an information limit of 1.1 Å is attainable.

2.3   Residual lens aberrations

So far, only defocus and spherical aberration have been considered as aberrations affecting the image contrast. Both depend only on the magnitude of the spatial frequency $|g|$, but not on the diffraction direction, thus resulting in rotationally symmetric phase shifts (Equation 11). However, the objective lens may exhibit further aberrations resulting in additional phase shifts, which are not necessarily rotational symmetric. The most important of these additional aberrations are astigmatism and coma.

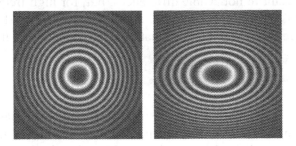

Figure 3. Two-dimensioal CTFs (at relatively high defocus): The left CTF only exhibits the effects of spherical aberration and defocus, whereas the right CTF also exhibits 2-fold astigmatism.

Coma is due to the electron beam being tilted away from the optical axis of the objective lens, the coma-free axis. The coma results in lattice fringes related to $+g$ and $-g$ being shifted by the objective lens. Lattice fringes belonging to different beam pairs, $+g$ and $-g$, are shifted differently resulting in an asymmetry of the HRTEM image.

Two-fold astigmatism means that the phase shifts result in different defoci along perpendicular directions, leading to a two-dimensional CTF as indicated in Figure 3b. Similarly, higher order astigmatism produces directionally dependent contrast transfer along more than two axes. As the phase shift induced by n-fold astigmatism depends on g^n, higher order astigmatism becomes increasingly important with increasing resolution; currently, mainly 2-fold and 3-fold astigmatism are relevant. The effect of 3-fold astigmatism on a HRTEM image is illustrated in Figure 4 for 'diamond-dumbbells' (O'Keefe, 2001).

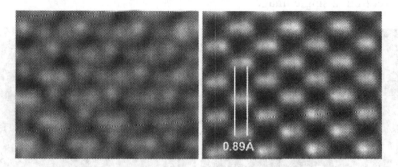

Figure 4. Influence of three-fold astigmatism, A_3, on HRTEM images of diamond in [110] orientation (left: $A_3 = 2250$ nm; right: $A_3 < 50$ nm). When the 3-fold astigmatism is corrected, the HRTEM image shows the 'dumbbell-structure' of diamond, whereas strong 3-fold astigmatism results in an image that cannot be directly interpreted in terms of the atomic structure (O'Keefe 2001).

The above mentioned aberrations can be measured (Zemlin et al. 1977) and, in principle, corrected for in a TEM. However, in practice, small residual aberrations are left due to the limited accuracy of the measurement and the stability of the microscope. Nonetheless, these residual aberrations have to be taken into account for a quantitative image analysis.

2.4 Contrast delocalization

Another effect of the spherical aberration of the objective lens is the lateral displacement of spatial frequencies in the image. This displacement increases strongly with spatial frequency. The effect is rarely a problem on LaB_6 instruments due to their limited coherence. However, on FEG instruments, with their high coherence resulting in a strong contribution of the high spatial frequencies, this effect is quite pronounced. Depending on the spatial frequencies transferred and the defocus, the delocalization can be up to several nanometers. The delocalization, R, is defined as the maximum gradient of the aberration function, χ, occurring between 0 and g_{max}, where g_{max} is the highest transferred frequency.

$$(17) \quad R(z,C_s) = \left| C_s \lambda^3 g^3 + z \lambda g \right|_{max}$$

One typical delocalization effect is that lattice fringes appear to continue well beyond the edge of a specimen into the vacuum or support film. In general, the delocalization becomes apparent only for non-periodic object features, e.g. interfaces or defects, which are broadened up to several nanometers (Figure 5). This means that a defect or interface, often the most interesting part of the image, appears blurred and is almost impossible to interpret from a single image.

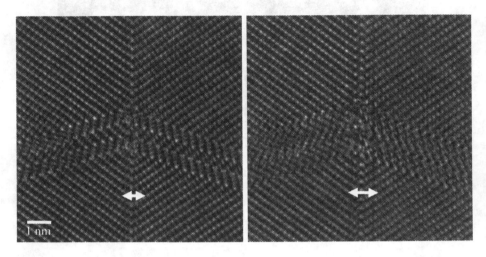

Figure 5. HRTEM images (left: $z = -123$ nm, right: $z = -195$ nm) of $\Sigma3$ twin boundaries in $BaTiO_3$. The contrast delocalization at the interface is indicated by arrows (Jia et al. 1999).

2.5   Non-linear image contributions

The above discussion of the image formation process was based on the weak phase object approximation in combination with linear imaging theory. In deriving equation 6, it was assumed that the object is only weakly scattering, so that the observed image intensities are dominated by the linear term in equation 5, whereas the quadratic term has been neglected. However, in practice, non-linear image contributions are always present and the effect they have on the HRTEM image of a real specimen is best discussed in Fourier space.

In Fourier space, the linear imaging process can be understood as the interference between the zero beam and a diffracted beam, whereas non-linear imaging can be expressed as interference between any two diffracted beams. As for the linear terms, the non-linear terms result in lattice fringes with a distance corresponding to their difference in scattering angle. For example, the linear interference of the zero beam and a $+g$ beam results in a certain spatial frequency, g, and, accordingly, the non-linear interference of the corresponding $+g$ and $-g$ beams results in twice that frequency 2g. This means that a HRTEM image can exhibit lattice fringes with half the distance of the aperture used for the image acquisition, which is an artifact with respect to the object structure.

Figure 6 illustrates the effect of linear and non-linear image contributions. The interference of the zero beam with n diffracted beams of the object results in 2n linear contributions to the image spectrum (indicated blue in Figure 6). For the non-linear image contributions, the interference between any two diffracted beams results in n^2 non-linear contributions to the image spectrum (some of them are indicated red in Figure 6). The relative importance of the linear vs. non-linear image contributions depends on the relative strength of the zero and the diffracted beams, and, in general, the non-linear terms become increasingly important with increasing sample thickness, t, (Equation 5).

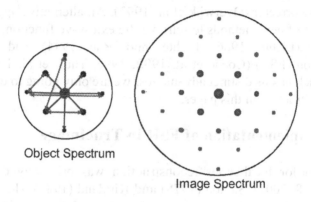

Object Spectrum

Image Spectrum

Figure 6. Schematic illustrating the effect of linear (blue) and selected non-linear (red) contributions to the image spectrum.

3. FOCAL-SERIES RECONSTRUCTION

3.1 Motivation

With the improved resolution in electron microscopy, especially with the introduction of FEG intruments, the complexity of HRTEM images became apparent as discussed in the previous chapter. It is no longer possible to directly interpret these high-resolution images, but a careful comparison with simulated images (EMS: Stadelman 1987, MacTempas: O'Keefe and Kilaas 1988) is necessary. The microscopist would 'guess' a model structure and simulate the HRTEM image while taking into account the microscope and specimen parameters. If the specimen structure, thickness, orientation and all microscope parameters including residual aberrations such as coma (Smith et al. 1983) and 3-fold astigmatism (Ishizuka 1994, Krivanek 1994, Krivanek and Stadelmann 1995) are considered, this results in a large number of variables required for image simulation. The main limitation of this approach lies in developing a good model for complex structures consisting of a large number of atoms in the aperiodic unit. Especially for defects and interface structures, a quantitative analysis becomes almost impossible, as contrast delocalization at the core area of interest limits the directly available information to develop a model.

Several other approaches have been followed towards quantitative HRTEM imaging. One approach is the development of new hardware to correct for or alleviate some of the aberrations in the image, e.g. spherical aberration corrector (Rose 1990, Haider et al. 1995) and three-fold astigmatism corrector (Overwijk et al. 1997). An alternative approach is the development of new methods to retrieve the exit wave function, e.g. off-axis holography (Lichte 1986, Lichte and Rau 1994) and focal-series reconstruction (FSR) (Coene et al. 1992, 1996, Thust et al. 1996a). While each approach has its distinct advantages, we are only going to discuss focal-series reconstruction in this paper.

3.2   Implementation of FSR in TrueImage

The idea for focal-series reconstruction was originally developed by Schiske (1968, 2002), Saxton (1978) and Kirkland (1984) who realized that although amplitude and phase cannot be retrieved from a single HRTEM image, they can be obtained by combining several HRTEM images recorded at different defocus settings. This idea has become the basis for various focal-series reconstruction schemes, one of which is the PAM-MAL method developed by Coene, Thust and coworkers (Coene et al. 1992, 1996, Thust et al. 1996a), which is implemented in TrueImage software package. The method has been further developed by Thust to allow, amongst other improvements, the determination and correction of residual aberrations. These improvements are also implemented in the TrueImage software package.

The reconstruction of the exit wave occurs in four steps. It starts with an alignment step where the images of the focal-series are aligned with respect to each other. In the second step, an analytical inversion of the linear imaging problem is achieved by using the paraboloid method (PAM) to generate a first approximation to the exit wave function. This approximated exit wave function is then refined in the third step by a maximum likelihood (MAL) approach that accounts for the non-linear image contributions. Finally, the exit wave is corrected for residual aberrations of the microscope.

With the microscope aberrations eliminated, the exit wave function can be directly understood only in terms of the electron beam – specimen interaction.

3.2.1   Image alignment

The image alignment is necessary prior to reconstruction because no microscope can record a series of 10 to 20 HRTEM images without mechanical drift between images. The initial alignment of the images is performed by cross-correlation between successive images, which is only a rough alignment as the contrast of the images in the focal-series is changing due to the defocus change. Therefore, the alignment is refined in each iteration step of the PAM and MAL reconstruction by cross-correlation between the experimental images and images simulated on the basis of the already available wave function.

3.2.2   Paraboloid reconstruction method (PAM)

For an intuitive explanation of the idea behind focal-series recon-struction, consider each point of the exit wave function as a vector with a real and an imaginary part. Only considering linear image contributions, we can describe the HRTEM image formation as a projection of this complex quantity. In the projection process, the convolution with the transfer function, T, of the microscope results in a transfer of real/imaginary part information of the exit-wave function into the imaginary/real part of the wave function in the image plane. The real part of the wave function in the image plane is recorded as the HRTEM image (Equation 6) and, as the transfer function changes (with defocus), different projections are observed. In principle, two projections are sufficient to invert the linear imaging process and solve for the real/imaginary part of the wave function. In practice, 10 to 20 images are used in order to overcome transfer gaps (zero-transitions in the transfer function).

However, even in the WPOA, the image formation is not a purely linear process as the diffracted beams will also interfere resulting in a non-linear contribution to the image, which cannot be directly reconstructed. It is not possible to distinguish the linear and non-linear contributions from a single image, but the defocus behavior of the non-linear terms is, in general, different from the linear terms. The defocus-induced phase shift of the linear term is proportional to g^2 (Equation 11), which means that all linear terms lie on two paraboloids in 3D Fourier space (Figure 7). Thus, by only using the components on the surface of the paraboloids $\xi = \pm 1/2\, \lambda g^2$, it is possible to select the linear image contributions and suppress most of the non-linear contributions, which are more evenly distributed in 3D Fourier space. Furthermore, the two paraboloids in Figure 7 represent the wave function and its complex conjugate (Op de Beeck et al. 1996), so by using only one of the paraboloids, it is possible to overcome the Friedel symmetry of the original HRTEM images and resolve the intensity for each +g and −g beam separately in the exit wave function. This approach is implemented in PAM as back-propagation of the images onto the paraboloid surfaces by means of a filter function.

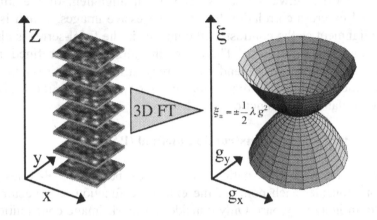

Figure 7. The linear image contributions of a focal-series are located on the surface of two paraboloids obtained by 3D Fourier transformation of the focal series. The two paraboloids correspond to the electron wave function and its complex conjugate.

3.2.3   Maximum likelihood method (MAL)

The result of the PAM reconstruction is, in general, only an approximation of the exit wave function. Some non-linear terms may be present exactly on the paraboloid surfaces, and, thus result in artifacts for the PAM reconstruction. However, the PAM result is a good approximation to the exit wave function, which, in the present implementation, is used as a starting point for a maximum likelihood (MAL) reconstruction that takes the non-linear image contributions fully into account (Coene et al. 1996, Thust et al. 1996a).

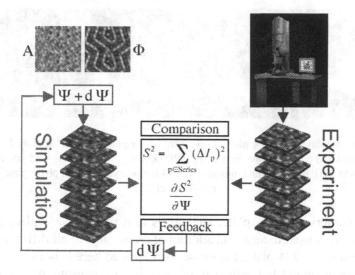

Figure 8. Schematic showing the iterative, nonlinear maximum-likelihood reconstruction scheme.

In the MAL approach, the approximated exit wave function, Ψ, obtained in the previous iteration step is used to simulate the images of the focal-series. These simulated images are quantitatively compared with the original HRTEM images and a correction to the exit wave function, $d\Psi$, is calculated to minimize the difference between the experimental data and the simulation. The corrected exit wave function is then used as the basis to simulate the images of the focal series and the whole process is repeated iteratively until the difference S^2 between simulation and experiment is sufficiently small (Figure 8).

3.2.4 Residual aberration determination and correction

With the paraboloid method followed by the maximum-likelihood refinement of the exit-wave function, the inherent effects of the microscope on the exit wave function due to spherical aberration and defocus are eliminated resulting in a complex-valued wave function with the delocalization removed. However, the electron wave function frequently suffers from residual aberrations due to insufficient microscope alignment. In a single image, it is not possible to remove these aberrations, but, with the reconstructed complex wave function, one can use a numerical phase plate to compensate the effect of aberrations by applying appropriate phase shifts (Thust et al. 1996b).

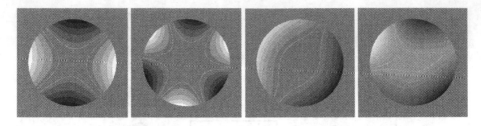

Figure 9. Numerical phase plates to compensate the phase shift induced by 2-, 3-fold astigmatism and coma – shown in $\pi/2$ steps (200 kV, $g_{max} = 8$ nm^{-1}, $A_2 = 15$ nm, $A_3 = 1100$ nm, b = 488 nm); the white line indicates the $\pi/4$ boundary. The right phase plate shows the combined effect.

The numerical phase plate, analogous to a physical $\lambda/4$-plate in optical microscopy, is used to apply an additional phase shift to all diffracted beams. In contrast to the $\lambda/4$-plate, the phase shift applied here is not constant for all spatial frequencies, but adjusted to exactly compensate the phase shifts induced by the residual aberrations of the microscope. This is illustrated in Figure 9 for different aberrations. The $\pi/4$ boundary (white lines in Figure 9) indicates the spatial frequency up to which the exit-wave function (or image) can be analyzed without explicitly taking the aberrations into account. In contrast, phase shifts higher than $\pi/4$ result in a strong change or inversion of the contribution of a diffracted beam to the exit-wave function preventing a direct interpretation. However, after numerical aberration correction, the exit-wave function can be directly interpreted up to the information limit of the microscope only by taking the electron beam-specimen interaction into account.

The only practical difficulty comes from determining the appropriate aberration correction. In TrueImage, this has been automated, e.g. for the defocus and the 2-fold astigmatism. For a weak phase object, the real part of exit wave function should be constant (Equation 2). Thus, the defocus and 2-fold astigmatism can be determined by minimizing the contrast of the real part of the exit-wave function.

4. APPLICATION EXAMPLES

Due to its unique ability to directly image the local structure of a thin object with atomic resolution, HRTEM is an extremely powerful tool for materials research. Metals, ceramics, and semiconductors are some examples of prominent materials of interest. HRTEM imaging used to be a high-end research tool mostly used in academia, but has now become standard for a wide variety of applications from materials science research to defect analysis in industrial semiconductor fabrication lines.

Focal-series reconstruction is used to reach the ultimate limits in high-resolution TEM imaging, either for quantitative structural analysis and/or to remove delocalization for interface and defect characterization. A number of academic and industrial research labs are using TrueImage and we are presenting two of their results as an example here. For a broader overview, there are several excellent articles featuring some recent work with focal-series reconstruction (Rosenfeld et al. 1998, Jia and Thust 1999, Zandbergen et al. 1999, C. Kisielowski et al. 2001, Ziegler et al. 2002, Shao-Horn et al. 2003)

4.1 Σ=3, {111} Twin boundary in a BaTiO₃ thin film

Single crystal and bulk BaTiO$_3$ exhibits a sharp paraelectric-to-ferroelectric transition at 393K. In the presence of submicron grains, the transition becomes diffuse and can be absent for polycrystalline BaTiO$_3$. Twin boundaries along the four crystallographically equivalent {111} planes constitute the main lattice defects. Junctions between such twin boundaries can be frequently observed within a grain. The local atomic arrangement of the core of twin intersections was studied by focal-series reconstruction (Jia et al. 1999).

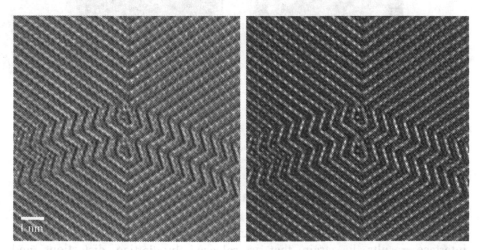

Figure 10. Reconstructed exit wave function (phase) of the same area as shown in Figure 5. The left image shows the phase of the exit wave function before aberration correction and the right image shows the phase after correction for residual 2-, 3-fold astigmatism, and coma. The difference clearly illustrates that numerical correction of residual aberrations is crucial for a quantitative analysis.

Figure 5 shows two high-resolution images representative of a focal series recorded of junctions between twin boundaries in BaTiO$_3$. However, in the focal series, the structure of the central junction can only be recognized roughly due to contrast delocalization in HRTEM imaging. After focal-series reconstruction, the resulting exit wave function reveals a sharp phase image of the central twin junction with two distinct types of phase maxima (Figure 10) corresponding to the positions of the atomic colums. The exit wave function can be interpreted almost intuitively, but simulations of the exit wave function were used to confirm the structure. Based on simulations of the single-crystalline areas, the sample thickness was estimated to be 9 nm and this thickness was used to simulate the central twin junction. A good agreement of the experimental exit wave function with simulations was obtained for the structural model shown in Figure 11. The TiO_2 columns correspond to the bright maxima in the phase image, while the BaO columns appear as darker maxima, except for the central polygon where the BaO columns appear bright due to an occupancy of 50%.

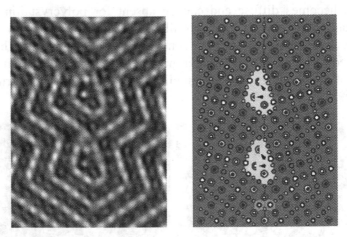

Figure 11. Phase of the reconstructed exit wave function and structure model of the same area (Ba: green, oxygen: blue, titanium: red).

4.2   Lattice distortions at a GaN/ sapphire interface

A mismatch in the lattice parameters of materials on both sides of an interface results in strain that is released by lattice distortions and dislocations next to the interface. Understanding the release mechanisms of the strain is crucial in order to tune the properties of the interface.

In a single HRTEM image, it is difficult to measure atomic positions with a high accuracy especially next to an interface due to contrast delocalization. However, the delocalization is removed in the exit wave function and the maxima in the reconstructed phase directly correspond to the center of the atomic positions. The atomic positions can, therefore, be measured with a very high accuracy enabling an accurate measurement of the lattice distortions. An example is shown in Figure 12 for a GaN/sapphire interface (C. Kisielowski et al. 2001), where it was possible to measure the in- and out-of-plane lattice parameters (perpendicular and parallel to the interface) with an accuracy of a few percent, corresponding to lattice distortions of a few picometers.

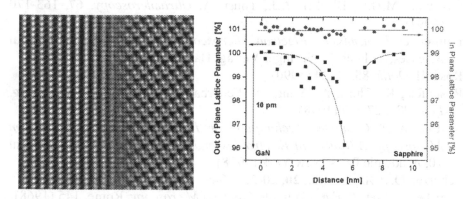

Figure 12. Reconstructed exit wave function (phase) of a GaN/Sapphire interface. The lattice parameters next to the interface were determined with very high accuracy. (C. Kisielowski et al. 2001)

References

Bethe, H. *Ann. Phys.* **87**, 55 (1928).

Coene, W.M.J., Janssen, A.J.E.M., Op de Beeck, M., Van Dyck, D. *Phys. Rev. Lett.*, **69**, 3743-3746 (1992).

Coene, W.M.J., Thust, A., Op de Beeck, M., Van Dyck, D. *Ultramicroscopy*, **64**, 109-135 (1996).

Cowley, J.M. and Moodie, A.F. *Acta Cryst.* **10**, 609-619 (1957).

Goodman, P. and Moodie, A.F. *Acta Cryst.* **A30**, 280 (1974).

Haider, M. Braunshausen, G., Schwan, E. *Optik*, **99**, 167-179 (1995).

Humphreys, C.J. *Rep. Prog. Phys.* **42**, 1864 (1979).

Ishizuka *Ultramicroscopy*, **55**, 407-418 (1994).

Jia, C.L., Rosenfeld, R., Thust, A., Urban, K. *Phil. Mag. Let.*, **79**, 99-106 (1999)

Jia, C.L., Thust, A. *Phys. Rev. Let.*, **82**, 5052-5055 (1999).

Kisielowski, C., Hetherington, C.J.D., Wang, Y.C., Kilaas, R., O'Keefe, M.A., Thust, A. *Ultramicroscopy*, **89**, 243-263 (2001).

Kirkland, E.J. *Ultramicroscopy*, **15**, 151-172 (1984).

Krivanek, O.L. *Ultramicroscopy*, **55**, 419-433 (1994).

Krivanek, O.L. and Stadelmann, P.A. *Ultramicroscopy*, **60**, 310-316 (1995).

Lichte H. *Ultramicroscopy*, **20**, 293-304 (1986).

Lichte, H. and Rau, W.-D. *Ultramicroscopy*, **54**, 310-316 (1994).

O'Keefe, M.A. and Kilaas, R. *Scanning Microscopy Supplement*, **2**, 225-244 (1988).

Op de Beeck, M., Van Dyck, D., Coene, W. *Ultramicroscopy*, **64**, 167-183 (1996)

Overwijk, M.H.F., Bleeker, A.J., Thust, A. *Ultramicroscopy*, **67**, 163-170 (1997).

Reimer, L. *Transmission Electron Microscopy*, Springer Series in Optical Sciences, Volume 36, Springer Verlag, Hamburg (1984).

Rose, H. *Optik*, **85**, 19-24 (1990).

Rosenfeld, R., Thust, A., Yang, W. Feuerbacher, A. Urban, K. *Phil. Mag. Lett.*, **78**, 127-137 (1998)

Saxton, W.O. *Computer Techniques for Image Processing in Electron Microscopy. Advances in Electronics and Electron Physics.* Supplement 10, Academic Press, New York (1978).

Scherzer, O. *J. Appl. Phys.*, **20**, 20-29 (1949).

Schiske, P. *Proc. 4th Eur. Conf. On Electron Microscopy*, Rome, 145 (1968).

Schiske, P. *J. Microscopy*, **207**, 154 (2002).

Shao-Horn, Y., Croguennec, L., De Delmas, C., Nelson, E.C., O'Keefe, M.A. *Nature Materials*, **2**, 464-466 (2003).

Smith, D.J., Saxton, W.O., O'Keffe, M.A., Wood, G.J., Stobss, W.M. *Ultramicroscopy*, **11**, 263-281 (1983).

Spence, J.C.H. *Experimental High-Resolution Electron Microscopy*, Oxford University Press, New York, Oxford (1988).

Stadelman, P. *Ultramicroscopy*, **21**, 131 (1987).

Thust, A., Coene, W.M.J., Op de Beeck, M., Van Dyck, D. *Ultramicroscopy*, **64**, 211-230 (1996a).

Thust, A., Overwijk, M.H.F., Coene, W.M.J., Lentzen, M. *Ultramicroscopy*, **64**, 249-261 (1996b).

Williams, D.B. and Carter, C. B. *Transmission Electron Microscopy*, Plenum Press, New York (1996).

Zandbergen, H.W., Bokel, R., Connolly, E. Jansen, J. *Micron*, **30**, 395-416 (1999).

Zemlin F., Weiss P., Schiske P., Kunath W. and Herrmann K.H.: *Ultramicroscopy*, **49**, 3 (1977).

Ziegler, A., Kisielowski, C., Ritchie, R.O. *Acta Materialia*, **50**, 565-574 (2002).

Applications

NEW INSIGHTS INTO THE NANOWORLD OF FUNCTIONAL MATERIALS

Gerhard Cox, Thomas Breiner, Hartmut Hibst & Ulrich Müller
BASF-AG, D-67056 Ludwigshafen, Germany

Key words: nanosized materials, catalysts, structure determination, HRTEM, HAADF

1.   INTRODUCTION

Functional materials have dedicated adjusted chemical and physical properties which are often determined by their microstructure. Such materials with custom tailored properties are the result of an interdisciplinary research with contributions from chemistry, physics and engineering. In the chemical industry functional materials are made in the range from several 100 kilograms to thousands of tons in complicated processes involving often several synthesis steps. At BASF these materials are produced in a production "Verbund" starting with nine raw materials. From these about 200 major basic products and intermediates and over 8000 commercial products are made (fig. 1). Verbund means that the production plants are linked with other plants by at least one product or process stage thereby saving energy costs and leftover material. BASF operates six Verbund sites at present (one starting in China in 2005) in combination with several local production sites in Europe, USA and Asia to ensure global presence in all major markets.

T.E. Weirich et al. (eds.), Electron Crystallography, 395–407.
© 2006 *Springer. Printed in the Netherlands.*

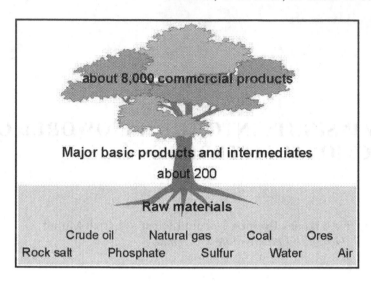

Figure 1. Production "Verbund" at BASF

2.   NANOSIZED MATERIALS

The properties of materials often change dramatically when their size reaches values below 100 nm. Especially the ratio of surface to diameter increases dramatically. Let us consider e.g. a quartz cube with an edge length of 27 cm weighting 50 kg (Fig 2). Its surface area measures less than half a square meter. If we now divide this cube into a lots of small cubes with an edge length of 1mm the surface area of all cubes together would amount to the size of a classroom. If we shrink the dimensions to 1 nm the surface area would increase to about 12 square km, so the specific surface area i.e. the area related to the weight increases by a factor of 30 million from its initial state. This effect plays an important role in the field of heterogeneous catalysis, where the chemical reactions take place at the surface of solid matter. The larger the catalytic surface area, the more active the catalyst. Hydrogenation catalysts for example often consist of metallic nanoparticles dispersed on an oxide carrier material. The nanoparticles have a high number of steps and kink sites on the surface where the hydogenation reaction can occur (fig 3).

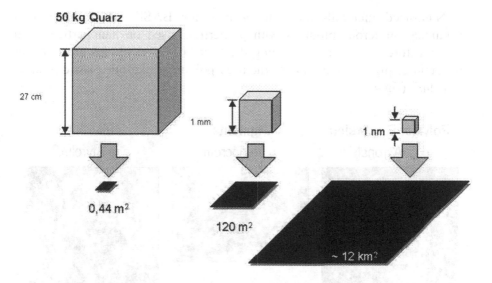

Figure 2 Dimensions from the Macro- to the Nanoworld

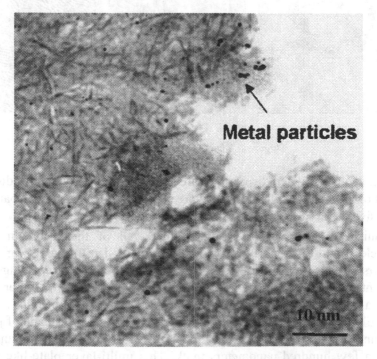

Figure 3. Hydrogenation catalyst with nm-size metallic particles on an oxide carrier
(inverted STEM dark field image)

3.   NANOMATERIALS MADE BY BASF

Nanosized materials are nothing new for BASF, which have been producing numerous products with properties based on nanoparticles and nanostructures for decades. Among the best known examples are polymer dispersions, pigments and nanostructured polymers like our plastic product Styrolux® (fig 4).

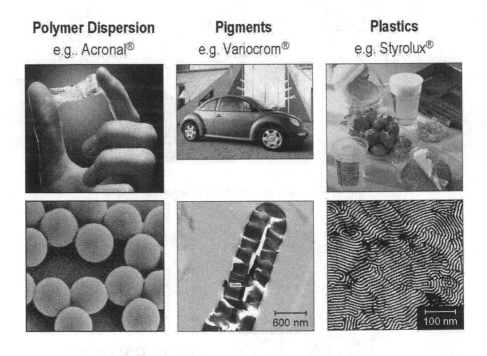

Figure 4. Nanomaterials in BASF products

Aqueous polymer dispersions are one of BASF biggest products with sales of around 1.5 billion €. They are found in exterior paints, coatings and adhesives or are used in

finishing papers, textiles and leather. All of them contain polymer particles ranging from 10 to several 100 nm in size. In paints e.g. polymer dispersions use water as a dispersing agent thereby replacing organic solvents in many applications. When the paint dries the polymer particles form a fine film thereby improving adhesion on various substrates .

Colour variable pigments in our Variocrom® range are made of µm-large aluminium flakes which are covered by several layers of SiO_2 and Fe_2O_3, each a few hundred nanometers thick. This multi-layer plate-like structure produces an angle dependent, iridescent colour play. The SiO_2-layer with a low refractive index produces strong angle dependent interference colours like in soap bubbles. But the tone is weak due to the fact that only 4% of the incoming light is reflected. In colour variable pigments the weak colour impression is intensified through a second Fe_2O_3 layer with a high reflection index.

Styrolux® is an example of a nanostructured polymer which is used in food packaging. It is a polystyrene-polybutadiene block copolymer where polymer chains are build up of alternating polystyrene and polybutadiene blocks. These blocks appear as dark lamellae in the TEM image due to the staining of the polybutadiene with OsO_4. This structured nanoscale architecture of the polymer, which can be controlled during manufacture, allows the optimum combination of impact resistance and transparency.

4. NANOMATERIALS IN NOVEL BASF BUSINESSES

In the examples shown before BASF has been exploiting the special properties of nanosized and nanostructured materials for many years, often unconsciously. But for BASF nanomaterials are not an end in themselves. Their incorporation in products such as coatings, dispersions, fibres and foams often first create the improved or even novel properties of materials with an added value for our customers and the consumer. The research and development of nanomaterials is therefore driven by market and customer needs. To span the wide range of possible applications of nanomaterials and to fully exploit their potential in possible applications, BASF uses its Research "Verbund" with its central technology platforms Chemical Research and Engineering, Speciality Chemicals Research and Polymer Research with over 6000 employees in Ludwigshafen and approximately 8000 worldwide. It is often the combined effort and knowledge of experts of various fields within the three central technology platforms, which is crucial for the successful introduction of new products into the market. The following chapters describe a few examples of the latest BASF innovations in the nanomaterials field.

4.1 Metal oxide frameworks for hydrogen storage

Fuel cells are an attractive alternative for replacing single-use and rechargeable batteries in mobile communication gadgets like cell phones, PDAs and laptops. Particularly the Proton Exchange Membrane (PEM) cell and the Direct Methanol (DM) cell, which operate at ambient temperature and in the lower power range, are suited for these purposes. The PEM cell uses hydrogen as a fuel which dissociates into H^+-ions and electrons at the cathode using an appropriate catalyst. The H^+ ions diffuse through the polymer membrane to the cathode where they react with oxygen to form water – the only by-product in this reaction. The electrons travel through an external circuit producing an electrical current, which can be used powering an electrical device. An important requirement for the PEM cell is the safe transportation and storage of hydrogen, particularly in the mobile use of small devices where liquefaction (at $-235°C$) or compressed gas storage (at 200-300bar) is not an option.

The task, therefore is, to find a sort of effective storage medium for hydrogen, which is light, compact and cost competitive for the practical operation of so-called mini fuel cells. What is needed is a material – a sort of nanocube –with a huge internal surface to store the hydrogen. Such materials have been developed by Omar Yaghi and coworkers at the University of Michigan [1,2]. They use organic units of terephthalic acid which are linked with inorganic zinc oxide via carboxylate bonds. Periodically consecutive structural units of terephthale ligands and Zn_4O clusters produce a highly porous space lattice and form a metal organic framework (MOF) (fig.5). MOFs with specific surface areas greater than 4000 m^2/g have been reported, which means that two grams of this new material could cover the area equivalent of a soccer field. BASF is currently able to produce nanocubes of this material with around 1 μm in primary particle size in kg quantities. Hydrogen uptake measurements showed a recharge capacity of about 1,85 weight percent hydrogen at ten bar for certain MOF types, and we expect this figure to be doubled in future with the development of new MOF types with even higher specific surface areas.

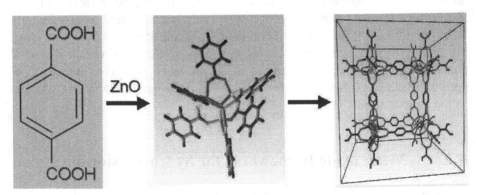

Figure 5. Building of a metal oxide framework.

4.2   Polymer nanocomposites

Polymer Nanocomposites are novel plastic compounds with a filler having dimensions between 1 and 100 nm. They have attracted much attention in the past because nanocomposites exhibit markedly improved properties like stiffness, thermal flammability, improved barrier properties and others compared to the unfilled matrices [3]. Among all potential fillers, those based on easily available clay and layered silicates have been more widely investigated for some time now.

The ideal layered silicate is the mineral montmorrillonite, which is main component of the clay bentonite (fig 6). Montmorrillonite is a so called 2:1 layered silicate. Each layer consists of two sheets of silica tetrahedrae which are sandwiched by an alumina octahedrae sheet. The layers are only weakly bound, often by hydrogen bonding from water. The structure has been extensively studied by transmission electron diffraction especially by Zyvagin and coworkers [4].

Figure 6. Structure of montmorrillonite (SiO_4 tetraedra red, $Al(OH,O)_6$ octahedra green, $Mg(O,OH)_6$ octahedra grey and Na atoms yellow (after [4])

In Montmorrilonite part of the Al^{3+} in the octahedrae is missing or replaced by Mg^{2+} resulting in a net negative electrical charge of the layers which is compensated by Na^+- or Ca^{2+} ions intercalated between the silicate layers. The interlayer distance can vary between 1-2 nm depending on the amount of water present in the mineral. Due to its polar structure montmorrillonite is good dispersable in aqueous media and is therefore often used as a thickener. Polymers are less polar and therefore the inorganic cations have to be replaced by organophilic molecules like quaternary alkylamines. This exchange reaction makes the montmorrillonite hydrophobic and it increases the interlayer distance. The montmorrillonite "swells" which increases its ability to exfoliate in the polymer. This is important because it is often easier to bring in the layered silicate particles via compoundation rather than during the polymerisation. But the shear energy in the extrusion process is often to low to generate single layer particles. Therefore additional measures have to be taken by adjusting the hydrophobisation to the polymeric system to exfoliate the layered silicates into primary platelets and to homogeneously distribute them throughout the polymer matrix (fig. 7).

One of the advantages of layered silicates when compared with ordinary fillers like glass fibres and standard minerals is their higher impact on the stiffness per weight fraction. For example the E-modulus reached by adding 15% of a 10 µm thick glass fibre is also gained with 5% layered silicate (fig 6). This effect is due to the much higher aspect ratio of the layered silicate particles in comparison to conventional fillers. Nanocomposites based on layered silicates have interesting and sometimes even fascinating properties. This is due to fact that the nanofiller does not accomplish an additive contribution to the properties like classical fillers but changes through intensive interaction the properties of the matrix material itself.

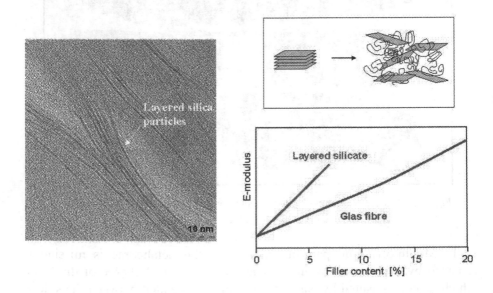

Figure 7. Layered silicate particles in a polymer and their effect on the E-modulus

Nevertheless the euphoric optimism where these materials were allowed a huge potential in material applications has given way to a more realistic view. Nanocomposites are not a universal solution for reinforced materials. Their full potential can only be realised if every step in the added value chain is taken into account during the whole development process. From todays perspective nanocomposite materials with an improved thermal flammability resistance or improved barrier properties have the best chances to fulfil these requirements.

4.3   Heterogeneous catalysis

Catalysis is a nanoscale phenomenon that has been the subject of research and development for about hundred years in the chemical industry. It is normally not associated with the prefix "nano", however catalysts are typical nanomaterials.

The contribution from Catalysis to the economics is remarkable. Based on estimates from the North American Chemical Society between 15 and 20% of the world gross net product are provided by catalytic processes [5]. Thereby the catalysis costs are much less than 1% of the sales revenues from the products, which they help to create, making catalysis a key technology to the sustainable and cost effective production of chemicals. Numerous things of our daily life like gasoline, plastics, cars, computers or drugs would not exist at all or at least not be available in the today's quality without catalysis. At BASF for example over 80% of the 8000 products see at least once a catalyst during their production cycle.

The development of heterogeneous catalysts has changed essentially over the last two decades away from an empirical strategy let by trail and error towards a more rational design approach. This is due to the fact, that analytical tools in combination with powerful modelling equipment allow us nowadays to identify structural key elements of a catalyst. The deeper knowledge on the structure-property relations permits a well-aimed and targeted development of heterogeneous catalysts.

The important role Transmission Electron Microscopy (TEM) can play in this process is demonstrated on the development of an oxidation catalyst for the production of acrylic acid. Acrylic acid is produced by BASF in quantities of several 100.000 tons per year in a two step gas phase oxidation process starting from propene, which is oxidised to acrolein in the first step and then further oxidised to acrylic acid in a second step, each step requiring a special developed catalyst. Acrylic acid is used as a base material for the production of superabsorbents for nappies, dispersions and emulsions for adhesives and construction materials.

A possible economically attractive alternative would be the production of acrylic acid in a single step process starting from the cheaper base material propane. In the nineteen nineties the Mitsubishi Chemical cooperation published a MoVTeNb-oxide, which could directly oxidise propane to acrylic acid in one step [6]. Own preparations of this material yielded a highly crystalline substance. Careful analysis of single crystal electron diffraction patterns revealed that the MoVTeNb-oxide consists of two crystalline phases- a hexagonal so called K-Phase and an orthorhombic I-phase, which is the actual active catalyst phase, as could be shown by preparing the pure phases and testing them separately.

Despite an intensive search in the data base of the international centre of diffraction data (ICDD, Newton Square, USA) no structure could be assigned to the I-phase. Solving the structure by direct XRD methods proved difficult due to the large unit cell. We were able to solve the structure by High Angle Annular Dark Field (HAADF)-STEM imaging of single crystals, which crystallised with the [001] direction parallel to the particle base plane. Even though HAADF imaging is slower and beam damage is more severe when compared with conventional high resolution TEM imaging, it has the big advantage, that it can be applied to thicker samples and that image interpretation is much easier. This is due to the incoherent spot illumination and the strong Z dependence of the HAADF signal [7]. Metal atoms in oxides therefore always appear as bright spots in HAADF images.

The orthorhombic structure of the I-phase in the (001) plane consists of edge- and corner connected octahedrae, which form pentagonal bipyramids and six- and seven sided channels, where the six-sided channels are occupied and the seven sided channels are empty (fig 8). Most of the particles crystallise with the (001)-plane being perpendicular to the particle base plane like in the MoVW-oxide system. Experiments showed that the activity increased by grinding the particles. This strongly suggests that the (001)-plane is the catalytic active surface for the oxidation of propane to acrylic acid [8]. Further XPS measurements and diffraction experiments using anomalous dispersion conditions at a synchroton allowed a consistent Rietveld refinement of the diffraction data starting with the structural model from the HAADF-TEM images. Nb occupies as Nb^{5+} the pentagonal bipyramids which are surrounded by five Mo^{6+} octahedral. The six- and seven sided channels are statistically occupied by Mo^{6+} and V^{5+} ions and Te is placed as Te^{4+} off center in the hexagonal channels.

A literature survey showed that the structure of I-phase was not new in fact, but had been discovered earlier in the Cs-Nb-W oxide system by Lundberg and Sundberg using high resolution TEM imaging [9]. We were able to prepare a single phase substance with the composition

$Cs_{0,5}[Nb_{2,5}W_{2,5}O_{14}]$, which was very similar to the one described by Lundberg et. al. Single electron diffraction patterns showed, that the particles now crystallised with the (001) plane being parallel to base plane making HAADF imaging much easier (fig.9).

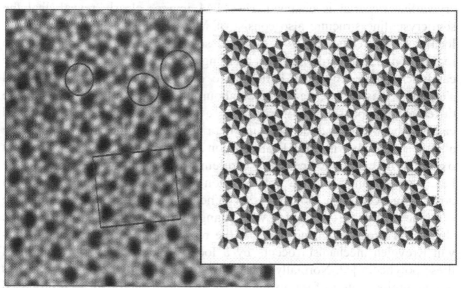

Figure 8. HR-HAADF-STEM image of a MVTeNb oxide along the [001] direction. The bright spots correspond to single metal atoms, which are coordinated to six oxygen atoms. These octahdrae are connected via their edges and corners to form this complex bronze structure. The rectangle markes the unit cell in the (001)-plane. The circles marke the pentagonal bypiramid and the sic- and seven-sided channels.

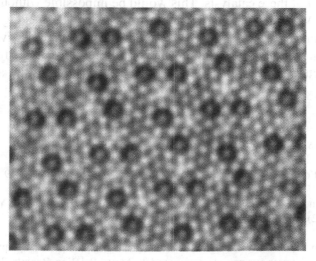

Figure 9. HR-HAADF-STEM image of a $Cs_{0,5}[Nb_{2,5}W_{2,5}O_{14}]$

Here Cs stabilizes the structure by occupying the six-sided channels and partially occupying the seven sided channels. The HAADF image in fig.9 shows very nicely, that Cs is placed off center in the seven sided channels.

When the $Cs_{0,5}[Nb_{2,5}W_{2,5}O_{14}]$ is only heated up to 700°C, a semi crystalline material is obtained. HR-TEM images clearly reveal, that the semicrystalline structure also consist of MO_6 octahedrae, which are only locally connected to the five-, six and seven-sided channels but with no regular longe range ordering. This ordering seems to be a prerequisite for the catalytic alkane activation. Furthermore different catalytic active isolated sites must exist in close atomic neighbourhood in a regular fashion for the different reaction steps to occur on one surface. Based on the observed catalytic behaviour and the structural data we presume that in a first step the alkane activation with the H^+ abstraction occurs at the Nb^{5+} site. The first oxidation of the allylic spezies to acrolein proceeds at the Te^{4+} sites in the hexagonal channels and the final reaction to acryl acid and the subsequent desorption occur at a $Mo^{6+}-V^{4+}$ site. When oxygen is removed during the catalytic cycle the polyhedra will change from corner sharing to edge sharing. It is easily conceivable that the reoxidation via the so called Mars van Krevelen mechanism occurs by a local back and forward rotation of these polyhedra [5]. Normally one would expect that the oxygen vacancies would agglomerate to form extended shear defects as it had been proposed for other oxide reactions [9]. This would lead to extended strikes in single electron diffraction patterns, which were not observed in this catalyst.

The example above shows that heterogeneous catalysts are multifunctional materials not at the nanometer but even at the atomic scale. A detailed structural understanding is a prerequisite for a targeted development of these catalysts. This would be impossible without the help of modern TEM techniques. It can be expected, that the next generation of TEM instruments with Cs-corrected condensor- and objective lenses in combination with high resolution energy electron loss spectrometers can reveal structural and chemical details of catalysts even at the atomic scale [10,11].

References

[1] H.Li, M. Eddaoudi, M.O. Keefe. O.M. Yaghi, Nature, 402 (1999) 276-279

[2] M. Eddaoudi, J.Kim, N. Rosi, D. Vodak, J. Waechter, M.O. Keefe. O.M. Yaghi, Science, 295 (2002), 469-472

[3] M. Alexandre, Ph. Dubois, Polymer-layered silicate nanocomposites : preparation, properties and uses of a new class of materials, Mater. Sci. Eng. Reports, 28 (2000), 1-63

[4] For a review s. N. Güven, Smectites, in Reviews in Mineralogy Vol 19, Mineral Society of America, 1988

[5] J.M. Thomas, W.J. Thomas, Principle and Practise of Heterogeneous Catalysis, VCH, 1997

[6] H. Watanabe, Y. Koyasu, Mitsubishi Chemical Corporation, 4th World Congress on Oxidation Catalysis, Berlin 2001

[7] S.J. Pennycook, D.E. Jesson, Ultramicroscopy 37, (1991), 14-38

[8] R.K. Graselli et.al. Topics in Catalysis, Vol 23, No1-4, 2003, 5-22

[9] M.Lundberg, M. Sundberg, Ultramicroscopy 52, (1993), 429-435

[10] P.L. Gai, E.D. Boyes, Electron Microscopy in Heterogeneous Catalysis, Institute of Physics Publishing, 2003

[11] C.L.Jia, M. Lentzen, K.Urban, Science, Vol233, 870-873

[12] N. Shibata, S.J. Pennycook, T.R. Gosnell, G.S. Painter, W.A. Shelton, P.F. Becher, Nature 428, 730-733

ELECTRON CRYSTALLOGRAPHY ON POLYMORPHS

U. Kolb and T. Gorelik
Institute of Physical Chemistry, Johannes Gutenberg-University, Welderweg 11, 55099 Mainz, Germany

Abstract: Structure analysis via electron diffraction in combination with x-ray powder diffraction and simulation methods has been performed on single crystals of small organic molecules which form polymorphs. For two different examples of pigments and non-linear optical active material data collection, cell parameter and space group determination, structure analysis and refinement are discussed.

Key words: electron crystallography, polymorphs, solvate, computer simulation, x-ray powder

1. INTRODUCTION

A common tool to analyse crystal structures is single crystal x-ray diffraction, nowadays a well developed routine method. Nevertheless, many interesting materials are preferentially micro- or nano- crystalline such that no crystals of a size sufficient for single crystal x-ray diffraction can be obtained. Other methods, like single crystal electron diffraction (ED) or x-ray powder diffraction (XRD) still present a challenge, especially for organic compounds. X-ray powder diffraction delivers one-dimensional data, which suffers often from overlapping reflections, inadequate crystal quality, unknown impurities and preferred orientation [1, 2]. Electrons undergo coulombic interactions with matter which are 10^8 times stronger than electromagnetic interactions. Therefore, it is possible to investigate extremely small volumes. In comparison with powder data ED provides us with three dimensional data but suffers from an incomplete reciprocal space (missing cone problem), elongated reflections (spike function) and is

T.E. Weirich et al. (eds.), Electron Crystallography, 409–420.
© *2006 Springer. Printed in the Netherlands.*

affected by multiple scattering and dynamical scattering for thick samples [3, 4]. Because dynamical scattering is less serious for organic materials, structure determination by electron diffraction can be carried out based on a kinematical diffraction theory in the first step and the resulting model can be refined dynamically in a second step.

Properties of a material in the solid state are strongly governed by its molecular and crystal structure and therefore a big effort is put into structure determination. Dependent on the crystallisation parameters used one may derive even different crystal structures from the same molecule constituting different polymorphs. Since the definition of polymorphism can not be given in an overall manner McCrone's defintion as "the possibility of a compound to crystallize at least in two different arrangements of the molecules" is used here [5]. In the following three examples of nanocrystalline organic compounds will bee discussed where electron diffraction was used to support or even to perform the structure analysis of polymorphs.

The non-linear optical (NLO) active aromatic chalcone 1-(2-furyl)-3-(4-benzamidophenyl)-2-propene-1-one (FAPPO), shown in Fig.1, was found to give a rather big SHG activity of 1-2 order of magnitude higher than crystalline urea [6]. Due to their conjugated system all discussed NLO active molecules are approximately flat.

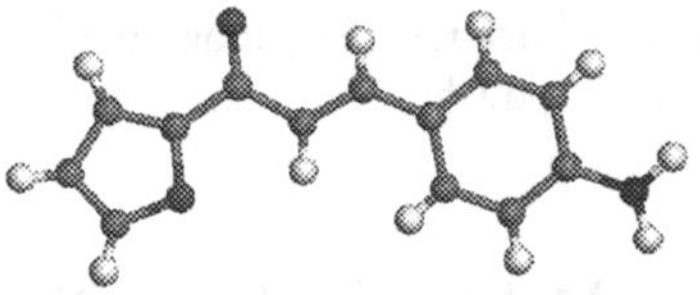

Figure 1: Molecular structure of FAPPO

It crystallizes from ethanol in two modifications which could not be separated from each other. Hence, a structure analysis from x-ray powder diffraction or even the indexing of the peaks turned out to be impossible.

Phthalocyanines exhibit high polymorphism, as well [7]. They are insoluble nano crystalline materials, which produce poor x-ray powder diagrams with high preferred orientation and have been intensely investigated by electron microscopy and diffraction [8]. From copper phthalocyanine (CuPc) (Fig.2) nine polymorphs (α, β, γ, δ, ε, π, ρ, ζ, σ and R) are known, but only the most stable β phase could be solved by single crystal x-ray diffraction [9] ($P2_1/c$: a=14.628 Å, b= 4.790 Å, c= 19.07 Å, b=120.93°).

Figure 2: Molecular structure of copper phthalocyanine (CuPc)

2. DATA COLLECTION

2.1 Sample preparation

From soluble compounds like FAPPO, crystals of approx. 100-200 Å thickness have been grown from thin solutions directly on a carbon coated copper grid (3mm/300 mesh) at room temperature by evaporation. With this method thin crystals with less dynamical effects are usually obtained which lie flat on the grid providing a view along the thinnest direction.

From insoluble materials, like CuPc, particles of high crystallinity have been selected through a light microscope and deposited onto a holey carbon film, after dispersion in ethanol by grinding and subsequent ultra sonication. To access other orientations, the particles can be embedded in epoxy raisin and the obtained pellets can be cutted by ultra microtomy with a diamond knife at room temperature. Generally, it is important to make sure that any influence on the crystal structure (e.g. through grinding or solvents used for embedding) is excluded as good as possible.

2.2 Electron diffraction

Electron diffraction for structure analysis is performed in a transmission electron microscope (TEM) usually at 100-400 kV acceleration voltage. Due to the small wavelength (usually 0.004 – 0.002 nm dependent on the acceleration voltage) of the electron beam, respectively the small scattering angle, a group of reflections (zone) will be in diffraction position for one crystal orientation (see Fig. 5 and 6). Imagine a Ewald sphere with a radius so huge that the surface is close to a straigth line leading to an almost planar cut through the reciprocal space. Through the tilt of the crystal, preferably about main axes, in respect to the electron beam, other zones can be derived. Suitable sample holders for this task need a high flexibility and it is important to use an appropriate holder.

Even for electron diffraction the Ewald sphere has a small curvature. Therefore, in the case of inorganic material one would be able to detect as well reflections from the next layers the "first/higher order laue zone" FOLZ/HOLZ shown in Fig. 3 as rings of reflections at higher angles. These rings can be used as a guide for the correct allignment of a low indexed zone and a double-tilt rotation holder can be used. Organic material diffracts weaker and therefore does not show HOLZ reflections. Hence, it is extremely difficult to find suitable zones for cell parameter and space group determination of an unknown cell. Here it is better to use a rotation-tilt holder where the crystal can be oriented with a chosen axis, preferably a main axis, and tilted around the goniometer axis of the TEM.

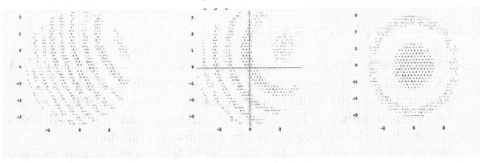

Figure 3: Simulated electron diffraction patterns of an inorganic compound with HOLZ reflections 6° mistilted (left hand side), 3° mistilted (middle) and in correct orientation (right hand side).

Compounds crystallized directly onto the carbon grid or with a defined orientation, due to other preparation methods, normally exhibit a suitable initial zone close to 0°. Samples from insoluble compounds are almost statistically oriented only biased by the particle shape. In this case, it is difficult to find a single crystalline part of appropriate thickness oriented with a suitable zone parallel to the surface. The best flexibility, and therefore the best possibility to orient a zone correctly, is given by a recently developed rotation-double tilt holder (Gatan Inc.). Through the combination of rotation and additional tilt (beta tilt) it is possible to orient the tilt axis exactly even if the crystal is not sitting flat on the support film (see Fig. 4). The tilt range, dependent on the pole piece distance of the objective lens, should be at least ±40°.

Figure 4: Movement steps using a double-tilt rotation holder: 1) Rotate the crystal until the chosen axis is oriented along the goniometer axis; 2) Tilt the crystal perpendicular to the goniometer axis until it is perpendicular to the electron beam; 3) Adjust eucentric height; 4) Tilt around the goniometer axis to reach different zones.

If the area of investigation needs to be reduced under the limit, given by the smallest available selected area diffraction apperture, nanodiffraction can be used. A C2 apperture of 10μm will provide us with an approximate parallel illumination at a beam diameter down to 5nm and reduce the applied electron dose significantly. Because the majority of organic samples are beam sensitive investigations should be carried out under low-dose conditions. In addition to that cryo techniques should be applied if no phase transition occurs. The diffraction patterns can be recorded on film and digitised by a scanner or directly on CCD.

2.3 Cell parameter determination

In the case of β Cu-phthalocyanine (space group P2₁/c) only 8 different zones were found at 100 kV via SAED using a rotation-tilt holder (a=14.50 Å, b= 4.70 Å, c= 19.32 Å, b=123°). With a rotation-double-tilt holder and in nano diffraction mode it was possible to detect 17 different zones at 300 kV (initial zone [001] and tilt about a* and b*) from one crystal at room temperature (a=14.50Å, b=4.84Å, c=19.41Å, β=120.3°). In Fig. 5 the initial zone (middle) and two other zones obtained through different tilt series (a*: left hand side; b*: right hand side) are shown exemplarily.

Figure 5: Selected electron diffraction patterns collected with CCD from β copper-phthalocyanine taken with a rotation double-tilt holder at 300 kV with FEI TECNAI F30 ST: zone [012] from tilt about a* (left hand side), initial zone [001] (middle), zone [-101] from tilt about b* (right hand side)

Based on a Vainshtein plot, where the d* values for the non tilt axis is plotted according to the tilt angle, the third axis and the one missing angle can be derived from each tilt series [10].

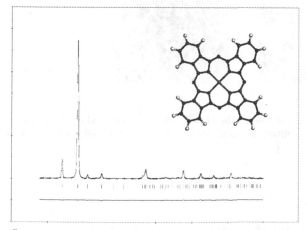

Figure 6: Pawley fit on x-ray powder data from β copper phthalocyanine based on cell parameters from electron diffraction (Rwp=9.46%, Rp=7.11%, a=14.606 Å, b= 4.799 Å, c= 19.477 Å, b=121.06° taking preferred orientation into account).

Experimental tilt angles have usually an accuracy of at best ±3°, leading to an error of about ±0.1 Å in cell axes. The calculated third cell axis will show a higher deviation. If possible an internal standard should be used for calibration purposes but a higher accuracy will be obtained with a Pawley fit (e.g. fit for β CuPc in Fig. 6) from x-ray powder diffraction data [11]. Especially for packing energy minimization used for simulation methods it is essential to determine the cell parameters as precise as possible. In the case of polymorphism, it is essential to use x-ray powder diffraction to ensure that bulk and investigated nano crystals represent the same modifications.

2.4   Space group determination

In the case of FAPPO, where two modifications are mixed, such a direct refinement was not possible. As given in Fig. 7, modification I exhibits platelets (0.5 x 2 µm) while modification II forms needle-like crystals of about 0.1 µm width.

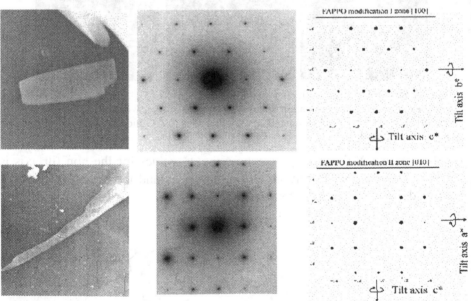

Figure 7: Platelets of modification I (top, left hand side) and needle-like crystal of modification II (bottom, left hand side). Experimental (middle) and kinematical calculated

(right hand side) electron diffraction patterns from start zone [100] for FAPPO modification I (top) and zone [010] for FAPPO modification II (bottom) at 100 kV using SAED.

Therefore the phases could be distinguished by electron microscopy. The start zone at $0°$ turned out to be [100] for modification I showing the parameters $b=8.9$ Å, $c=5.0$ Å and $\alpha=90°$ whereas modification II gave $a=10.1$ Å, $c=4.1$ Å and $\beta=90°$ from the [010] start zone.

During the tilt about the main axes, $90°$ angles were preserved and the tilt series turned out to be symmetrical, which implies an orthorhombic system in each case. The third axes were calculated as 23.5 Å and 25.3 Å, respectively. These preliminary cell parameters have been refined according to the ratio of the axes in every measured zone leading to a higher cell parameter precision for both modifications (Mod I: $a=23.83$ Å, $b=9.10$ Å, $c=5.03$ Å; Mod II: $a=10.22$ Å, $b=25.43$ Å, $c=4.18$ Å). With the refined cell parameters it was possible to index the two-phase diffraction diagram.

Based on the extinctions found from tilt experiments, the space group $Pna2_1$ (Pna2 or Pnam) were derived for both modifications. In zone [100] of FAPPO modification I the n-glide plane (0kl: k+l=2n) is clearly visible, whereas the a-glide plane (h0l: h=2n) can be recognized only during tilt by the appearance and disappearance of odd reflections perpendicular to the tilt axis b^*. Zone [010] of FAPPO modification II shows the a-glide plane which could be distinguished by the appearance of new lines during the tilt around the c^* axis. The n-glide plane can be recognized during tilt by the appearance and disappearance of odd reflections perpendicular to the tilt axis a^*. Experimental electron diffraction patterns of modification I are given exemplarily together with kinematical calculated diffraction patterns in Fig. 8 In $Pna2_1$ intensity for extincted reflections (k=2n+1) can be detected especially along the b-axis. These forbidden reflections were identified due to their sharp- and weakness and their dis- and reappearance during tilt as caused by dynamical scattering.

2.5 Collecting quantitative data

After the determination of cell parameters and space group, intensity data for structure determination must be collected. For each zone, exposure series have to be performed and assembled into a 3D data set, subsequently. To provide the CCD from damage by the main beam a beam stopper, recently developed by us, has been used. It is small enough not to cover important reflections and even suitable for a Gatan Imaging Filter (GIF) CCD. Intensities can be determined by fitting the peaks with a gaussian function using the program ELD [12]. For kinematical approach each zone needs to be oriented as correctly as possible and should yield an internal R-value R_{int} of max. 20 %.

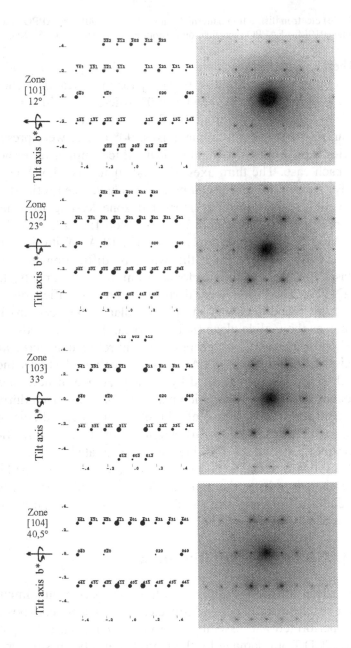

Figure 8: Selected experimental (right hand side) and kinematically calculated (left hand side) electron diffraction patterns from tilt about b* of FAPPO modification I

$$R_{\text{int}} = \sum \left| F_o^2 - F_o^2\,(mean)\ \right| / \sum \left[F_o^2 \right] \qquad (1)$$

These R-values have been calculated using SHELXL [13]. Different methods are discussed to determine a merging factor between single zones. In our case the factors were derived by minimization of the internal R-values for combination of zones. The merging procedure needs to be carried out with great care and step by step. One not suitable zone can spoil the whole data set. This holds especially for high tilt zones, which reach a high thickness (60° tilt causes double thickness) and therefore suffer from increasing dynamical effects.

3. STRUCTURE ANALYSIS AND REFINEMENT

Deficiencies in intensities, which occur in x-ray powder diffraction as well as in single crystal electron diffraction, may cause problems even in early stages of *ab initio* structure analysis. Nevertheless, examples for successful use of the tangent formula or Sayre equation for structure determination from ED data have been worked out [14]. Other direct methods, like maximum entropy can provide us with an envelope of the molecules in the cell, which delivers an idea of its orientation [15]. An alternative approach to "ab initio" structure determination is the calculation of the gas phase conformation of an initial model for subsequent refinement by energy minimization [16].

3.1 Simulation method

In order to get a good start model for packing energy minimization, the molecule geometry can be calculated in the gas phase by *ab initio* quantum-mechanical calculations or derived from structural data taken from the Cambridge Structural Data Base (CSD) [17]. It is essential to derive the molecule geometry as exact as possible because slight changes, especially in shape determining bond angles, may influence packing energies severely.

In a first step the rigid molecule should be packed according to all possible space groups. For FAPPO the space groups Pna2 and Pnam could be excluded because it was not possible to generate a reasonable packing. For minimization of packing energies a DREIDING2.2 force field was used with semi-empirically calculated charges from MOPAC6.0 using MNDO ESPs. If available, an allready solved crystal structure can be used to find the best force field and ESP charges [18]. From the initial model derived from packing energy minimizations single crystal electron diffraction pattern as well as x-ray powder pattern data can be calculated using Cerius2 [19] and compared with experimental data. Further packing energy minimizations should be performed with free torsion angles and restrained bond length and angles. Usually, three dimensional ED data does not cover the whole reciprocal space providing us with reflections which originate mainly from the short axes. Complementary to that XRD data will deliever long distances with little overlap. The simulation approach allows us to combine the available information, but unfortunately, not yet in an automated manner.

3.2   Structure refinement

After merging of the single zones, data sets of approx. 100-300 independent reflections can be obtained as described in chapter 2.5. In a first step a kinematical structure refinement should be performed using the program SHELXL [13]. The temperature factors for FAPPO were chosen as $U = 0.06$ Å^3 for C, N and O as $U = 0.10$ Å^3 for H atoms apart from H atoms situated at N with $U = 0.12$ Å^3 (Electron scattering factors [20]). To prevent the molecules from being distorted a refinement, where the whole molecule was kept rigid, was performed. This also improves the usually bad parameter/reflection ratio. In the case of modification I we obtained R-values of 31% (481 unique reflections with $I > 2\sigma$) for the 100 kV data and of 25% (385 unique reflections with $I > 2\sigma$). The sparse 100 kV data of modification II was not analysed quantitatively. From 300 kV data we obtained an R-value of 23% (226 unique reflections with $I > 2\sigma$).

In a second step dynamical effects must be taken into account. For dynamical refinement we used a multi-slice least squares (MSLS) procedure [21], where the centre of laue circle, crystal thickness and scaling factor were refined for each zone separately with fixed atom positions. In the case of non-centrosymmetric space groups the enantiomorph used for refinement can be chosen for each zone, additionally. After the basic parameters have been derived satisfactorily, the atom positions can be refined. The R-values (see Eq. 2) refined by MSLS usually dropped down to 6-13%.

$$R_{MSLS} = \sum \left\{ I_m^{obs} - I_m^{calc} \right\}^2 / \sum \left\{ I_m^{obs} \right\}^2$$

An advantage of this approach is that diffraction patterns, which are not suitable for kinematical procedures, can be used now. The disadvantage is that this approach needs a model and is therefore only suitable for structure refinement but not for structure solution.

The resulting structures of FAPPO, given in Fig. 9, exhibit hydrogen bonding and a herring bone like orientation but form layers which are tilted differently towards each other.

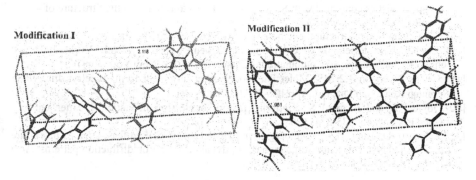

Figure 9: Refined structure for both modifications of FAPPO; distances of hydrogen bonding are indicated as dotted lines.

References

[1] Pawley, G.S.: Unit-cell refinement from powder diffraction scans. J.Appl. Cryst. 14: 357-361 (1981).

[2] David, W.I.F.; Shankland, K.; McCusker, L.B. and Baerlocher, Ch.: Structure Determination from Powder Diffraction Data, Oxford University Press, New York, 2002.

[3] Dorset, D.L.: Structural Electron Crystallography, New York: Plenum Press 1995.

[4] Williams, D.B. and Carter, C.B.: Transmission Electron Microscopy, Vol. II Diffraction, Plenum Press, New York, 1996.

[5] Bernstein, J.: Polymorphism in Molecular Crystals, Oxford University Press, Oxford, 2002.

[6] U. Kolb and G. Matveeva, Electron crystallography on polymorphic organics, Z. Krist. spezial issue: Electron crystallography, 218, 1-10 (2003).

[7] Horn, D. and Honigmann, B.: Polymorphie des Kupferphthalocyanins,181-189 (1978).

[8] Kolb, U.: Electron microscopy on pigments in Industrial applications on electron microscopy ed. Z.R. Li, Marcel Dekker, in press (2002).

[9] Brown, C.J.: Crystal structure of β-copper phthalocyanine. J. Chem. Soc. A:2488 – 2493, 1968.

[10] Vainshtein B.K. translated and edited by E. Feigl and J.A. Spink, Structure Analysis by Electron Diffraction, Pergamon Press Oxford 1964.

[11] Materials Studio™ Module Reflex, Accelrys Inc., 9685 Scranton Road, San Diego, CA 92121-3752, USA, 2002.

[12] S. Hovmöller, ELD – a program system for extracting intensities from electron diffraction patterns. Ultramicroscopy 49:147-158, 1993.

[13] G.M. Sheldrick, SHELXL, Program for crystal structure refinement. University of Göttingen, Germany 1999.

[14] Dorset, D. L.; Tivol, W.F. and Turner, J.N.: Electron crystallography at atomic resolution: ab initio structure analysis of copper perchlorophthalocyanine, Ultramicroscopy 38: 41, 1991.

[15] Gilmore, C.J.; Shankland, K. and Bricogne, G.: Application of the maximum entropy method to powder diffraction and electron crystallography. Proc. Roy. Soc. (London) A442:97–111 (1993).

[16] Voigt-Martin, I.G.; Zhang, Z.X.; Kolb, U. and Gilmore, C.J.: The Use of Maximum Entropy Statistics combined with Simulation Methods to Determine the Structure of 4-dimethylamino-3-cyanobiphenyl. Ultramicroscopy 68:43-59, 1997.

[17] CSD – Cambridge structural database, CCDC Software Ltd., Cambridge, GB.

[18] Cramer, C.J.: Essentials of Computational Chemistry, Wiley, New York, 2002 and Jensen, F.: Introduction to Computational Chemistry, Wiley, New York, 2001.

[19] Cerius 2 version 4.2 MS. Molecular modeling environment from Accelrys Inc., 9685 Scranton Road, San Diego, CA 92121-3752, USA.

[20] Doyle, P.A. and Turner, P.S.: Scattering factors for electron diffraction, J. Amer. Oil Chemists Soc. 45, 333-334, 1968.

[21] J. Jansen, D. Tang, H.W. Zandbergen and H. Schenk, MSLS, a least-squares procedure for accurate crystal structure refinement from dynamical electron diffraction patterns. Acta Cryst. A54:91-101, 1998.

ELECTRON CRYSTALLOGRAPHY IN MINERALOGY AND MATERIALS SCIENCE

D. Nihtianova, Jixue Li and U. Kolb

1 Central Laboratory of Mineralogy and Crystallography, Bulgarian Academy of Sciences, Acad. G. Bonchev Str., bl. 107, 1113 Sofia, Bulgaria: 2,3 Institute of Physical Chemistry, Johannes Gutenberg-University, Welderweg 11, 55099 Mainz, Germany

Abstract: The mineral aerinite is investigated by electron crystallography (NED and HREM). TEM results of aerinite are compared with X-ray and synchrotron powder diffraction data. Six selected area electron diffraction (SAED) patterns and two HREM images from Pb_5MoO_8 single crystals are used to solve their structure. The unit cell parameters of these crystals confirm the known powder diffraction data.

Key words: electron crystallography, aerinite, Pb_5MoO_8, SAED, HREM

1.   MINERAL DEFINITION AT THE MACROSCOPIC SCALE

Minerals are generally regarded as naturally occurring crystalline phases defined on the basis of their macroscopic physical properties [1]. As phases, or more specifically bulk phases in the "Gibbsian" or classical equilibrium thermodynamic sense, a mineral is said to be homogeneous with respect to its macroscopic physical properties and separable from other so-called phases (and external surroundings) by a physically distinct or discontinuous boundary.

Minerals have definite chemical compositions when analyzed at the macroscopic scale, but as exemplified by solid solutions, the local chemical composition may vary within fixed limits. As crystalline phases, whether chemical end members or solid solutions, minerals have internal atomic structures said to possess long-range order on average in three dimensions.

T.E. Weirich et al. (eds.), Electron Crystallography, 421–433.

The average structure is defined at the microscopic (near atomic) scale and represented by a unit cell that when repeated over an essentially infinite number of atomic distances generates microscopic crystallites, which in turn comprise macroscopic single crystals, polycrystalline aggregates, or powders.

The two most fundamental macroscopic properties of a mineral are its chemical composition and crystallographic symmetry. Together these serve as the basis for naming and classifying minerals. Other macroscopic properties (e. g., optical, electrical, magnetic, thermal, mechanical, etc.) can be considered manifestations of a mineral's average structure.

In practice, minerals are defined using so-called conventional methods to analyze their chemical compositions and determine their crystallographic properties. Chemical composition refers to the relative concentrations of different elements making up a mineral from which an average stoichiometry can be derived and expressed as a formula unit. Depending on the method used, a compositional analysis may also include a determination of oxidation states and/or atomic coordination environments. Crystallographic properties include the point group and space group symmetry of a mineral, its unit cell parameters, number of formula units per cell, average structure, etc.

A variety of instrumental methods apart from classical wet chemical methods, are available for analyzing a mineral's chemical composition. Probably the most common methods today are x-ray fluorescence analysis and electron probe microanalysis. Crystallographic properties are generally determined with x-ray diffraction (XRD) or neutron diffraction, using both powder and single crystal methods. It is common practice and generally a good idea to assess the "homogeneity" of a mineral specimen using some form of optical microscopy (e. g., reflected light or polarized transmitted light) before it is chemically analyzed or examined by powder or single crystal diffraction methods.

There are over 3,500 recognized mineral species and countless mineral varieties that have been classified as crystalline phases using the conventional methods listed above. Guidelines for naming minerals have been published by the Commission on New Minerals and Mineral Names (CNMMN) of the International Mineralogical Association (IMA).

Minerals are generally regarded as crystalline phases formed as a result of geological processes. As a (bulk) crystalline phase in the classical sense, a mineral must satisfy the conditions of long-range structural order in three dimensions, and homogeneity with respect to its macroscopic physical and chemical properties.

Presently, conventional microscopy, diffraction, and analytical methods can be used to study minerals as crystalline phases in the classical sense as long as a macroscopic specimen exists. Macroscopic refers to the quantity of material available, expressed in terms of either the size of a single crystal, mass (volume) of material as a polycrystalline aggregate or powder, number of crystallites, and so on. To put things into perspective, the unaided human eye can resolve objects as small as 10-100µm, which serves as a practical lower limit for the macroscopic scale with respect to the size of a specimen. A petrographic microscope and electron microprobe can be used to examine small grains (single crystals or crystal fragments) or small regions in a larger crystal with a diameter of a few micrometers, but generally a macroscopic quantity (statistically significant number) of grains or regions are sampled. For conventional powder XRD, using a flat-plate, or capillary specimen, at least 1 mg of crystalline material is needed to obtain measurable diffracted intensity (1 µg using a synchrotron source). For single-crystal XRD, a suitable crystal must be in the order of 100 µm (10 µm with a rotating anode or synchrotron source). Crystallites comprising a powder or single crystal specimen must consist of at least six unit cells in three dimensions to act as a coherent scattering domain. Depending on the unit cell size and scattering power of the atoms in the cell, coherent domains as small as 2 to 20 nm across can be detected by XRD if present in sufficient quantities. Coherent domains in minerals are more often in the order of 100 to 1000 nm (0.1 − 1 µm) and represent literally millions to billions of unit cells.

2. MINERAL DEFINITION USING HRTEM

High resolution transmission electron microscopy (HRTEM) offers the unique ability to observe minerals (or any solid material) directly in real space at or close to the atomic scale, i. e. the scale at which they are ultimately defined. Whereas conventional methods provide information about the average macroscopic structural and chemical properties of a mineral specimen, HRTEM can be used to describe the microstructural and microchemical properties. It combines various modes of imaging, electron diffraction, and chemical microanalysis in a manner that complements the conventional methods of optical microscopy, x-ray and neutron diffraction, and instrumental chemical analysis. With modern HRTEM instruments, lattice or structure images of very small crystals (crystallites) or very small regions in larger crystals can be obtained with 0.2 − 0.3 nm (2 − 3 Å) resolution (less than 0.2 nm with dedicated intermediate − or high-voltage HRTEM instruments). Diffraction patterns (SAED, CBED) and chemical analyses (EDS, EELS) can be taken from areas of a few micrometers to a few nanometers across (and less than 1 nm on dedicated STEM instruments).

Basically two types of HRTEM studies involving minerals have been reported in literature. One type may be considered as "microcrystallographic" in nature, dealing preferably with mineral properties, another as "micropetrographic", focussing on mineral behavior. HRTEM and associated TEM methods have been used to confirm and refine existing structure models of known minerals, to propose new structure models, to define new minerals, and even to discredit minerals. HRTEM information becomes even more useful when combined with data from other sources. Expect to see more HRTEM studies combined with Rietveld powder x-ray diffraction studies and synchrotron powder diffraction studies involving small volumes of sample or small single crystals ($1 - 10$ μm across). Minerals are often fine-grained, exist in small quantities, and are insufficiently ordered. The two nanocrystalline examples (aerinite $(Ca_{5.1}Na_{0.5})(Fe^{3+}AlFe^{2+}_{1.7}Mg_{0.3})(Al_{5.1}Mg_{0.7})[Si_{12}O_{36}(OH)_{12}H]$ $[(CO_3)_{1.2}(H_2O)_{12}]$ and Pb_5MoO_8) which are discussed below represent ideal samples for HRTEM studies [1].

3. CELL PARAMETER DETERMINATION ON AERINITE

Aerinite, a blue fibrous aluminosilicate mineral associated with the alteration of ophitic rocks, is found in the southern Pyrenees or Marocco and was commonly used as a blue pigment in most catalanic roman paintings between the XI – XV centuries. It was first described in 1876 and added to the mineralogical tables in 1898. Structure solution of aerinite is complicated due to its extremely small crystal size which makes it impossible to isolate single crystals for structural studies. Furthermore, aerinite is always found in mixtures with some other minerals and obtaining pure specimens is extremely difficult. Early attempts to analyse the structure lead to a proposal of a monoclinic unit cell (a = 14.690, b = 16.872, c = 5.170, β = 94.45°, Z=2) with $(Ca_{4.0}Na_{0.2})(Fe^{3+}_{1.7}Fe^{2+}_{0.9}Mg_{1.5}Al_{6.1})_{\Sigma=10}$ $(Si_{1.6}P_{0.2}Al_{0.3})_{\Sigma=12}$ $O_{36}(OH)_{12}12 H_2O + CO_2]$ for aerinite from Saint-Pandelon (France) [2]. The sample used for the present study is found in Camporrells – Estopanyà area (Huesca – Lleida, Spain) and consists of long bright blue fibres of a diameter of $\sim 0.2 - 0.4$ μm as shown in Fig. 1. For preparation fibres were preselected by a light microscope, crushed in a mortar and dispersed in water via ultrasonic treatment.

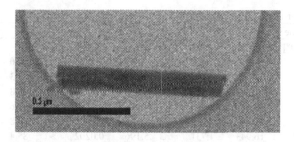

Figure 1. TEM image of an aerinite fibre on quantifoil support (scale bar 0.5 μm).

Based on lab x-ray powder diffraction Rius et.al.determined cell parameters as a = b = 16.872(1) Å, c = 5.2256(5) Å, α=β=90°, γ=120° with the most likely space group of P3c1 (No. 158) [3]. In order to investigate the structure of aerinite via TEM we performed electron diffraction using a FEI Tecnai F30 ST at 300 kV with a beam diameter of approx. 20 nm in nanodiffraction mode and 1k x 1k CCD camera. To allow the highest possible movement we used a double tilt–rotation holder and performed a tilt around two axes of the crystal [4]. For strong scatterers like inorganics we can observe often the high order Laue zone (HOLZ) reflections (see Fig 2. right hand side) and use those as a guide to the next zone. Therefore, it is as well convenient for inorganic samples to use a double-tilt holder. Starting from the initial zone [100], given in Fig. 2 (left hand side), diffraction patterns were taken through tilts around the axes b* and c*; distances and angles were calculated by ELD and are given in Tables 1, 2 (mean values: a= 16.67 Å; b = 16.92 Å; c = 5.22 Å, α=90.1°; β=92.0°, γ=119.7°)[5].

Table 1. Data for tilt series of aerinite: Tilt about c*

tilt	in-zone									
angle[°]	angle[°]	c[Å]	b'[Å]	h	b[Å]	a*[Å⁻¹]	k	[°]	a[Å]	zone
33.0	89.9	5.205	4.060	4	17.008	0.0522	-3	59.6	16.565	[430]
30.0	89.8	5.209	2.341	7	16.910	0.0490	-5	59.9	16.820	[750]
28.0	89.8	5.169	5.488	3	16.850	0.0468	-2	60.2	16.414	[320]
23.0	89.9	5.221	3.345	5	16.861	0.0415	-3	60.1	16.674	[530]
17.0	89.9	5.220	8.404	2	16.812	0.0353	-1	60.4	16.358	[210]
10.0	89.9	5.232	3.346	5	16.814	0.0280	-2	60.4	16.495	[520]
6.0	89.9	5.231	5.507	3	16.761	0.0237	-1	60.8	16.216	[310]
0.0	89.9	5.223	4.051	4	16.833	0.0172	-1	60.2	16.757	[410]
-14.0	89.9	5.222	14.612	0	16.824	0.0000	1			[100]
-28.0	89.9	5.229	4.064	3	16.868	0.0228	1	60.7	16.893	[3-10]
-33.0	90.0	5.223	5.533	2	16.772	0.0340	1	60.0	16.971	[2-10]
-38.0	90.0	5.215	3.350	3	16.982	0.0463	2	61.0	16.613	[3-20]
	89.9	**5.217**			**16.858**			**60.3**	**16.616**	

Table 2. Data for tilt series of aerinite: tilt about b*

tilt angle[°]	in-zone angle[°]	b[Å]	c'[Å]	h	c[Å]	a*[Å⁻¹]	l	[°]	a[Å]	zone
25.0	85.5	16.951	2.369	2	5.133	0.0895	-3	89.0	16.771	[203]
22.0	96.4	17.029	1.615	3	5.126	0.0778	-4	88.3	17.141	[304]
17.0	90.2	16.884	4.995	1	5.172	0.0585	-1	88.8	17.091	[101]
12.0	96.7	17.070	1.710	3	5.174	0.0408	-2	87.6	16.330	[302]
9.0	95.2	16.989	2.583	2	5.184	0.0305	-1	86.8	16.428	[201]
6.0	93.5	16.943	1.731	3	5.196	0.0202	-1	85.4	16.487	[301]
0.0	90.0	16.896	2.628	1	5.259	0.0000	1			[100]
-9.0	95.2	16.946	2.586	2	5.246	0.0304	1	87.2	16.453	[20-1]
-17.0	90.1	16.869	4.993	1	5.279	0.0586	1	88.8	17.083	[10-1]
-25.0	94.8	17.155	2.400	2	5.366	0.0884	3	89.5	16.986	[20-3]
-32.0	90.3	17.097	4.500	1	5.425	0.1178	2	89.1	16.991	[10-2]
		16.985			**5.233**			**87.9**	**16.777**	

Calculation given for tilt about b^*: $\gamma = \arcsin\left(\dfrac{|\sin\Theta|}{c'a^*}\right)$ with

$$c' = c * \cos(90 - \gamma - \Theta) \text{ and } a^* = \sqrt{\left(\frac{\sin\Theta}{c'}\right)^2 + \left(\frac{1}{c_0'} - \frac{\cos(\Theta)}{c'}\right)^2} \text{ where } \Theta$$

is the tilt angle.

Due to the high flexibility of the rotation-double tilt sample holder it was possible to collect 24 zones in a reasonable orientation. In addition the use of a small C2 aperture of 10 nm diameter delivered an intensity low enough to avoid significant beam damage so that all diffraction patterns could be collected from one crystal. During the tilt around c* the in-zone angle stayed 90° indicating that β must be 90° as well whereas the in-zone angle through the b* tilt changes leaving γ different from 90° (compare Fig. 2 and 3). At this point we do not need to know in advance which reflections to use for the correct definition of the zone axes, which can be quite problematic. The in-zone angle can be calculated in later for correct indexing of the reflections.

The next step was to identify the zones correctly. For this task one can use "ghost reflections". Due to the thin specimen the reflections are no sharp spots but elongated in the direction of the beam. This is the reason that we find some weak reflections, which do not belong to the actual but to adjacent zones. For example, zone [410] (Fig.2, right hand side) exhibits 3 weak reflections between the allowed ones indicating an index of h=4. More striking is the comparison of the zones [20-1] and [10-2] given in Fig. 3. Due to the ratio of the cell axes a/c (long/short) the zone [20-1] passes adjacent reflections closer than zone [10-2]. Please note the difference between an extinction in zone [100], where every odd line in c* direction is missing and no ghost reflections are found (first reflection is indexed as (002)), and zone

[20-1] where we find one "ghost line" (first strong reflection is indexed as (102)).

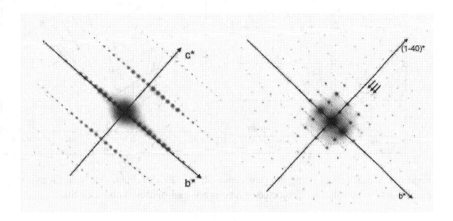

Figure 2. Initial zone [100] of aerinite (left hand side) with 1/b* = 14.612 Å / α= 90° and 1/c* = 2.611 Å * 2 = 5.222 Å where every odd line in c* direction is extincted; Zone [410] at tilt about c* of 14° (right hand side) exhibits "ghost reflections" indicated by black arrows.

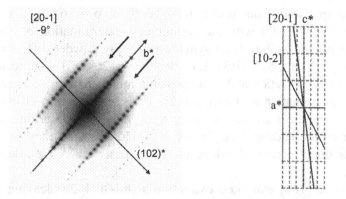

Figure 3. Zone [20-1] of aerinite with an in-zone angle of 87° (left hand side) the "ghost reflections" are indicated by blach arrows; Zone [10-2] with an in-zone angle of 73° (middle) and a scheme of the reciprocal space view down titl axis b (right hand side).

To test the indexing, one should always produce a Vainshtein diagram by plotting the changing reciprocal distance along the axis defined by the tilt angle [6]. It will lead to a regular pattern of reflections and the projected unit will be visible as given in Fig. 4.

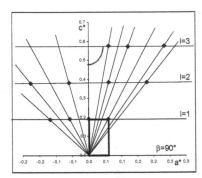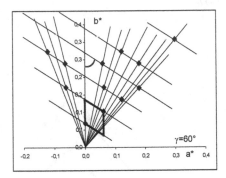

Figure 4. Vainshtein drawing for view down tilt axis b* (left hand side) and tilt axis c* (right hand side) for aerinite; the projected unit cells are highlighted with black lines.

The Vainshtein diagram is the geometrical basis of the calculations used to produce Tables 1, 2 and delivers the missing values of a and γ as well. Please note the asymmetric indices of zones for the tilt about c* (γ ≠ 90°). As an example we can compare the zones [410]/[3-10], [310]/[2-10] and [520]/[3-20] which have comparable tilt angle resp. d-values and show the same pattern.

The comparison of our results (a = 16.70 Å, b = 16.92(1) Å, c = 5.22 Å, α=β=90°, γ=119.7°) with the former cell determination of Rius and a recent one refined by him from synchrotron x-ray powder diffraction (a = b = 16.8820(9) Å, c = 5.2251(3) Å, α=β=90°, γ=120°) [7] is quite satisfying. Due to the spike function and some possible misalignment the uncertainty in tilt angle determination is at least ± 0.5°. Therefore, electron diffraction data will never be as accurate as x-ray powder data. Assuming a correct alignment of the eucentric height we still find an inaccuracy of minimum ±0.2 Å for direct measured values and of at least ±0.5 Å for calculated cell distances.

Only zone [100] exhibited extinction's (h-h0l: l=2n) leaving over still four possible space groups (P3c1, P$\bar{3}$c1, P6₃cm, P6₃mcm). Unfortunately, due to the high beam sensitivity it was not possible to determine the existence of a centre of symmetry or a 6₃ screw axis by convergent beam electron diffraction. Recently, Rius et al. [7] have determined the crystal structure in space group P3c1 (No. 158) by applying the direct methods modulus sum function to synchrotron powder diffraction data. As shown in Fig. 5, the framework of aerinite is formed by units of three pyroxene chains ($Si_2O_6^{4-}$) placed around one three-fold rotation axis, laterally linked through [001] chains of edge-sharing AlO_6 octahedra to other similar units. In each unit, the apical O atoms of the three pyroxene chains point inwards thus coordinating Fe, Al and Mg atoms, and giving rise to columns of face-sharing octahedra. Two neighboring units, i. e., two units connected by Al chains, are not symmetry-related: called units A and B. In fact, both are mutually shifted by 0.93 Å along c. The large [001]-channel formed by the pyroxene and AlO_6 chains are filled with Ca, H_2O and CO_3 groups. The carbonate group is on the third three-fold axis. The framework of aerinite also contains zigzag chains of edge-sharing AlO_6 octahedra that join the cylindrical units A and B. Each AlO_6 octahedron consists of four hydroxyl groups (OHA, OHA', OHB, OHB') and of the two unshared basal O atoms of the next pyroxene chains (02A and 02B). Ca (also occupied by a small amount of Na^+ and some vacancies) is at the centre of a distorted O octahedron. The atomic arrangement formed by one zigzag chain of AlO_6 octahedra and the adjacent Ca coordination polyhedra may be described as forming a four-row wide slab of a brucite-like layer. The two inner octahedra are filled with Al (and Mg) atoms and the two outer with Ca (and Na). The internal O atoms of the brucite-like layer are hydroxyl groups, the intermediate are unshared basal O atoms of the neighboring pyroxene chains, while the external ones are water molecules forming relatively strong H-bridges with the partially disordered CO_3 groups. Presumably, the hydroxyl groups in the brucite-like layer also form H-bridges with the apical O atoms of the neighboring pyroxene chains to compensate for the defect of charge caused by the presence of divalent cations in the face-sharing octahedra.

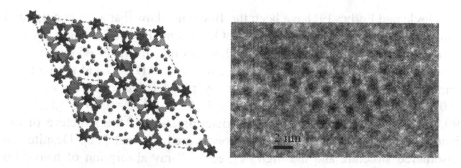

Figure 5. Crystal structure of aerinite view down c (left hand side) and corresponding HRTEM [001] image (right hand side).

In addition, from ultrathin slices of the embedded material HRTEM images were obtained in different crystallographic orientations ([001] (see Fig.5), [101], [110], [-111], [102], [120], [201], [122], [203], [301], [1-19], [-124]). Using the structure model of Rius et al. the comparison of kinematically electron diffraction patterns have been compared with experimental data (Cerius 4.2 [8]) as shown in Fig. 6. The basic course of intensity is distinguishable but clearly the patterns deviate from each other. For quantitative analysis significantly thinner crystals are needed.

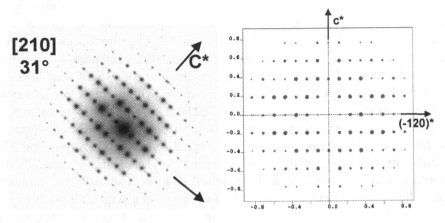

Figure 6. Zone [210] of aerinite from tilt about c* (left hand side and the kinematically calculated data for comparison.

4.   STRUCTURE INVESTIGATION OF PB₅MoO₈ SINGLE CRYSTALS BY SELECTED AREA ELECTRON DIFFRACTION.

Doyle and Forbes [9] have been the first who show that a new compound Pb_5MoO_8 is formed in the $PbO - MoO_3$ system. Pb_5MoO_8 single crystals were obtained from the melt under slow cooling using Pt-wire and were investigated by single crystal structure analysis (CAD – 4) and the unit cell parameters were confirmed through x-ray powder data by Mentzen et al. [10] (PDF 37-1086: $a = 15.316(4)$ Å, $b = 11.827(1)$ Å, $c = 11.6387(3)$ Å, $\beta = 90.200°$, $Z = 8$, SG P2₁/c. A partial determination of the structure of the isomorphous compound Pb_5SO_8 was performed by Sahl [11]. Despite the disordered structure and the high degree of X-ray absorption of heavy Pb atoms he succeeded to determine the Pb atomic positions. We try to solve this problem by the use of transmission electron microscopy (EM 420 – T/EDAX, Philips at 120 kV) because the electron diffraction allows us to distinguish more precisely the superstructural details. Six ([100], [010], [001], [011], [101], [110]) selected area electron diffraction (SAED) patterns and two ([001], [021]) HREM images from several particles were obtained and analysed. The diffraction patterns have been photographed on AGFA EM films (6.5 x 9 cm) with exposure times in the interval between 0.1 s and 99 s. The negatives were digitized by a Bosch video-camera of high linearity, connected to the specialized image processing system of Kontron – electronics group. The system is linked to a PC/486 computer, where the calculations have been performed.

Since the diffraction patterns consist of strong α - PbO reflections and essentially weaker superstructure reflections, which contain the main information of the Pb_5MoO_8 superstructure, the dynamical range of the electron beam intensities incident on the photoplate is extremely wide and overpasses the linear region of the sensitivity of the photographic material. To overcome this problem a multi-exposure technique was used. The diffraction patterns were photographed in a broad interval of exposure times (0.1 s, 0.2 s, 0.5 s, 1.0 s, 2.0 s, 5.0 s, 10.0 s, 20.0 s, 50.0 s, 99.0 s), so that all of the reflections lie in the linear region of a photoplate of a certain exposure time. Collecting the intensity data at all exposure times, a map of the relative intensities of the reflections is obtained for each EDP. After a detailed analysis of the measured diffraction intensities a principle structure model of the positions of the heavy Pb and Mo atoms of Pb_5MoO_8 was constructed and tested by a comparison between the theoretical intensities, calculated by multislice dynamic calculations [12], and the observed ones in all six measured EDP. HREM image simulations were performed in [001] and [021] zone axes and compared to the experimental ones.

The SAED patterns consist of an intense base set of α - PbO subcell reflections and weak superstructure reflections h'/4, k'/3, l'/2 referring to the α - PbO cell due to a modulation of the structure. To determine the Pb and Mo positions in Pb_5MoO_8 on the base of the relative intensities of the SAED reflections a multislice dynamic calculation [12] was performed. The calculations were performed separately for each of the 6th experimental microdiffraction patterns. The thickness of the specimen was determined independently for each of the patterns. The performed procedure is close to the published idea by Bing – Dong et al. [13].

The most effective starting model happens to be the one presented by Sahl [11] for Pb_5SO_8, the modulation of which is changed from 2 : 3 : 2 to 4 : 3 : 2 by doubling the *a*-parameter of the cell. The procedure of comparison was first applied to the zones [001], [010] and [011] where the (100) reflection appears because of the quadrupole modulation of structure along the *a*-direction. The best similarity for [001], [010], [011] zones is obtained when Mo atoms replaces 2/3 of Pb atoms in one of four sectional layers perpendicular to the *a*-axis and forming the fourfold superstructure along the *a*-direction. The 2/3 Mo substitution in the Mo – Pb layer is responsible for the threefold superstucture along the *b*-direction. The presented data in this work is not enough to explain the doubling of the *c*-axis. It only confirms the existence of such twofold modulation along *c*. Another possibility for the Mo-substitution of Pb atoms is that Mo atoms replace 1/3 of Pb atoms in one of two sectional layers perpendicular to the *a*-axis forming a twofold superstructure along the *a*-direction. In this case the observed fourfold superstructure must be due to some disturbances of one of two Mo – Pb layers or Pb – Pb layers. The analysis of the structure types give essentially less intensities for (2h + 1, 0, 0) reflections compared to (2h, 0, 0) reflections than the measured in all directions of [001], [010], [011].

Anyway, due to the limited precision of the EDP – intensity measurement, especially for the (100) reflection, which is too close to the central beam - this argument is not very strong and cannot reject the 1/3 substitution at all. To confirm the hypotheses of the proposed 2/3 substitution HREM observation in [001] and [021] zones were performed in parallel with multislice calculations, based on the 2/3 substitution. These projections are used because the fourfold modulation due to the Mo atoms is well expressed in them. The observed images are enhanced by a Fourier filter with Bragg mask to eliminate the noise frequencies.

The determination of *y* and *z* coordinates of Pb and Mo atoms in the layer distinguished by its Mo – content is made by a fit of the [100] zone intensities, which give the best information about the [100] Mo – Pb layer projection. Two other zones [101] and [110] were also used to verify the atomic positions.

The dynamic calculations include all beams with interplanar distances d_{hkl} larger than 0.75 Å at 120 kV acceleration voltage and thickness between 100 Å and 300 Å for the different zones. The structure factors have been calculated on the basis of the relativistic Hartree – Fock electron scattering factors [14]. The thermal diffuse scattering is calculated with the Debye temperature of α-PbO: 481 K [15] at 293 K with mean-square vibrational amplitude $<U_s^2> = 0.0013$ Å^2 following the techniques of Radi [16]. The inelastic scattering due to single-electron excitation (SEE) is introduced on the base of real space SEE atomic absorption potentials [17]. All calculations are carried out in zero order Laue zone approximation (ZOLZ).

HREM image simulations are performed using the optical parameters of EM 420-T/EDAX Philips microscope with $C_S = 1.0$ mm at Scherzer defocus = -57.6 nm at a thickness of 23.3 Å along [001] direction and 105.4 Å along [021] direction.

The structural peculiarities of the whole class of isomorphous compounds Pb_5AO_8 (A = S, Se, Cr) are up to now investigated only by single crystal and powder X-ray methods. Thus two types of superstructures obtained are: 4 : 3 : 2 from single crystal diffractometry and 2 : 3 : 2 from powder diffractometry. This contradiction between the two methods can be solved by using the SAED method as it is done here with Pb_5MoO_8. SAED investigations undoubtedly confirmed that 4 : 3 : 2 superstructure of deformed α-PbO is more probable to this class of compounds [18].

References:

[1]. Fred M. Allen, Mineral definition by HRTEM: Problems and opportunities, ch. 8, 289-333 in: Reviews in Mineralogy, Vol. **27**, 1992, Minerals and Reactions at the Atomic Scale: Transmission Electron Microscopy, Ed. P. R. Buseck.

[2]. B. Azambre and P. Monchoux, Précisions minéralogiques sur l'aérinite: nouvelle occurrence à Saint-Pandelon (Landes France), Bull. Minéral., 111, 39-47 (1988).

[3]. J. Rius, F. Plana, I. Queralt, D. Louër. Preliminary structure type determination of the fibrous aluminosilicate 'aerinite' from powder x-ray diffraction data. Anales de Quimica Int. Ed., **94**, 101-106, 1998.

[4]. U. Kolb and G. Matveeva, Electron Crystallography on polymorphic systems, Z. Krist., **218**, 259-268, 2003.

[5]. S. Hovmöller, ELD – a program system for extracting intensities from electron diffraction patterns, Ultramicroscopy, **49**, 147-158, 1993.

[6]. Vainshtein B.K. translated and edited by E. Feigl and J.A. Spink, Structure Analysis by Electron Diffraction, Pergamon Press Oxford 1964.

[7]. J. Rius, E. Elkaim and X. Torrelles, Structure determination of the blue mineral pigment aerinite from synchrotron powder diffraction data: The solution of an old riddle, European Journal of Mineralogy, Vol. **16**, No 1, 127-134, 2004.

[8]. Cerius 2 version 4.2 MS. Molecular modeling environment from Accelrys Inc., 9685 Scranton Road, San Diego, CA 92121-3752, USA.

[9]. W. P. Doyle, F. Forbs, Determination by Diffuse Reflectance of the Stoichiometry of Solid Products in Solid – Solid Additive Reactions, J. Inorg. Nucl. Chem., **27**, 1271-1280, 1965.

[10]. B. F. Mentzen, A. Latrach, J. Bouix, Mise en évidence d'une surstructure dans le pentaplomb (II) octaoxomolybdate (VI) Pb_5MoO_8 et la solution solide $Pb_5S_xMo_{1-x}O_8$, CR. Acad. Sci. Paris, **297**, 887-889, 1983.

[11]. K. Sahl, Zur Kristallstruktur von 4PbO: $PbSO_4$, Z. Kristallogr., **141**, 145-150, 1975.

[12]. J. M. Cowley, Diffraction Physics, Amsterdam, North Holland, 1975.

[13]. S. Bing-Dong, F. Hai-Fu, L. Fang-Hua, Correction for the Dynamical Electron Diffraction Effect in Crystal Structure Analysis, Acta Crystallogr., **A49**, 877-880, 1993.

[14]. P. A. Doyle, P. S. Turner, Relativistic Hartree-Fock X-ray and Electron Scattering Factors, Acta Cryst., **A24**, 390-397, 1968.

[15]. Gmelins Handbuch der Anorganischen Chemie. Verlag Chemie, Weinheim/Bergstr., 1969.

[16]. G. Radi, Complex Lattice Potentials in Electron Diffraction Calculated for a number of Crystals, Acta Cryst., **A26**, 41-56, 1970.

[17]. W. Coene, D. Van Dyck, Inelastic Scattering of High-Energy Electrons in Real Space, Ultramicroscopy, **33**, 261-267, 1990.

[18]. D. D. Nihtianova, V. T. Ivanov and V. I. Yamakov, Structure investigation of Pb_5MoO_8 single crystals by selected area electron diffraction, Zeitschrift für Kristallographie, **212**, 191-196, 1997.

References

STRUCTURES OF ZEOLITES AND MESOPOROUS CRYSTALS DETERMINED BY ELECTRON DIFFRACTION AND HIGH-RESOLUTION ELECTRON MICROSCOPY

Osamu Terasaki
Structural Chemistry, Arrhenius Laboratory, Stockholm University, Stockholm, Sweden n

Abstract: The present article is a brief introduction to structure determination of meso- and microporous materials (zeolites) by electron crystallography. The used approaches are illustrated by several examples.

Key words: microporous materials, zeolites, electron microscopy, structure determination

Porous materials are classified into three different categories based on the size of pores r_{pore}, that is, microporous for $r_{pore} < 20$ Å, mesoporous for 20 Å$< r_{pore} < 500$ Å, and macroporous for $r_{pore} > 500$ Å. Here I will discuss micro- and meso-porous materials. Microporous crystals (hereafter called zeolites) with more than 150 different framework-type structures have been reported.

Zeolites are crystalline alumino-silicate materials with $M_{x/m}[Al_xSi_{1-x}O_2]$ $\times n$ H_2O, where M is a cation of valence m. Their frameworks are built from corner-shared TO_4-tetrahedra (T stands for Si, Al, P, Ga, Ge, etc.) to produce channels or cavities (spaces) of molecular dimensions. Based on the characteristic structures, zeolites have attracted much attention not only for catalytic applications but also as containers in which materials are confined. It is becoming more and more important to understand the fine structures of zeolites in order to synthesize high quality or novel zeolites not only for producing better catalysts but also for better containers of confined materials.

T.E. Weirich et al. (eds.), Electron Crystallography, 435–442.

Zeolite Framework Type: Channel & Cage

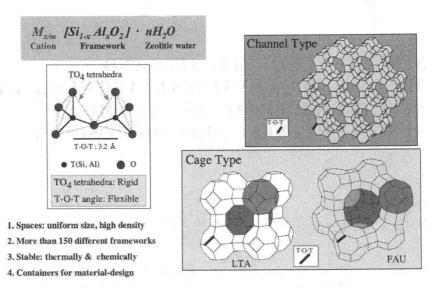

$$M_{x/m} \; [Si_{1-x} Al_x O_2] \cdot nH_2O$$

Cation   Framework   Zeolitic water

TO₄ tetrahedra

T-O-T : 3.2 Å

● T(Si, Al)   ⬤ O

TO₄ tetrahedra: Rigid
T-O-T angle: Flexible

1. Spaces: uniform size, high density
2. More than 150 different frameworks
3. Stable: thermally & chemically
4. Containers for material-design

Channel Type

T-O-T

Cage Type

LTA   T-O-T   FAU

Structural characterisation of zeolites by HRTEM
-through arrangement of building blocks (not atom)-

Structural Units: **D4R, D6R, TO₆, etc.**

— *Sheet:*

ABC-6(SOD, ERI, OFF,..),

Faujasite sheet (FAU/EMT)

— *Rod:*

Pentasil(MFI, MEL),

CAN+D6R(LTL),

TO₆+3R(ETS-10)

Defects

Block channels

LTL, ABC-6

Make different channels

EMT

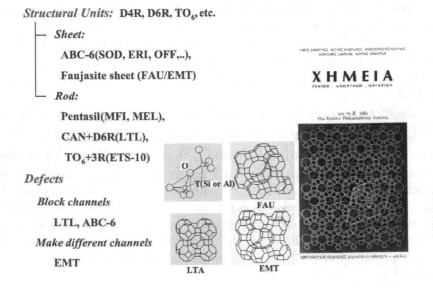

X H M E I A

O   T(Si or Al)

FAU

LTA   EMT

Zeolites form many different families in which common structure building units for the frameworks are found. They have a tendency to form intergrowths between different structures within the same families.

Many structures of zeolites can be described by certain structural units called building units, such as double 4 or 6 membered rings (D4R or D6R), structural sheets and rods. The main requirement for structure analysis of zeolites at present is not the refinement of electron charge distribution in the unit cell but the determination of the framework type structures, manner of arrangements of secondary building units or characterizing their lattice defects.

Recently, new ordered mesoporous silicas have also been synthesized by using self-organization of amphiphilic molecules, surfactants and polymers either in acidic or basic condition. A schematic phase diagram of water-surfactant is shown in the figure.

A Silica-network is formed at the boundary between water and surfactant and later surfactant will be removed by calcination.

Transmission electron microscopy (TEM) can provide detailed structure of zeolites. I use the word "characterize or characterization" for structural study on a unit cell scale, such as various kind of structural defects and basic structural units, and "determine or determination" for obtaining atomic coordinates within the unit cell for all the atoms of a crystal. A simple text or reviews for structural characterization of porous materials can be found in a book or review articles [1-6]. Now, we are in a new era, that is, we can determine new structures of micro- and mesoporous materials only by electron microscopy(EM), an area called electron crystallography (EC) [7-11].

TEM has several advantages for structural studies of porous materials over X-ray diffraction (XRD) and can overcome the following problems;

(i)   Most porous materials synthesized are microcrystalline with particle dimensions of 1 micron or less, and are too small for structural determination by single-crystal XRD experiment.

(ii)   Microporous and mesoporous crystals may often contain different kind of faults and tend to form intergrowths. Therefore, the first task is to characterize structural units followed by analysis of the defects or intergrowths and then to derive/speculate ideal structure(s) from the observations.

Self-organisation of surfactant molecules in a water-surfactant system

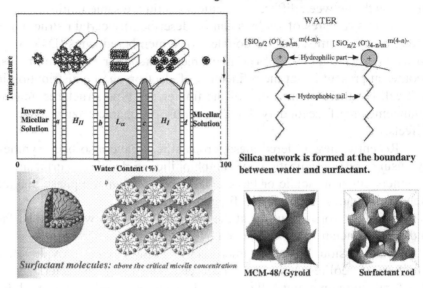

Silica network is formed at the boundary between water and surfactant.

Surfactant molecules: *above the critical micelle concentration*

MCM-48/ Gyroid   Surfactant rod

Length scale of our interest from structural view points

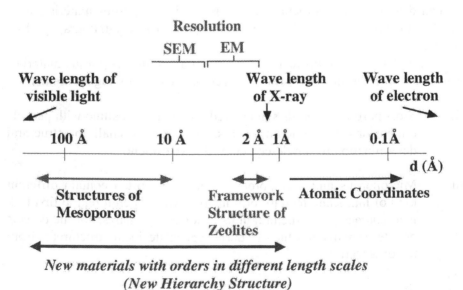

New materials with orders in different length scales (New Hierarchy Structure)

(iii)   In case of microporous crystals, the difficulties in solving crystal structures from powder XRD are that the three dimensional structural information is projected to one dimension (d-spacing). Without having a reasonable initial structure model, it is hard to decompose the heavily overlapped powder XRD profile into individual reflections and to refine atomic coordinates.

(iv) In case of mesoporous materials, it is hard to synthesize a single phase. Even for a single phase material it is difficult to determine the crystal class, let alone determine the structure by powder XRD experiment. This is because only a few reflections with large broadening are observed at the small scattering angles in powder XRD pattern. Mesoporous materials show local structural variations caused by local fluctuations in synthesis conditions.

References

1. O. Terasaki. Molecular Sieves, vol. 2, Heidelberg; Springer-Verlag,1999, pp 71-112.
2. O. Terasaki. J. Electron Microscopy 43: 337-346, 1994.
3. O. Terasaki, T. Ohsuna, N. Ohnishi, K. Hiraga. Current Opinion in Solid State & Materials Science 2: 94-100, 1997.
4. O. Terasaki, T. Ohsuna. Studies in Surface Science and Catalysis 135 Elsevier, (2001), eds. by A. Galarneau, F. Di Renzo, F. Fajula and J. Vedrine, pp. 61-71.
5. J.M. Thomas, O. Terasaki, P.L. Gai, W. Zhou, J. Gonzalez-Calbet, Account of Chemical Research 34: 583-594, 2001.
6. O. Terasaki, T. Ohsuna, Handbook of Zeolite Science and Technology, Marcel Dekker, 2003, pp 291-315.
7. P. Wagner, O. Terasaki, A. Ritsch, SI. Zones, ME. Davis, K. Hiraga. J Phys Chem B103: 8245-8250, 1999.
8. A. Carlsson, M. Kaneda, Y. Sakamoto, O. Terasaki, R. Ryoo, SH. Joo. J Electron Microscopy 48: 795-798, 1999.
9. Y. Sakamoto, M. Kaneda, O. Terasaki, DY. Zhao, JM. Kim, G. Stucky, HJ. Shin, R. Ryoo, Nature 408: 449-453, 2000.
10. T. Ohsuna, Z. Liu, O. Terasaki, K. Hiraga, M A. Camblor. J. Phys. Chem. B106: 5673-5678, 2002.
11. S. Che, Z. Liu, T. Ohsuna, K. Sakamoto, O. Terasaki, T. Tatsumi, Nature **429**(2004), 281-284.

Case study 1: Fine structures of zeolites, intergrowths and surface
structures.

Schematic drawings of ABC-6 Frameworks

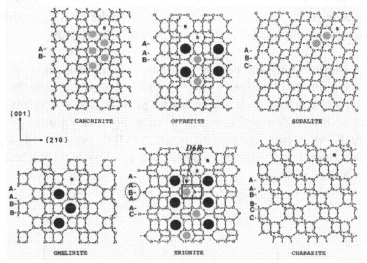

Structures are described by the stacking sequences of the sheet along the unique axis like AABAAC.
Successive stack of the same type forms D6R and 8-rings(next to D6R at the same stacking level).

FAU & EMT, Basic structural unit and rule for connectivity

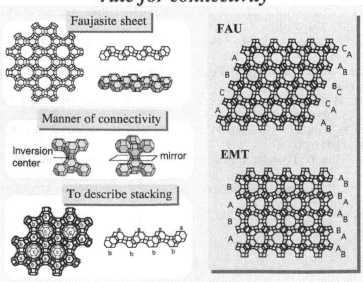

Case study 2: Structural solution of zeolite from electron diffraction data, with a help of (a) Direct method, and (b) Patterson Map

Using Patterson map-Enhancement of atomic positions-

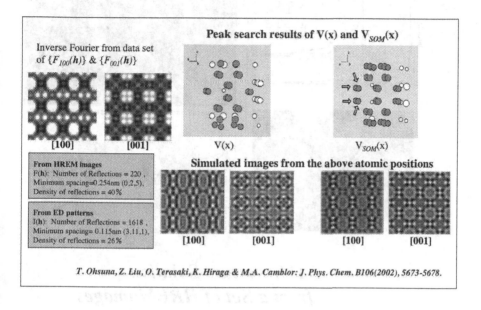

Peak search results of V(x) and $V_{SOM}(x)$

Inverse Fourier from data set of $\{F_{100}(h)\}$ & $\{F_{001}(h)\}$

[100]   [001]   V(x) $V_{SOM}(x)$

From HREM images
F(h): Number of Reflections = 220 ,
Minimum spacing=0.254nm (0,2,5),
Density of reflections = 40%

From ED patterns
I(h): Number of Reflections = 1618 ,
Minimum spacing= 0.115nm (3,11,1),
Density of reflections = 26%

Simulated images from the above atomic positions

[100]   [001]   [100]   [001]

T. Ohsuna, Z. Liu, O. Terasaki, K. Hiraga & M.A. Camblor: J. Phys. Chem. B106(2002), 5673-5678.

Case study 3: Three dimensional structural solution of mesoporous crystals.

Silica mesoporous material is crystalline !

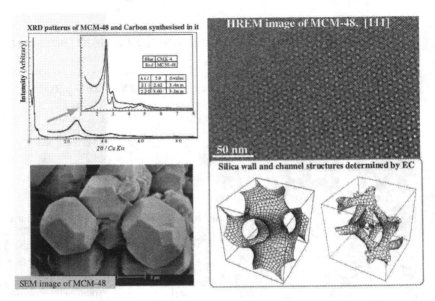

3-D Structure Solution from a Set of HREM images

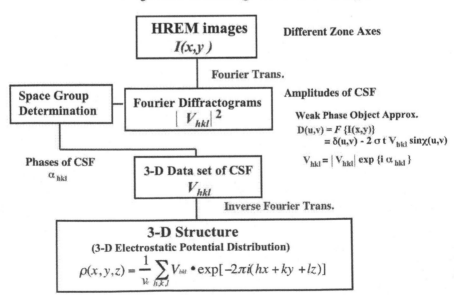

HRTEM INVESTIGATION OF NANOCRYSTALLINE MATERIALS

Andjelka M. Tonejc

Department of Physics, Faculty of Science, Bijenička 32, Zagreb, Croatia

Abstract: The microstructural and structural investigation performed on nanocrystalline (NC) materials will be reported using TEM, SAED, EDS, HRTEM and XRD measurements. The aim of this investigation was to characterize ZrO_2 and ZrO_2 -Y_2O_3 NC materials prepared by ball-milling. To elucidate the process of aloying of ZrO_2 -Y_2O_3 NC material the HRTEM image processing was applied which gave a strong support to the methods mentioned. The methods gave the evidence into a particular process on the atomic level.

Key words: nanocrystalline materials, ZrO_2, ZrO_2 -Y_2O_3, HRTEM, IP, Fourier filtering

1.   INTRODUCTION

Nanocrystalline (NC) materials exhibit many physical and chemical properties that are found to be fundamentally different and superior to those of conventional   coarse-grained crystalline or amorphous materials. For such properties, the deviation from conventional behaviour is a sensitive function of the average grain size and its distribution [1-3]. The understanding of physical properties of such materials is not possible without the detailed knowledge of their structure properties: nanocrystalline sizes, the shape of crystallites, the structure of grain boundaries, internal strains, influence of defects, appearance of phases. This information can be obtained mostly only by using X-ray and electron diffraction (XRD and ED) and high   resolution electron microscopy (HRTEM) together with specialised computing methods which permit to perform the investigation at the cutting edge. In the structural investigation of NC materials, the experimental results show the grain   size dependence   of some structure

T.E. Weirich et al. (eds.), Electron Crystallography, 443–453.

parameters, such as   lattice parameters, microstrains and Debye-Waller parameter B [4,5,6,7].

In this article we report on the initial stage of the mechanical alloying process in ZrO_2-10mol%Y_2O_3 nanocrystalline powders using HRTEM image processing performed at the Department of Physics, Faculty of Science, University of Zagreb.

In pure zirconia ZrO_2, having monoclinic structure at room temperatures, the high temperature cubic phase is stable above 2370^0C and cannot be quenched to room temperature [8]. By alloying zirconia with other oxides such as MnO, NiO, Cr_2O_3, Fe_2O_3, Y_2O_3 or Ce_2O_3 the stabilisation of the cubic phase   at room temperature   is possible. However, a very high temperature (over 1000^0C) of calcination or sintering is required for preparation of solid solutions of zirconia with these oxides [9,10,11]. Y. L. Chen and D. Z. Yang showed [12] that mechanical alloying (MA) at room temperature could also be used for alloying ceramic materials (ZrO_2-CeO_2 system), and we reported [13] that it is possible to synthesize ZrO_2-10mol.% of Y_2O_3 cubic solid solutions as the end product although in first stages of mechanical alloying the tetragonal phase appeared.

2.   EXPERIMENTAL TECHNIQUES

Powders of zirconia and yttria were alloyed in a Fritsch planetary micro-ball mill Pulverisette 7, as reported previously [13,14]. The resulting specimens were examined by XRD, TEM, SAED and HRTEM with EDS.

TEM and HRTEM investigations were carried out by using JEOL JEM 2010 200 kV microscope, Cs=0.5 mm, point resolution of 0.19 nm with a beryllium window energy-dispersive (EDS) detector. For each sample the dark-field (DF) image, bright-field (BF) and HRTEM image were recorded and the average grain size was determined by measuring individual grains from captured micrographs. The XRD measurements of the samples were carried out at room temperature on a Philips powder diffractometer (PW 1820) with monochromatized CuKα radiation.

2.1   High Resolution Imaging and HRTEM image processing analysis

In order to understand the image processing (IP) of HRTEM images, the image formation process [15], when the electron beam passes through the microscope, will be described. We distinguish the following phenomena (see Fig. 1(a)): diffraction phenomena in the plane of the object; the image formation in the back focal plane of the objective lens; the interference, of diffracted beams, in the image plane of the objective lens.

The wave function ψ_{ex} (r) of electrons at the exit face of the object can be considered as a planar source of spherical waves according to the Huygens principle. The amplitude of diffracted wave in the direction given by the reciprocal vector **g** is given by the Fourier transformation of the object function, i.e.

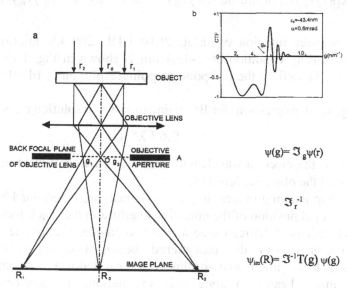

$\psi(g) = \Im_g \psi(r)$

$\Im_r^{-1}$

$\psi_{im}(R) = \Im^{-1} T(g) \psi(g)$

Figure 1. (a) Image formation in an electron microscope: R_o - undiffracted beam; O - objective aperture **A** as placed in (a); (b) Contrast transfer function of JEOL JEM 2010 200 kV electron microscope: Scherzer underfocus $\varepsilon = -43.4$ nm, $\alpha = 0.6$ mrad, $g_s = 5$ nm^{-1}.

$$\psi(g) = \Im_g \psi(r)$$

The intensity distribution in the diffraction pattern is given by $|\psi(g)|^2$ in the back focal plane of the objective lens.

$$|\psi(g)|^2 = |\Im_g \psi(r)|^2$$

During the second step in the image formation, which is described by the inverse Fourier transform ($\Im^{-1}$), the electron beam undergoes a phase shift $\chi(g)$ with respect to the central beam. The phase shift is caused by spherical aberration and defocus and damped by incoherent damping function $D(\alpha,\Delta,g)$, so that the wave function $\psi_{im}(R)$ at the image plane is finally given by

$$\psi_{im}(R) = \Im^{-1} T(g) \psi(g)$$

where T(**g**) is the contrast transfer function (CTF) of thin phase object. T(**g**) includes damping envelope $D(\alpha,\Delta,g)$ and phase shift $\chi(g)$:

$$\chi(\mathbf{g}) = \pi\varepsilon\,\lambda\mathbf{g}^2 + \pi\,C_s\lambda^3\,\mathbf{g}^4/2 \qquad T(\mathbf{g}) = D(\alpha,\Delta,\mathbf{g})\,\exp[i\,\chi(\mathbf{g})]$$

α is the convergent angle of the incident electron beam and Δ is the half-width of the defocus spread ε due to chromatic aberration. For the details consult references [15].

In the weak-phase object approximation contrast transfer function $T(\mathbf{g})$ is:

$T(\mathbf{g}) = \exp(i\chi(\mathbf{g}))$, should be $\sin\chi(\mathbf{g}) = 1$ and $\cos\chi(\mathbf{g}) = 0$, $\chi(\mathbf{g}) = n\pi/2$

for all $\mathbf{g}$.

Contrast transfer function of JEOL 2010 UHR 200 kV microscope, for Scherzer focusing condition of -43.4 nm is shown in Fig. 1(b). The first zero g_s of CTF defines the EM point-to-point resolution d_s ($d_s = 1/g_s = 0.197$ nm).

The general expression for BF point-to-point resolution d_s, is

$$d_s = 0.67\,\lambda^{3/4}C_s^{1/4}, \tag{1}$$

where λ is the electron wavelength and C_s is the coefficient of spherical aberration of the objective lens [15].

The different imaging modes, in phase contrast, are determined by the size and geometrical position of the objective aperture in the back focal plane of the objective lens. A lattice fringe image is obtained, if only one diffracted beam interferes with the unscattered beam. The period of fringes corresponds to the interplanar spacing of the excited beams. Using into an aperture many beams in zone axis orientation, a many-beam image (structure image) will be observed. A dark-field lattice image is formed if particular diffracted beams of interest interfere and all other beams are excluded.

The HRTEM image processing is inverse procedure from the image formation procedure. The Fourier transform of selected area of HRTEM image is obtained, diffracted beams assigned, filtering in Fourier space using particular diffracted spots is performed. After performing inverse FT, the images of the filtered planes in the real space are obtained.

HRTEM images were digitised by scanner and analysed with CRISP programme [16]. The programme calculates Fourier transform which provides the information on the present phase composition and the diffracted family of planes (d-values). With different sizes of filtering masks (1-5 pixels) one can select some spot in FT image and create inverse FT which gives the image of diffracted planes; or if one chooses all diffracted spots, the refined reconstructed image of the selected area is obtained. The great advantage of the image processing is the approach in fine sample details such as: defects, dislocations, grain boundaries, grain sizes, pore sizes, lattice and phase analysis [14,17,18], which gives the results usually inaccessible with the other methods of analysis of nanocrystalline materials.

3. RESULTS AND DISCUSSION

The decrease of the crystallite size to nano-size dimensions has become a common features of the mechanically alloying process. From previous XRD, TEM, HRTEM and ED measurements the evidence of the refinement of the crystalline size down to nanometer-sizes was confirmed in ball-milled ZrO_2-10mol%Y_2O_3 powders (Fig. 2) [13,14]. The cubic ZrO_2-10mol.% of Y_2O_3 cubic solid solutions appeared as the end product although the tetragonal zirconia solid solution phase (t-ZrO_2 SS) was detected in first stages of mechanical alloying (10 minutes of milling). These first grains of t-ZrO_2 solid solution nucleated at m-ZrO_2 layers (Fig. 2(b)).

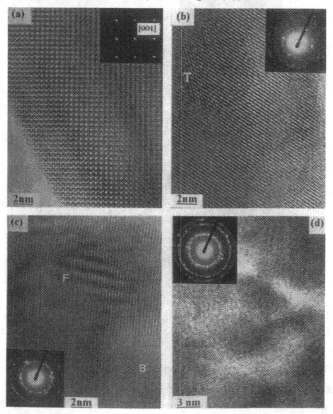

Figure 2. HRTEM lattice image of: (a) monoclinic ZrO_2 viewed along [001] direction (crystallite size is 200 nm); (b) d_{111} lattice image of a sample of the m-ZrO_2-10mol% Y_2O_3 powder mixture milled for 10 minutes. Tetragonal crystallite T is stuck on large monoclinic ZrO_2 oxide grain having layered structure, deformed and ruptured lattice planes; (c) a sample of the m-ZrO_2-10mol% Y_2O_3 powder mixture milled for 1 h. Stacking faults or antiphase boundary F and crystallite B stuck on larger crystallite; (d) a sample of the m-ZrO_2-10mol% Y_2O_3 powder mixture milled for 3 h. Crystallite sizes from 2 to 12 nm. Corresponding SAD patterns are shown as the inserts.

A further insight in the early stages of the alloying process was obtained by image processing of HRTEM photographs. One example from the early stages of the milling process will be given. During milling, fragmentation and adhesion of the grains is a steady process. The newly formed interfaces between monoclinic zirconia grains and Y_2O_3 grains can be brought into the intimate contact by subsequent ball collision forming an agglomeration, composed of grains G, C, M and region D (Fig. 3(a)). Fig. 3(a) also shows a large monoclinic grain, M, nearly in the [0$\bar{1}$ $\bar{1}$] orientation. From this region two selected area diffraction photographs (SAD), shown in Figs. 3(b) and 3(c), were taken. In the SAD pattern the strongest reflections belong to the m-ZrO_2 grain, M, and the weak reflections belong to the Y_2O_3 layer, L in

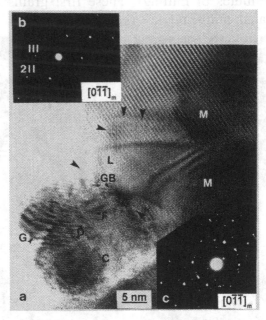

Figure 3. (a) HRTEM photograph of mixture of m-ZrO_2 and Y_2O_3 powders ball milled for 10 min; (b) ED pattern of (a) [0 1 1] m-ZrO_2 ; (c) ED pattern of (a) nearly [0 1 1] m-$ZrO_2 \parallel$ [01 1] Y_2O_3 zone. The meaning of lettering explained in the text.

[01$\bar{1}$] orientation. The region marked L is composed of m-ZrO_2 grain overlapped with the thin layer of Y_2O_3 in [01$\bar{1}$] orientation, which may serve as an evidence that alloying takes place in layers. Some other nanocrystalline Y_2O_3 grains distributed in the aggregate give additional spots, which form very sparse, spotty rings in the SAD pattern shown in Fig. 3(c).

To elucidate the process of the solid-state reaction during ball-milling one has wanted to find out how Y_2O_3 penetrates into the m-ZrO_2 lattice and how this results in the formation of a tetragonal ZrO_2 solid solution (t-ZrO_2 SS).

The Fourier transforms (diffraction patterns) of several regions in Fig. 3(a) were calculated. The regions selected for study are marked in Fig. 3(a) as: GB - the grain boundary; L - a region, of overlapping layers of zirconia and yttria; F - a region containing stacking faults.

The Fourier transforms of regions of size 256 x 256 pixels (corresponding to around 6x6 nm² in the specimen) were calculated. The effect of hole size was investigated within a range from 1 to 5 pixels, the largest size being chosen in order to ensure that very large reflections from the ZrO_2 were completely encircled by the mask hole.

1. Grain boundary region: The Fourier transform of the grain boundary region is shown in Fig. 4(d) and contains the reflections from the ED pattern

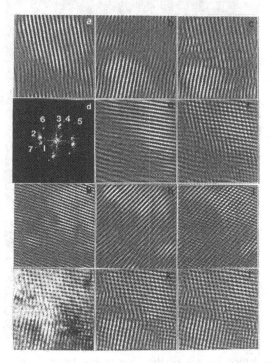

Figure 4. The filtered lattice images from the grain boundary GB region obtained with particular diffraction spots from FT of Fig. 4(d): (a) 1 - ($\bar{1}$ 11)m-ZrO_2 , d=0.316 nm; (b) 2 - (111)m-ZrO_2 and (222)Y_2O_3 ; (c) superposition of lattice images given in (a) and (b); (d) FT of GB region; (e) 3 - ($\bar{1}$ 11) and (222)Y_2O_3 , d=0.3 nm; (f) superposition of images in reflections 1, 2 and 3 from (a) , (b) and (e); (g) 4 - (332)Y_2O_3 , d=0.226 nm; (h) 6 - (400)Y_2O_3 , d=0.265 nm; (i) superposition of lattice images of Y_2O_3 reflections 4 and 6 from (g) and (h); (j) unfiltered original image; (k) all reflections from FT (d); (l) all marked reflections from FT.

shown in Fig. 3(b). The reflections in the FT are marked by numbers 1 to 7, and are assigned to m-ZrO$_2$ and Y$_2$O$_3$. The results obtained after filtering using particular reflections are displayed in Figs. 4(a) to 4(l).

Two sets of planes from Y$_2$O$_3$ are shown in Figs. 4(g) and 4(h), corresponding to reflections 4 and 6, reveal the appearance of yttria in the grain boundary region. We can follow successive formation of the final filtered HRTEM image by making the comparison of Figs. 4(f) and 4(i) to Fig. 4(l) and to the original image given in Fig. 4(j). Here using a HRTEM image processing of original image we revealed and proved how "alloying" took place in the GB.

2. Overlapping layers L of zirconia and yttria: The electron diffraction pattern shown in Figs. 3(b) and 3(c) is close to the $[0\bar{1}\bar{1}]$ zone of m-ZrO$_2$ lattice. The faint reflections are due to Y$_2$O$_3$. It is not obvious whether the large grain M has a thin layer of Y$_2$O$_3$ overlapping it. All Fourier transforms

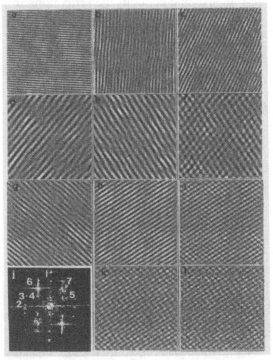

Figure 5. Filtered lattice images of the overlapping layers L obtained with particular diffraction spots from FT of Fig. 5(j): (a) 1 - (440)Y$_2$O$_3$, d=0.187 nm; (b) 2 - (200)m-ZrO$_2$, d=0.262 nm; (c) 3 - (400)Y$_2$O$_3$, d= 0.265 nm; (d) 4 - ($\bar{2}$11)Y$_2$O$_3$, d=0.434 nm; (e) 5 - (211)Y$_2$O$_3$, d=0.434 nm; (f) superposition of images of yttria from (d) and (e); (g) 7$_1$, 7$_2$, 7$_3$ - ($\bar{1}$11)m-ZrO$_2$, d=0.316 nm, (222)Y$_2$O$_3$, d=0.306 nm, ($\bar{1}$11)m-ZrO$_2$, d=0,316 nm; (h) 6 - (1$\bar{1}$1)m-ZrO$_2$, d=0.316 nm; (i) the superposition of lattice images of m-ZrO$_2$ 6 and 7 from (g) and (h): (j) FT of the region L; (k) reconstructed image with superposition of (f) and (i) images; (l) reconstructed original image with all reflections from FT (j).

from this region had the same appearance (Fig. 5(j)), showing that the two zones, one from yttria and one from zirconia, are parallel, and that the layers are parallel. The reflections in the FT are marked by numbers from 1 to 7, and are assigned to m-ZrO_2 and Y_2O_3 .

The results obtained after filtering using particular reflections from FT of Fig. 5(j) are displayed in Figs. 5(a) to 5(l).

Fig. 5(k) shows the reconstruction obtained when using contributions both from yttria and from m-ZrO_2, reflections 4, 5, 6 and 7_1, 7_2, 7_3. Finally, Fig. 5(l) shows the reconstruction obtained with all the reflections identified in the FT (Fig. 5(j)). One can see that this image reveals stripped details from the original HRTEM photograph of Fig. 3 in the region marked L.

Figure 6. The filtered lattice images from the stacking fault region obtained with particular diffraction spots from FT of Fig. 6(g): (a) 1_1 - ($\bar{1}11$)m-ZrO_2 , d=0.316 nm; (b) 1_1+1_2 - ($\bar{1}11$)m-ZrO_2 , (200)m-ZrO_2 and (400)Y_2O_3; (c) 4 - (431)Y_2O_3, d=0.208 nm ; (d) 2 - (110)m-ZrO_2 , d=0.369 nm; (e) 3 - m-ZrO_2 + Y_2O_3 , d=0.18 nm; (f) superposition of images (d) and (e); (g) FT of stacking fault region; (h) stacking fault - original image; (i) reconstruction of original image with all reflections from FT (g).

Fig. 5(l) shows a periodicity of 0.74 nm, arising from the superposition of the images in yttria and zirconia reflections. The Fourier transform and the ED pattern of this region have shown the strong streaked reflection from the $[0\bar{1}\bar{1}]$ zone of m-ZrO_2, while the faint reflections come from the $[01\bar{1}]$ zone of Y_2O_3 (see Fig. 3(c)). This zone is parallel to the $(0\bar{1}\bar{1})$ of m-ZrO_2 (see Fig. 3(c)), which means that the layers of zirconia and yttria are also parallel.

3. Stacking fault region: The results of filtering analysis of stacking fault region F are displayed in Figs. 6(a) to 6(i), while the *d*-values assignment obtained from FT of Fig. 6(g) are given in figure caption. Elongated spot ("streak") in FT of Fig. 6(g) cut in the Fourier space region, giving in real space the *d*-values from 0.316 to 0.266 and it could be regarded as a complex broadened spot containing at least two components 1_1 and 1_2, of the planes having meeting point at the stacking fault. As the lattice image in the reflection 3 was assigned as a mixture of m-ZrO_2 and Y_2O_3 reflections, one can see in the Fig. 6(e) that corresponding planes are introduced as the dislocations in Fig. 6(e). In Fig. 6(f) the superposition of images given in Figs. 6(d) and 6(e) are shown revealing interwoven m-ZrO_2 and Y_2O_3 planes originated from reflections 2 and 3.

As there is a change in the contrast at stacking fault, it is probable that the nucleation of new t-ZrO_2 SS phase took place at stacking fault F, so the Y_2O_3 segregates on defects in m-ZrO_2 lattice that is revealed in the filtering analysis as penetration of as assigned Y_2O_3 and ZrO_2 planes and revealed in Fig. 6(f). The presence of Y_2O_3 (400) planes in Fig. 6(b) and (440) planes in Fig. 6(e) at stacking faults means that the process of segregation of yttria on stacking faults in m-ZrO_2 occurred!

4.   CONCLUSION

Using the HRTEM image processing analysis, it was possible to deduce that the alloying of monoclinic zirconia and yttria and the formation of a new tetragonal ZrO_2 solid solution occur simultaneously.

The formation of dislocations (ZrO_2 or Y_2O_3 planes) in GB region represents one of the tentative mechanisms of alloying, reported recently [19]. The other mechanism of alloying proceeds via the segregation of Y_2O_3 on the stacking faults in m-ZrO_2 [20].

The decrease of grain sizes is accompanied by stresses in ball milling procedure. The defects and stacking faults are formed. By Fourier filtering analysis inserted planes are assigned to particular planes according to the corresponding FT of the region. Alloying was observed by HRTEM as interpenetration of ZrO_2 and Y_2O_3 planes.

The HRTEM IP analysis of ball milled monoclinic m-ZrO_2 and m-$ZrO_2 + Y_2O_3$ with different oxides revealed transition to tetragonal t-ZrO_2

and t-ZrO_2 solid solution, respectively. The appearance of high-temperature phase at room temperature as result of milling treatment is explained by thermodynamic consideration about the role of the defects and the value of the grain size in the transition from m-ZrO_2 to t-ZrO_2 (and appearance of high temperature phase at room temperature as a result of ball milling). It is now generally accepted that the defects (dislocations, vacancies, stacking faults, grain boundaries, interfaces) initiate these transitions. Those defects give the contributions to the free energy of defects ΔG_d. When free energy of monoclinic phase G_m and defects (strain energy) G_d are higher than free energy of tetragonal phase G_t, tetragonal phase can be formed even at room temperature [21].

References

[1]   H. Gleiter, *Europhys. News* 20 (1989) 130.
[2]   R.W. Siegel, *J. Phys. Chem. Solids* 55 (1994) 1097.
[3]   G. Herzer, *Phys. Scripta* T49 (1993) 307.
[4]   K. Lu and Y.H. Zhao, *Nanostruct. Mater.* 12 (1999) 559.
[5]   Y.H. Zhao, K.Zhang and K. Lu, *Phys. Rev.* B56 (1997) 14322.
[6]   J.A. Eastman and M.R: Fitzsimmons, *J. Appl. Phys.* 77 (1995) 522.
[7]   A.M. Tonejc, I. Djerdj and A. Tonejc, *Mat. Sci. Eng. C* 19 (2002) 85.
[8]   E.C. Subbaro, H.S. Maiti and K. K. Srivastava, *Phys. Stat. Sol. (a)* 21 (1974) 9.
[9]   A. Keshavaraja, A.V. Ramaswamy, *J. Mater. Res.* 9 (1994) 837.
[10] M.C. Caracoche, P. C. Rivas, A. P. Pasquevich, A. R. Lopez Garcia, *J. Mater. Res.* 8 (1993) 605.
[11] H.J. Fecht, *Nature* 356 (1992) 133.
[12] Y.L. Chen and D.Z. Yang, *Scripta Met.* 29 (1993) 1349.
[13] A.M. Tonejc, A. Tonejc, *Mater. Sci. Forum* 225-227 (1996) 497.
[14] A.M. Tonejc, A. Tonejc, G. W. Farrants, S. Hovmöller, *Croat. Chem. Acta* 72 (1999) 311.
[15] J.M. Cowley, *Electron Diffraction Techniques, High Resolution Imaging*; International Union of Crystallography, Oxford University Press, Oxford, 1993, p. 131.
[16] S. Hovmöller, *Ultramicroscopy* 41 (1992) 121.
[17] A.M. Tonejc, *Acta Chim. Slov.* 46 (1999) 435.
[18] A.M. Tonejc, I. Djerdj and A. Tonejc, *Mat. Sci. Eng.* B 85 (2001) 55.
[19] R.B. Schwartz, *Mater. Sci. Forum* 269-272 (1998) 665.
[20] R.D. Doherty, *Diffusive Phase Transformations in the Solid State*, in Physical Metallurgy, fourth revised editions, editors R.W. Cahn and P. Haasen, Elsevier Sci. BV, Amsterdam, 1996, p. 1385.
[21] M. Qi and H.J. Fecht, , *Mater. Sci. Forum* 269-272 (1998) 187.

ELECTRON CRYSTALLOGRAPHY ON BEAM SENSITIVE MATERIALS
— Electron Microscopy and Electron Diffraction of Polymers —

Masaki Tsuji

Laboratory of Chemistry of Polymeric Functionality Materials, Division of Materials Chemistry, Institute for Chemical Research, Kyoto University, Uji, Kyoto-fu 611-0011, Japan

Key words: electron diffraction, beam sensitive materials, polymers, sample preparation

1.   INTRODUCTION

Under a "microscope", a tiny object can be observed as its magnified image. Although the resolving power of an optical microscope is of the order of the wavelength of the employed light [1], a transmission electron microscope (TEM), which utilizes high-speed electrons having a much smaller wavelength, has a higher resolving power of the molecular or atomic level (a point-resolution of 0.2—0.3 nm and a line-resolution of 0.14nm or smaller even with 100—200kV TEMs).

Polymers (here, the term "polymer" refers to a material consisting of linear-chain macromolecules) are typical examples of radiation-sensitive materials against electron beams [2,3]. Therefore, linear polymers are utilized here as examples of beam sensitive materials. In the field of polymer science, TEM has been restricted in principle to morphological observation of polymer solids at low direct-magnifications and to selected-area electron diffraction (SAED) of them, because of their electron irradiation damage [4]. Recently, however, high-resolution observation with TEMs is attempted extensively to investigate molecular-level structures in polymer solids, especially of crystalline polymers, owing to the improved performance of the

455

TEM instruments and the improved operation procedures using devices such as a Minimum Dose System (MDS; JEOL Ltd.) or a Low Dose Unit (LDU; Philips) [2,3].

2.   PROCEDURE FOR TEM STUDY ON BEAM SENSITIVE MATERIALS [7]

Operation procedures for SAED, dark-field (DF) imaging and high-resolution observation [5,6] are briefly described here, which are utilized in our laboratory for structural investigation of "crystalline polymers". As for conventional bright-field (BF) imaging, any text books [5,6] can be referred to.

2.1   Selected-area electron diffraction (SAED)

Electron beams are spread out as widely as possible by controlling the 2nd condenser lens (CL2). In specimen-searching, a caustic image [6] (or caustic pattern) is made in the diffraction mode but without using any selected-area aperture (SAA). In this case, by setting the diffraction lens (DL) more or less out of focus, the outline of the specimen can be seen in high contrast even under a fairly weak illumination condition (owing to "defocus contrast": see Section 2.3.2). SAA having a desired opening is introduced, and then electron beams are focused on the fluorescent screen by adjusting DL. Desired brightness is made by adjusting CL2, and then an SAED pattern is recorded onto a photographic film with high sensitivity (or onto an Imaging Plate, or with a CCD camera) with fairly weak illumination [7]. After recording an SAED pattern, DL is set again out of focus and then a double-exposed photo with and without SAA is taken: this procedure is useful to know the specimen position/orientation giving the SAED pattern in question. For beam sensitive materials, an SAED pattern is recorded before taking an image of the target specimen-portion.

2.2   Dark-field (DF) imaging

There are two methods for DF imaging [5]: 1) tilted illumination with an objective lens aperture (OLA) set symmetrically on the optic axis, and 2) un-tilted illumination with OLA set off from the optic axis. Method-2 is frequently adopted for morphological observation of polymers [7]. Direct magnification is at most 10,000, and that in a range of 2,000—5,000 is often utilized for polymers. If several DF images, each of which is made by using a different reflection, are desired from a same specimen-area, then the first image should be recorded by using the weakest reflection. A moiré pattern [5,6] can be obtained even in BF imaging, but usually DF imaging is utilized to obtain the pattern. For obtaining a moiré pattern from a polymer single crystal, a bi-layered crystal must be prepared [8].

2.3 High-resolution TEM

2.3.1 High-voltage TEM (HVTEM)

Here, instruments operated at 300kV or higher are termed HVTEM. The advantages of HVTEM are as follows: 1) it enables us to observe rather thick specimens; 2) it suppresses electron irradiation damage of specimens. On the other hand, the disadvantages of high voltage are: 1) the higher the voltage, the lower the contrast; 2) the higher the voltage, the lower the sensitivity of image recording systems such as photographic films. It seems that 200—300 kV is adequate for polymer crystals. Examples of our work were carried out mostly with 200kV TEMs.

2.3.2 Phase contrast and high-resolution observation

When a thin specimen regarded as a "phase object" is examined in the BF imaging mode by using OLA having a considerably large opening, appropriate contrast is not expected in the image taken at the just focus (Gaussian focus). If phase modulation is not so large and accordingly an image with appropriate contrast is not obtained even by using a small OLA, then satisfactorily better contrast for such a phase object can be obtained in BF imaging, by applying a certain amount of underfocus and making resultant proper phase difference between scattered (or diffracted) and transmitted waves. This is termed "defocus contrast imaging" [2,9], which visualizes the stacked-lamellar structure in an oriented thin film of polyethylene (PE) and the 2-dimensional spherulites in a thin film of isotactic polystyrene (i-PS). Defocus contrast imaging and the high-resolution observation are both based on the phase contrast. Here, the word, "high-resolution", refers to the resolution which is of the order of a distance between adjacent chain-stems in a polymer crystal.

2.4 Electron irradiation damage of crystalline specimens

The electron dose needed for complete disappearance of all crystalline reflections in the SAED pattern is termed the total end-point dose (TEPD), and is a measure of the durability of the specimen under consideration against electron irradiation [2,3,7]. The greater the TEPD value, the more stable the specimen against electron irradiation [7]. From the TEPD value, the resolution limit for the specimen can be predicted [2,3]. The TEPD value of a material at 4.2K is more than 10 times that at room temperature, though some differences are recognized among materials [3]. Cooling a specimen down to such a low temperature to suppress electron irradiation damage is termed "cryo-protection". The resulting images of a given specimen obtained at a low temperature, however, may or may not reflect the structure at room temperature. A specimen can be mounted on a conventional copper (Cu) grid in usual cases, but a silver (Ag) grid is utilized when thermal conduction must be taken into consideration. When a specimen, which has already been cooled down (*viz.*, frozen) outside the TEM column, is desired to bring with no frost to the specimen stage in the column, a cryo-transfer device is indispensable.

3.   SAMPLE PREPARATION FOR TEM STUDY ON POLYMERS [7]

3.1   How to support a specimen

In the case of conventional TEMs of a 100—200kV class, the specimen thickness is desired to be smaller than 100nm even for organic materials. A thin film-like specimen can be mounted directly on a specimen grid (Cu: 150—400 mesh) for TEM. The grid may have a right side and a wrong one. The right side has smaller openings than the other, and therefore gives a wider contact between the grid and a specimen film or support film. Practically there is no big difference between the sides except for a special purpose such as high-resolution observation. In the case of specimen tilting, however, the specimen should be mounted on the right side of grid; if it is mounted on a wrong side, the edge of the opening will interrupt the field of view by tilting and accordingly the range of available tilt-angle becomes small.   Special adhesive, "Mesh-cement" (a 0.2—0.5% solution of chloroprene rubber in toluene), is used to fix a support film or a thin film-shaped specimen on a grid

3.1.1 Support films

Granular specimens such as polymer single crystals (see Section 3.2.1) are mounted on a support film which has been put on a grid in advance, if the specimens are finer than the opening of the grid. A "Collodion" (cellulose acetate) film, a "Formvar" (polyvinyl formal) film, a vapor-deposited carbon (C) film, and so on are used as a support film. Onto a plastic film, C is vapor-deposited lightly under vacuum in order to reinforce the film mechanically and/or to improve chemical stability. For example, in the case of PE single crystals which are suspended in xylene, a plastic film (*e.g.*, Formvar) must not be used because the film is dissolved in xylene, and therefore, for example a C support-film should be used. When granular specimens dispersed in water are desired to mount onto a specimen grid coated with a support film, hydrophilization treatment [7] might be done to recover hydrophilicty of the support film just before use. In particular, the treatment is indispensable for electron staining with, *e.g.* uranyl acetate.

Depending on purposes, a plastic film under the vapor-deposited C is dissolved away with a solvent. By this method, a very thin C support-film of smaller than 10nm in thickness can be made easily. An aluminum (Al) support-film can be made similarly, by vapor-deposition of Al onto a plastic film put on a grid and then by dissolving away the plastic film. The reflection rings from the Al support-film can be used as an internal reference to calibrate the camera length of SAED pattern, and this support film does give no amorphous halo. When an ultra-thin C support-film (less than 5nm in thickness) is desired, a "microgrid" (MG: see Section 3.1.2) should be used on which an ultra-thin film made by indirect vapor-deposition of C has been put in advance.

When the fine specimens are suspended in a volatile solvent such as ethanol, it is recommended that a drop of the suspension is added with a pipette onto a grid from right above. (The grid coated with a support film is placed on a piece of filter paper.) With this procedure, excess liquid which oozes out of the grid can be immediately absorbed by the filter paper.

Crystalline clumps in suspension are ultra-sonicated into thin plates/fragments suitable for TEM investigation [10], but such ultra-sonication may damage the polymer single crystals.

3.1.2 Microgrid (MG)

A thin net-like film, namely "MG", which has much smaller openings (0.1—10 □m in diameter) than those in a specimen grid, is put on the grid, and an ultra-thin support-film or a film-shaped specimen is mounted on MG, which is, for example, made of cellulose acetobutyrate ("Triafol"). At least, C is vapor-deposited onto a plastic MG for reinforcement, and in addition Au should be vapor-deposited to improve stability. Alumina super-MG of ultra-fine openings (20—100 nm in diameter) has been devised.

3.2   Preparation of thin specimens of polymers for TEM

3.2.1   Preparation of solution-grown crystals (including so-called "single crystals") of linear polymers [2-4, 8, 10-13]

"Single crystals" of linear flexible-chain polymers are grown isothermally from dilute solution (0.01—0.1 wt%). They are thin platelets, namely lamellar crystals in which the polymer chains are folded back and force and the chain-stems are set perpendicularly to the wide basal surface in principle, and they have a thickness of several nanometers to some ten nanometers and a lateral size of micrometer order. Single crystals of various polymers and preparation procedure for them are compiled in the books by Geil [12], Wooward [4] and Wunderlich [8], and also in the review article by Lotz and Wittmann [13]. When thin platelets of such single crystals are mounted onto a support film, they are to be settled there in a "flat-on" fashion. If electron beams are desired to be incident onto them in the direction perpendicular to the chain axis (the c-axis of polymer crystals), then for example, "edge-on" lamellar crystals should be prepared by epitaxial growth (see Section 3.2.2.5). To prepare lamellar crystals of polymers, the "confined thin film melt polymerization method" is developed [42]. Single-crystal-like lamellae can be grown in a polymer thin film (see Section 3.2.2.1) on a substrate (*e.g.*, glass slide) by melt crystallization [33,34]. Procedures to prepare crystals of low-molecular-weight materials and are compiled in the book by Dorset [10].

3.2.2   Preparation of thin film specimens of polymers

3.2.2.1   Solution-cast method [14]

A hot solution (0.1—1 %) is added drop-wise onto a pre-heated substrate such as a glass slide or a piece of mica sheet (the substrate may have been coated with vapor-deposited C). After being dried and solidified, the resulting thin film specimen is mounted onto a specimen grid with the aid of water surface if the specimen is insoluble in water. When a glass slide is used as a substrate, a dilute aqueous solution (<0.02M) of hydrofluoric acid (HF) might be used for easy detachment. The resulting thin film can be mounted onto a specimen grid with the aid of polyacrylic acid and of water surface. A variation of the solution-cast method is as follows: a substrate is soaked in a large quantity of a given solution and slowly pulled up vertically from the solution. Another variation of the method is used to make a uniaxially oriented thin film by sandwiching a hot polymer solution between two glass slides and then by displacing one of the slides quickly just after evaporation of the solvent.

3.2.2.2　Thin polymer film spread on the water surface

For example, by adding a drop of a *ca.* 2wt% solution of natural rubber (NR) in benzene onto the water surface, a thin amorphous film of NR can be made [15]. This procedure is similar to that for making the support films of Collodion or Formbar. When xylene is used as a solvent, hot water near the boiling point (*viz.*, close to 100°C) must be utilized. If, for example, a drop of a PE solution in xylene is added onto the surface of hot water, a crystalline thin film having 2-dimensional spherulites is prepared. By stretching this PE film on the hot water surface, a uniaxially oriented thin film having fiber structure can be made [2,3]. A thin film of i-PS can be made by casting a hot solution in xylene onto the surface of hot water, but the resulting film is amorphous. The i-PS film which has been mounted on a specimen grid can be crystallized by heating (for example, at 160—200 °C) [2]. Ortho-phosphoric acid (H_3PO_4) might be utilized in the place of water if temperatures higher than 100°C are desired [16]. Glycerin or fusible salt (or fusible alloy) is also used [16].

3.2.2.3　Preparation of highly oriented thin films of polymers: film stretching method [2, 3, 16, 17]

For preparing highly oriented thin films [17], one should use as high degree of polymerization as possible. Polymers such as PE, i-PS, isotactic polypropylene, poly(1-butene), poly(4-methyl-1-pentene) are dissolved in *e.g.* xylene (about 1 wt%). Some drops of a hot solution of each material are added and spread on the glass plate which is thermostated at the desired temperature for the material. A thin molten film is left after evaporation of the solvent. A highly oriented thin film can be made, by pinching the rim of the molten film with a pair of tweezers and then by stretching the film upwards from the plate.

To prepare nano-fibers of polymers, "electrospinning" is developed [43].

3.2.2.4   Vacuum evaporation

Most of the inorganic materials such as metals and of the organic compounds of "low molecular-weight (MW)" can be vaporized under vacuum to make respective thin films. When a polymer is heated under vacuum, generally it decomposes into lower MW entities but its vapor-deposited film can be made on a substrate. A newly cleaved (001) face of a certain alkali halide is used as a substrate for epitaxial growth (see Section 3.2.2.5) of vapor-depositing polymers. In the case of polymers which are polymerized from vapor phase, they can be polymerized/crystallized directly on an appropriate alkali halide [18].

An application of vacuum evaporation to polymers is the surface decoration of polymer crystals such as a single crystal [19,20]: for example, a TEM photograph of the PE single crystal surface-decorated with vapor-deposited PE shows that a lozenge-shaped single-crystal platelet of PE has four {110} sectors and that the folding direction of PE chains in the platelet is parallel to the lateral growth surface in each sector .

3.2.2.5   Epitaxy (or epitaxial growth) [8, 20-22]

Epitaxially grown thin films of vapor-deposited organic compounds and defect structures in these films have been studied by electron diffraction and high-resolution imaging with TEM [23]. Here, epitaxial crystallization of polymers is briefly described.

An alkali halide (NaCl, KCl, KBr, *etc.*) which has a freshly cleaved (001) face is introduced into a polymer solution and then the polymer is to be crystallized epitaxially onto the face of the alkali halide. In general, flexible-chain polymers are apt to be crystallized as rod-like crystals on such an alkali halide: these rod-like crystals are "edge-on" lamellae with their lateral surface being mostly in contact with the (001) face of the alkali halide. A given polymer deposited on an alkali halide can be melted and subsequently crystallized there in an epitaxial fashion.

When a glass plate which is heated *e.g.* at 130°C is rubbed with a polytetrafluoroethylene (PTFE) briquette, a highly oriented thin film (2—100 nm thick) of PTFE can be made on the plate: such a film is termed a friction-transfer layer [24]. This method is applicable to make thin oriented layers of other polymers. Various kinds of organic compounds can be oriented on the PTFE friction-transfer layer from vapor phase, from solutions and from the melts [24].

As for epitaxy of polymers by using various kinds of organic compounds including polymers as substrates, the review article by Wittmann and Lotz [21] should be referred to. In particular, crystallization of polymers can be well controlled with the following procedure proposed by them: A thin film of polymer is prepared by solution-casting onto a glass slide (or a piece of freshly cleaved mica). A layer of the appropriate nucleating agent (*e.g.*, naphthalene, benzoic acid, *etc.*) is spread over the dried polymer film and these two materials are melted together and then cooled by displacing the slide on a "Kofler bench" having a temperature gradient. The crystallized eutectic solid can be examined by optical microscopy. Then its appropriate portion is mounted onto a TEM grid, after the nucleating agent is washed away from the solid with a proper solvent and the remaining polymer specimens are reinforced with vapor-deposited C. This procedure is cited in the book by Dorset [10] as "growth from a co-melt" to orient the polymer crystals.

3.2.3   Ultra-microtoming [5]

In order to observe a structure inside a bulk sample with TEM, one should prepare a replica of the "fracture surface" of the sample, or must make an ultra-thin section whose thickness is small enough for electron beams to pass through the section. Unfortunately, however, orientation of crystallites might be changed by microtoming.

4.   STRUCTURAL ELECTRON CRYSTALLOGRAPHY FOR BEAM SENSITIVE MATERIALS [10]

TEM as a crystallographic instrument and crystal symmetry are described in detail in the book written by Dorset [10]. Examples of crystal structure analysis for various organic molecules including linear polymers are well compiled in his book. Most of the beam sensitive materials are organic ones and "most probable space groups for organic molecules" are summarized in Table 2.6 of his book.

5.   **EXAMPLES OF OUR STRUCTURAL ANALYSIS
        FOR POLYMER CRYSTAL BY ELECTRON
        DIFFRACTION COMBINED WITH HIGH-
        RESOLUTION TEM**

5.1   **Crystal structure analysis of poly(*p*-xylylene) (PPX)
         β-form [2, 25]**

5.1.1   **Historical background**

The β-form crystal was already analyzed by Iwamoto and Wuderlich [26] by X-ray diffraction, and their value for the *c*-axis (chain axis), namely the identity period is 0.655nm corresponding to the extended trans-zigzag conformation. However, two crystal systems, monoclinic and hexagonal (or trigonal), were proposed for the PPX β-form crystal. Niegisch [27] studied optical properties of the β-form, and observed the β-form single crystals grown from β-chloronaphthalene solution with TEM. Taking into account his results and our electron/X-ray diffraction data as well as the already analyzed crystal structure of the β-form (especially the *c*-dimension), we determined the crystal system and unit cell dimensions of the β-form to be hexagonal or trigonal; $a=b=2.052$nm, c(chain axis)=0.655nm, and $\gamma=120°$.

From the highest observed density (1.15g/cm^3) of the PPX β-form and its lattice dimensions, the unit cell must contain 16 monomer units, *i.e.*, 16 polymer chain-segments (the length of a "chain-segment" is the fiber period) ought to be contained in the unit cell. Prior to the crystal structure analysis of β-form, the polymer crystals belonging to the hexagonal or trigonal system were classified into three groups:

Group 1: Each chain itself has 3- or 6-fold axis or screw axis (The number (N) of chains in a unit cell is one or an integral multiple of 3).

Group 2: Each chain has cylindrical or nearly cylindrical symmetry (A unit cell must be made up of only one chain, and the symmetry is restricted to be hexagonal or pseudo-hexagonal).

Group 3: A set of identical chains produces the 3- or 6-fold symmetry, though one chain itself has no symmetry features of Group 1 or 2 (A unit cell consists of such sets, and N is an integral multiple of 3).

5.1.2 Determination of the β-form crystal structure in the projection on the *ab*-plane

The high-resolution TEM image of the β-form single crystal was successfully obtained at room temperature with JEOL JEM-500 (500kV, point resolution =0.14nm) and then was image-processed to improve the S/N ratio. Each black spot in the image was to represent the projection of one molecular chain-stem on the *ab*-plane. Thus we have first succeeded in direct imaging of the individual chain-stems of polymer viewed along the chain axis (*c*-axis). Some of the spots in this micrograph appeared to be elliptic. The mutual positions of the molecular centers in the projection on the *ab*-plane were recognized. The feature of the arrangement of molecular centers seemed to be represented in terms of the 2-dimensional pattern (space group = *p*6), which is well represented by "Chidori" that is one of the repeating patterns in dyeing and industrial arts in Japan. Thus the 2-dimensional space group of the molecular arrangement in the β-form crystal was approximated to be *p*6 which has no conditions limiting possible reflections. Indeed, the SAED pattern of β-form single crystal has neither systematic absences of reflections nor symmetry planes, and only shows 6-fold rotational symmetry. We first assumed a 2-dimensional structure model of β-form so as to satisfy the symmetry of *p*6, taking van der Waals' radii and the TEM image into account. The mutual positions and orientations of 16 molecules were changed so as to be consistent with the SAED pattern. In computation, the orientation of the molecular chain at the origin of a unit cell was assigned statistically to one of three equivalent orientations to satisfy the symmetry of *p*6.

Reflection intensity in the SAED negatives was measured with a microdensitometer. The refinement of the structure analysis was performed by the least square method over the intensity data (25 reflections) thus obtained. A PPX single-crystal is a mosaic crystal which gives an "N-pattern". Therefore we used the $1/d_{hk0}$ as the Lorentz correction factor [28], where d_{hk0} is the ($hk0$) spacing of the crystal. In this case, the reliability factor R was 31%, and the isotropic temperature factor B was 0.076nm^2. The molecular conformation of the β-form took after that of the β-form since R was minimized with this conformation: benzene rings are perpendicular to the trans-zigzag plane of $-CH_2-CH_2-$.

5.1.3 Crystal structure analysis of the β-form by X-ray diffraction

In the case of linear polymers, it is almost impossible to make a large single crystal which can be utilized for X-ray crystal structure analysis.

Therefore, uniaxially oriented samples should be prepared for this purpose, which give so-called fiber pattern in X-ray diffraction. The diffraction intensities from the PPX specimen of β-form, which had been elongated 6 times at 285°C, were measured by an ordinary photographic method. The reflections were indexed on the basis of the lattice constants: $a=b=2.052$nm, c(chain axis)$=0.655$nm, $\alpha=\beta=90°$, and $\gamma=120°$. Inseparable reflections were used in the lump in the computation by the least square method.

5.1.3.1   Two-dimensional analysis using X-ray equatorial reflections

First of all, we analyzed the 2-dimensional structure projected on the *ab*-plane in order to confirm the result of the analysis by electron diffraction. The 17 independent reflections on the equator were used (R=35%, B=0.075nm^2 (isotropic)). The molecular conformation of the β-form was adopted because it lowered R. The mutual positions and orientations of molecular chains are almost identical to those analyzed by electron diffraction. In the 3-dimensional analysis, therefore, this structure in the *ab*-plane projection was basically fixed.

5.1.3.2   Three-dimensional structure analysis

The two molecular orientations (these coincide with each other by 180° rotation about the chain axis) can give the identical projection on the *ab*-plane. (Such a problem took place also in the case of the β-form crystal structure analysis, and the orientation was determined using the reflections on the first layer line [26].) The 3-dimensional space group of the β-form crystal is to give approximately the 2-dimensional space group, *p*6, in the projection on the *ab*-plane, so that it should be either *P*3 or *P*3. We determined the 3-dimensional structure whose space group is *P*3 (trigonal), because the structure having the *P*3 symmetry lowered R than *P*3. In this case, R=21% and B=0.09nm^2 (isotropic). The orientation of the molecular stem at the origin of a unit cell was assigned statistically to one of three equivalent orientations.

We computer-simulated the through-focusing images to be taken with a JEOL JEM-500 (500kV, spherical aberration coefficient (Cs) =1.06mm, resolving power =0.14nm). In the calculation, an imaginary crystal with the final structure parameters was used. The computer simulation was performed according to the kinematical imaging theory, and confirmed that the obtained micrograph is nearly the best one that can be expected with the JEM-500 [25]. By the way, edge-dislocations were first observed at molecular level resolution in the high-resolution images of both β- and β-form single crystals [2].

5.2 Stacking faults in syndiotactic polystyrene (s-PS) β-form crystals [3, 29-34]

5.2.1 5.2.1 Solution-grown single crystal (β''-modification) [3,29-32].

Various types of morphologies and crystalline modifications of s-PS have been reported [44]. One of these modifications has an orthorhombic form (space group = $P2_12_12_1$) with a planar zigzag conformation of back-bone chain. This form corresponds to the α-phase found by Kobayashi et al. [35] and to the β-phase named by Guerra et al. [36]. The crystal structure of this form was analyzed and the plane of phenyl rings within this form is normal to the chain axis (c-axis) [37-39].

s-PS was dissolved in 2:1 (v/v) *n*-tetradecane/decahydronaphthalene, and single crystals were grown isothermally in a temperature range of 160°C through 200°C from a 0.01—0.02wt% solution. A mono-layered single crystal was grown isothermally at 200°C and the corresponding SAED pattern was obtained. The a^*- and b^*-axes are perpendicular to each other. Because the $h00$ and $0k0$ reflections with h and k being odd are absent, the extinction rule for reflection suggests that the 2-dimensional space group is *pgg* in the *ab*-plane projection. The cell dimensions of the *a*- and *b*-axes were estimated at 2.87nm and 0.88nm, respectively. The c-dimension was estimated at 0.51nm by X-ray diffraction, *i.e.* by introducing X-ray beams parallel to the surface of a sedimented mat of single crystals. From these results, it was concluded that the s-PS single crystals have an orthorhombic form identical with those reported by Chatani et al. [37,38] and De Rosa et al. [39]. The *a*- and *b*-axes, however, were identical with *b*- and *a*-axes of the unit cells defined by both of them, respectively. Thus, the s-PS single crystals have an orthorhombic unit cell ($P2_12_12_1$) with a planar-zigzag molecular conformation; $a = 2.87$nm, $b = 0.88$nm and c (chain axis) = 0.51nm

The striking features found in the SAED pattern are as follows: (1) the $h00$ and $0k0$ reflections with h and k being odd are not observed (the extinction rule of reflection); (2) the reflections corresponding to $h+k$=even are spot-like, but others streak along the a^*-axis (the degree of streaking is independent of the order of reflection); and (3) diffuse scattering is not observed. The 2nd feature strongly suggests that the single crystal contains some kind of lattice defects, and other features impose the following structural restrictions on the defects:

(a) Even if the defects are introduced in the crystal, projected inner-potentials of the crystal onto the *a*-axis and *b*-axis must be identical respectively with those of the regular crystal, *e.g.* the 2_1 screw symmetry must be retained along the *b*-axis.

(b) The defects cause a subdivision of crystallite-size along the *a*-axis and its effect upon the reflections of *h+k*=odd only. Accordingly, the defects are deduced to be stacking faults of molecular layers parallel to the *bc*-plane when they stack in the direction normal to the plane.

(c) A structural model, in which one of the two possible molecular arrangements is statistically chosen in individual molecular layers, is not adequate, because such a model ought to give diffuse scattering and the 2nd feature mentioned above cannot be explained with the model.

To meet these features in the SAED pattern, a model for the stacking fault was made with the following procedure. A pair of neighboring molecular layers extended parallel to the *bc*-plane was considered to be a basic structural unit, "motif", of s-PS crystal. By dividing regular crystal lattice into such bi-layered motifs every other layer, we get two types of motifs A and B: motif B corresponds to the mirror image of motif A with respect to the *bc*-plane (here, A_2 and B_2 correspond respectively to A_1 and B_1 shifted by half a period along the *b*-axis). Thus, the regular crystal is expressed by a stacking --$A_1B_2A_1B_2$-- or --$A_2B_1A_2B_1$--, and a sequence such as --$A_1B_2B_1A_2$-- or --$B_2A_1A_2B_1$-- forms a stacking fault. There are two ways of introducing stacking faults, which are closely related to the ways of making motifs: Case-1 and Case-2 [3,29]. Theoretical calculation of diffraction intensity based on both models well explained the features of the SAED pattern, but the calculation of interfacial energy at the boundary indicated that Case-1 is more plausible than Case-2 [29].

The lattice image of an s-PS single crystal was successfully obtained at 4.2K with JEOL JEM-2000SCM (160kV), showing (200) lattice fringes running in the vertical direction and also oblique (110) and (210) fringes. The (210) fringes are not perfectly straight, but kinked or shifted by half of the (210) spacing at the places corresponding each to the stacking fault in question, although no shift of (110) fringes was recognized. For the (210) planes, accordingly, introducing such stacking faults in a crystallite corresponds to cutting the crystallite perpendicularly to the *a*-axis into some slices. This is the geometrical reason why the *hk*0 reflections are streaking along the *a**-axis when *h+k*=odd and are spot-like when *h+k*=even. De Rosa et al. [39] and Chatani et al. [40] have also proposed structure models of disorder in the s-PS orthorhombic crystal, which are the same as our plausible model of stacking fault (Case-1).

5.2.2 Quantification of stacking faults in the solution-grown single crystals [31]

On the basis of the model (Case-1), diffraction intensity from an infinite 2-dimensional crystal was mathematically formulated by incorporating the probability, p, of occurrence of a stacking fault (-AA- or -BB-) in the crystal [30-32]. The result clearly showed that the SAED pattern is well reproduced by this formulation with p<0.5.

The probability, p, of the presence of stacking faults in the β-form single crystals was estimated from the mean half-breadth of the streaking reflections in the SAED pattern. The p values were also estimated in the DF and the high-resolution images, by counting the number of the faults in a unit length along the *a*-axis of the crystal. The p values derived by three different methods were in good agreement with one another. These p values showed weak dependence on crystallization temperature (Tc), with a maximum value at 165°C. When the single crystals grown at 165°C were dried and then annealed isothermally at 200°C, only a small decrease in the p value was detected. These results suggest that it is difficult to grow the fault-free crystal of β-form [31].

5.2.3 Energetic analysis of the incorporation mechanism of stacking faults into the crystal [29, 32]

The Tc-dependence of p in the solution-grown single crystal of β-form was theoretically formulated on the basis of the growth theory of polymer crystal [32]. By comparing the experimentally obtained Tc-dependence of p with theoretical one, it is deduced that the relative energy difference, $\Delta E/\Delta h_f$ ($=\xi$), between the regular and the faulted structures decreased linearly with increasing Tc: here, (Δh_f - ΔE) and Δh_f are the heat of fusion for faulted structures and that for the regular one, respectively. The mathematical formulation was made for estimating ξ from p at any Tc. These results lead to an expectation that the faulted structure will be dominant when Tc is higher than a certain critical temperature at which $\xi=0$. Such a fault-rich crystal (p>0.5) is to give the *hk*0-diffraction pattern in which the reflections with *h*+*k*=odd are absent [31]. This feature has been really found in the X-ray diffraction pattern from the β'-form crystal [39], and also in the SAED pattern of melt-crystallized thin film (see Section 5.2.4) [34].

5.2.4   Melt-crystallized β'-modification [33, 34]

The faulted structure was found to be more stable ($\xi < 0$) than the regular one at temperatures higher than a certain critical temperature, as predicted in Section 5.2.3. In this case, the faulted structure is expected to become dominant (p>0.5). When a thin film of s-PS was melt-crystallized above the critical temperature, single-crystal-like lamellae were successfully grown in the film at Tc's between 230°C and 250°C. Their SAED patterns showed another feature of streaking which implies that the stacking faults are dominant [31]. The structural model for the melt-grown lamellae is as follows: The succession of motifs of a same type makes a monoclinic unit cell, and comprises a domain elongated in the direction parallel to the *b*-axis. Such molecular arrangement was confirmed in the high-resolution TEM images of the melt-grown lamella, which were taken at 4.2K with a JEOL JEM-4000SFX (400kV) and in which each dark spot corresponded to the *ab*-plane projection of a molecular stem of s-PS. The domain structure was also detected in the DF images, in which irregularly spaced parallel striations were observed.

The p values were estimated from the high-resolution and DF images, and then these values were converted to ξ using the mathematical formula mentioned in Section 5.2.3. The Tc-dependence of ξ was, therefore, experimentally obtained from both solution-grown and melt-crystallized specimens. The straight line showing the Tc-dependence of ξ obtained from the solution-grown specimens may be reasonably extrapolated to 518K (=245°C) or higher. It is concluded that the structure in the s-PS β-form is affected by the Tc-dependence of ξ, which may hold between the solution-grown single crystals and the melt-grown lamellae.

From the relationship between ξ and Tc, the crystallization behavior of the s-PS β-form is summarized as follows:

1) At a sufficiently low Tc (below the critical temperature (around 220~230 °C) at which ξ =0), the alternation of two types of motifs (*viz.* "regular" structure) is favored. Occasional occurrence of the succession of motifs of a same type by kinetic reason results in the stacking fault, and gives the β"-modification.

2) With increasing Tc, the relative stability of the regular structure decreases, and accordingly the driving force to incorporate the stacking fault increases thermodynamically. On the other hand, the chance to incorporate the stacking fault decreases due to the slower growth rate. As a result of compensation between the thermodynamic and kinetic effects, the Tc-dependence of p becomes very weak.

3) Above the critical temperature, the stacking fault turns to be a favorable structure. A succession of motifs of a same type may be dominant

in the crystal grown in this temperature range, and accordingly the β'-modification is formed under this condition.

5.3 Other examples of electron crystallography on polymers

Crystal structure analysis for poly(*p*-benzoic acid) whisker [40] and structural study of stacking faults in syndiotactic polypropylene [41] were carried out.

References

1. M. Born and E. Wolf, *Principles of Optics*, 5th edn., Pergamon Press (1975).
2. M. Tsuji, In (Eds.; Sir G. Allen and J.C. Bevington) *Comprehensive Polymer Science*, Vol.1, (Vol. Eds; C. Booth and C. Price), Chap.34, pp.785-840, Pergamon Presss (1989).
3. M. Tsuji and S. Kohjiya, *Progr. Polym. Sci.*, **20**, 259 (1995).
4. A.E. Woodward, *Atlas of Polymer Morphology*, Hanser Pub. (1989).
5. D.B. Williams and C.B. Carter, *Transmission Electron Microscopy, I,II,III & IV*. Plenum Press (1996).
6. L. Reimer, *Transmission Electron Microscopy—Physics of Image Formation and Microanalysis*. 4th Edition, Springer (1997).
7. M. Tsuji and M. Fujita, In *Encyclopedia of Materials: Science and Technology*, Elsevier Sci., pp. 7654-7664 (2001).
8. B. Wunderlich, *Macromolecular Physics Vol.1 (Crystal Structure, Morphology, Defects)*, Academic Press (1973).
9. J. Petermann and H. Gleiter, *Phil. Mag.*, **31**, 929 (1975).
10. D.L. Dorset, *Structural Electron Crystallography*, Plenum Press (1995).
11. M. Tsuji, M. Tosaka, A. Kawaguchi, K. Katayama and M. Iwatsuki, *Bull. Inst. Chem. Res., Kyoto Univ.*, **69**, 117 (1991).
12. P.H. Geil, *Polymer Single Crystals*, John Wiley and Sons (1963)/ (reprint), R.E. Krieger Pub. (1973).
13. B. Lotz and J.C. Wittmann, In *Materials Science and Technology, "Structure and Properties of Polymers"*, Vol.12, Eds., R.W. Cahn, P. Haasen and E.J. Kramer, V.C.H., Chapt.3, pp.79-151 (1993).
14. H. Kawamura, M. Tsuji, A. Kawaguchi and K. Katayama, *Bull. Inst. Chem. Res., Kyoto Univ.*, **68**, 41 (1990)/ A. Magoshi, M. Yamakawa, M. Tsuji and S. Kohjiya, *Kobunshi Ronbunshu*, **53**, 375 (1996)/ T. Yoshioka, M. Tsuji, Y. Kawahara and S. Kohjiya, *Polymer*, **44**, 7997 (2003).
15. E.H. Andrews, *Proc. Roy. Soc.*, **A277**, 562 (1964)/ M. Tsuji, T. Shimizu and S. Kohjiya, *Polym. J.*, **31**, 784 (1999)/ idem, *ibid.*, **32**, 505 (2000).
16. J. Petermann, *Bull. Inst. Chem. Res., Kyoto Univ.*, **69**, 84 (1991).
17. J. Petermann and R.M. Gohil, *J. Mater. Sci. Lett.*, **14**, 2260 (1979).
18. S. Isoda, *Polymer*, **25**, 615 (1984).
19. J.C. Wittmann and B. Lotz, *J. Polym. Sci. Polym. Phys. Ed.*, **23**, 205 (1985).
20. M. Tsuji and K.J. Ihn, *Bull. Inst. Chem. Res., Kyoto Univ.*, **72**, 429 (1995).

21. J.C. Wittmann and B. Lotz, *Prog. Polym. Sci.*, **15**, 909 (1990).
22. M. Fujita, K.J. Ihn, M. Tsuji and S. Kohjiya, *Kobunshi Ronbunshu*, **56**, 786 (1999)/ M. Fujita, M. Tsuji and S. Kohjiya, *Polymer*, **40**, 2829 (1999)/ idem, *Macromolecules*, **34**, 7724 (2001).
23. T. Kobayashi, In *Crystals: Growth, Properties, and Applications, Vol.13, Organic Crystals I: Characterization*, Springer-Verlag, pp. 1-63 (1991).
24. J.C. Wittmann and P. Smith, *Nature*, **352**, 414 (1991).
25. M. Tsuji, S. Isoda, M. Ohara, A. Kawaguchi and K. Katayama, *Polymer*, **23**, 1569 (1982)/ S. Isoda, M. Tsuji, M. Ohara, A. Kawaguchi and K. Katayama, *Polymer*, **24**, 1155 (1983).
26. R. Iwamoto and B. Wunderlich, *J. Polym. Sci.-Phys.*, **11**, 2403 (1973).
27. W.D. Niegisch, *J. Appl. Phys.*, **37**, 4041 (1966).
28. B.K. Vainshtein, *Structure Anlysis by Electron Diffraction*, Pergamon-Press, Chapt.3, (1964).
29. M. Tsuji, T. Okihara, M. Tosaka, A. Kawaguchi and K. Katayama, *MSA Bull.*, **23**, 57 (1993).
30. M. Tosaka, N. Hamada, M. Tsuji, S. Kohjiya, T. Ogawa, S. Isoda and T. Kobayashi, *Macromolecule*, **30**, 4132 (1997).
31. N. Hamada, M. Tosaka, M. Tsuji, S. Kohjiya and K. Katayama, *Mcromolecules*, **30**, 6888 (1997).
32. M. Tosaka, N. Hamada, M. Tsuji and S. Kohjiya, *Macromolecules*, **30**, 6592 (1997).
33. M. Tosaka, M. Tsuji, L. Cartier, B. Lotz, S. Kohjiya, T. Ogawa, S. Isoda and T. Kobayashi, *Polymer*, **39**, 5273 (1997).
34. M. Tosaka, M. Tsuji, S. Kohjiya, L. Cartier and B. Lotz, *Macromolecules*, **32**, 4905 (1999).
35. M. Kobayashi, T. Nakaoki and N. Ishihara, *Macromolecules*, **22**, 4377 (1989).
36. G. Guerra, V.M. Vitagliano, C. De Rosa, V. Petraccone and P. Corradini, *Macromolecules*, **23**, 1539 (1990).
37. Y. Shimane, T. Ishihara, Y. Chatani and T. Ijitsu, *Polym. Prepr., Jpn.*, **37**, 2534 (1988).
38. Y. Chatani, Y. Shimane, T. Ijitsu and T. Yukinari, *Polymer*, **34**, 1625 (1993).
39. C. De Rosa, M. Rapacciuolo, G. Guerra, V. Petraccone and P. Corradini, *Polymer*, **33**, 1423 (1992).
40. M. Tosaka, M. Yamakawa, M. Tsuji, S. Kohjiya, T. Ogawa, S. Isoda and T. Kobayashi, *Microsc. Res. Technique*, **46**, 325 (1999)/ M. Tosaka, M. Yamakawa, M. Tsuji, S. Murakami and S. Kohjiya, *J. Polym. Sci.: Part B: Polym. Phys.*, **37**, 2456 (1999)/ M. Tosaka, N. Hamada, Y. Yamakawa, M. Tsuji and S. Kohjiya, *Compu. Theoretical Polym. Sci.*, **10**, 355 (2000).
41. M. Tosaka, Y. Endo, S. Murakami, M. Tsuji and S. Kohjiya, *Sen'i Gakkaishi*, **57**, 207 (2001)/ M. Tosaka, Y. Endo, S. Murakami, M. Tsuji, S. Kohjiya, T. Ogawa and S. Isoda, *ibid.*, **57**, 244 (2001)/ M. Tosaka, Y. Endo, M. Tsuji and S. Kohjiya, *J. Mocromol. Sci., Part B-Phys.*, **B42**, 559 (2003).
42. J. Liu and P.H. Geil, *J. Macromol. Sci.-Phys.*, **B36**, 61 (1997).
43. A. Formhals, *U.S. Patent No.1975504*, (1934)/ S. Koombhongse, W. Liu and D.H. Reneker, *J. Polym. Sci.: Part B: Polym. Phys.*, **39**, 2598 (2001)/ R. Dersch, T. Liu, A.K. Schaper, A. Greiner and J.H. Wendorff, *J. Polym. Sci.: Part A: Polym. Chem.*, **41**, 545 (2003).
44. E.M. Woo, Y.S. Sun and C.-P. Yang, *Progr. Polym. Sci.*, **26**, 945 (2001).

CHARACTERISATION OF CATALYSTS BY TRANSMISSION ELECTRON MICROSCOPY

Di Wang

Department of Inorganic Chemistry, Fritz Haber Institute of the Max Planck Society, Faradayweg 4-6, D-14195, Berlin, Germany

Abstract: Catalysts are one of the most important and the most widely used materials. Microstructures in heterogeneous catalysts are closely related to the catalytic properties. TEM and related microanalytic techniques are powerful tools in characterising catalysts at atomic level. The obtained structural information is essential to the understanding of correlations between microstructures and catalytic properties. In this lecture note, the general principle of characterization of catalysts by TEM is introduced and the applications on Pt/SiO$_2$ model system and on VPO catalysts are intensively described.

Key words: Heterogeneous catalysts; High-resolution transmission electron microscopy; Electron diffraction; Electron energy-loss spectroscopy

1. INTRODUCTION

Catalysis is the technology of modifying the rate of a desired chemical reaction by using catalyst materials. This technology represents a great deal of the present chemical, petrochemical, and life science industry [1]. It controls more than 90% of the world's chemical manufacturing processes [2]. The amount of catalysts applied in industry is very large. However, the cost of catalyst is relatively much smaller compared with the revenues. In addition, catalysis also plays important role in environmental protection, pharmacy, agriculture and biotechnology. Due to all these fact, catalyst is one of the most important and the most widely used materials.

In spite of the long history of catalysis application in chemical industry over 150 years, catalysis is largely an empirical science. The difficulty of a

T.E. Weirich et al. (eds.), Electron Crystallography, 473–487.

precise description comes from the enormous complexity of catalyst-reactant interaction in macro-, meso- and micro-dimensional levels and dynamics at different time scales [1]. Catalysis extends over surface science, solid state physics and chemistry, material synthesis and chemical engineering. Various techniques such as Raman spectroscopy, XPS, XAS, UPS, and AES are used to characterise the catalysts and to study the reaction mechanisms. Most of these techniques are surface sensitive. In the past a few decades, the development of electron microscopy and related microanalytical techniques makes it possible to obtain very local structural and chemical information from heterogeneous catalysts, thus helps the understanding of correlations between microstructure and catalytic properties [2–9], which is of great importance to the design of new catalysts.

Application of transmission electron microscopy (TEM) techniques on heterogeneous catalysis covers a wide range of solid catalysts, including supported metal particles, transition metal oxides, zeolites and carbon nanotubes and nanofibers etc.

In many catalytic systems, nanoscopic metallic particles are dispersed on ceramic supports and exhibit different structures and properties from bulk due to size effect and metal support interaction etc. For very small metal particles, particle size may influence both geometric and electronic structures. For example, gold particles may undergo a metal-semiconductor transition at the size of about 3.5 nm and become active in CO oxidation [10]. Lattice contractions have been observed in metals such as Pt and Pd, when the particle size is smaller than 2-3 nm [11, 12]. Metal support interaction may have drastic effects on the chemisorptive properties of the metal phase [13–15]. Therefore the structural features such as particles size and shape, surface structure and configuration of metal-substrate interface are of great importance since these features influence the electronic structures and hence the catalytic activities. Particle shapes and size distributions of supported metal catalysts were extensively studied by TEM [16–19]. Surface structures such as facets and steps were observed by high-resolution surface profile imaging [20–23]. Metal support interaction and other behaviours under various environments were discussed at atomic scale based on the relevant structural information accessible by means of TEM [24–29].

Transition metal oxides attract great interests mainly due to their redox nature, which is thought to be related with their flexible structure modification under reductive and oxidative conditions. Such structure modification takes place by forming so called crystallographic shear (CS) structures to accommodate anion vacancies in specific crystallographic planes by simultaneous shear displacement and crystal structural collapse [30–32]. High-resolution transmission electron microscopy (HRTEM) is a promising technique that reveals local structure of nonstoichiometry [33–39].

With rapid development of zeotypic materials and mesoporous solids and their application in heterogeneous catalysis, HRTEM shows its advantages in distinguishing the ultrastructural features [40, 41]. Carbon materials are used as support in catalytic reactions due to some of their specific characteristics and many publications report the TEM investigations on various forms of carbon related materials [42–48].

Electron energy-loss spectroscopy (EELS) is nowadays widely used to obtain the information with respect to chemical composition, oxidation state and electronic structure of solids. Since all catalytic processes concern the exchange of electrons between the reactants, EELS is extremely valuable in catalysts investigations [9, 49–57]. EELS in an electron microscope exhibits the advantage of high spatial resolution in area of interests with simultaneous structure determination by electron diffraction and imaging.

At the same time, one should notice that the real catalysts are applied in the gas/liquid environments at usually an increased temperature so that dynamic structural evolution of a real catalyst would not be probed in a conventional electron microscope. To bridge the gap, in situ environmental electron microscope is developed by placing a micoreactor inside the column of an electron microscope to follow catalytic reaction processes [58–62]. However, the specimen in an in situ TEM may suffer from interaction with ionised gas (plasma), making the interpretation of in situ TEM study of catalytic reaction more complicated. Characterisation of static, post-reaction catalysts is still the most commonly used. Well-designed model catalysts and reasonable interpretation of the results are essential to a successful study.

This lecture note includes the introduction to TEM techniques that are often used for structure characterisation in heterogeneous catalysis and subsequently some examples of application.

2. TEM TECHNIQUES IN HETEROGENEOUS CATALYSIS

Heterogeneous catalysts are usually complex in nanoscale and may have different structures from the average. TEM is a unique tool in investigation of local structure and chemistry of the catalysts. The principles of TEM can be found in many specialised books [63–67]. In this section, electron diffraction, HRTEM and EELS will be introduced concerning the practical application on heterogeneous catalysts.

2.1   Electron diffraction

Electron diffraction can be used for both single crystal and polycrystalline material to determine the phases and lattice parameters. In electron diffraction pattern, a diffraction spot arises from the reflection by a series of crystallographic planes (*hkl*) and the geometry is described by Bragg's law, $2d\sin\theta = \lambda$, where λ is the wavelength of the incident electrons, d the lattice planar spacing of *hkl* crystallographic planes and θ the scattering angle. When scattering angle is small, $2\sin\theta$ can be replaced by D/L, where D is the distance from the incident beam spot to the diffraction spot measured in negative film and L the camera length. The lattice planar spacing d_{hkl} therefore can be derived by $d_{hkl} = \lambda L/D$. For samples containing a known phase, a better method is to use the diffractions from the known phase for calibration and the lattice planar distance d values corresponding to the other diffractions in the same pattern can be calculated. Phase determination is achieved by comparing the d values with the crystallographic data of possible phases. Due to the complexity a diffraction pattern often exhibits, other knowledge about the sample is often helpful.

For structural study of small features in catalysts, e.g., nanoparticles with size of a few to several tens nanometers, electron microdiffraction is especially useful [65, 68]. By choosing small spot size, electron beam can be converged into a probe illuminating only the area of interests. With convergence angle smaller than Bragg diffraction angle, diffraction disks are resolved individually, i.e., so called Kossel-Möllenstedt pattern [69] can be obtained. Indexing the zero order Laue zone (ZOLZ) in such a pattern is similar to that of selected area electron diffraction (SAED) pattern. In addition, the dynamical contrast in diffraction disks can be used for symmetry determination [68].

2.2   HRTEM

HRTEM with resolution below 0.2 nm directly reveal the local structures of catalysts, which are highly relevant to special properties and offer information about reaction mechanisms. Crystallinity, lattice spacings and crystallographic orientations of the sample, as well as defect structures such as dislocation, planar defect, interface and cluster can be readily resolved. Surface facets and features such as roughness and decoration by other species can also be resolved by profile imaging. Care must be taken when interpreting HRTEM images. Only for the sample thin enough, satisfying weak-phase object approximation (WPOA) [63] or pseudo-WPOA [70], the contrast in a HRTEM image taken near Scherzer focus condition [71] corresponds to the positions of atomic columns. Some features that are not distinct in the image can be revealed by Fourier transform (FT) of a HRTEM image or part of it, e.g., overlapping of more than one phase and lattice fringes in a very noisy image.

2.3 EELS

Coulomb interaction between a fast incident electron and atomic electron may result in excitation of the atomic electron to an unoccupied electronic state at high energy level and at the same time a characteristic amount of energy loss of the incident electron. By detecting the energy distribution of electrons after passing through the sample, the chemical composition of a material can be known. In addition, energy-loss near edge structure (ELNES) reflects the unoccupied states of higher energy level and therefore can be used for probing the electronic configuration. ELNES on appropriate ionisation edges of oxygen, carbon and metals, etc., can serve as "fingerprints" reflecting changes in oxidation state, chemical bonding and coordination environment of the detected species.

The amount of each element can be deduced by measuring the area under the ionisation edge, after subtracting the background and removing the plural scattering [49]. In case ionisation edges of two elements are resolvable in one spectrum, the relative concentration ratio of these two elements can be deduced. The specific methods for EELS quantitative analysis are described elsewhere [49, 72]. The quantitative information helps the determination of phases, especially for unknown structures. By using electrons at an energy loss corresponding to a characteristic inner-shell excitation, it is possible to obtain image representing the distribution of the concerned element [73]. Such element mapping can reach the lateral resolution down to 1 nm.

3. TEM INVESTIGATIONS ON SOME HETEROGENEOUS CATALYSTS

3.1 Metal support interaction in Pt/SiO$_2$ model catalyst

In view of the complexity of "real" supported catalysts, consisting of randomly oriented and irregularly shaped metal particles on high surface area porous supports, well oriented and regularly shaped metal particles grown on planar thin supports are frequently used as model catalysts [19]. This facilitates the study by surface science and TEM techniques [11, 74, 75]. In the present work, Pt particles were grown at 623 K by electron beam evaporation of Pt at a pressure of 10^{-6} mbar on vacuum-cleaved (001) NaCl single crystal. They were then covered with a thin supporting film of amorphous silica (25 nm thick). Subsequently, the NaCl was dissolved in distilled water and the films were washed and mounted on gold grids. After an oxidising treatment in O_2 at 673 K for 1 h, the reduction was performed with 1 bar hydrogen gas at 873 K for 1 h. The effects of metal support interaction were then examined ex-situ in a Philips CM200 FEG microscope using SAED, microdiffraction, HRTEM and EELS.

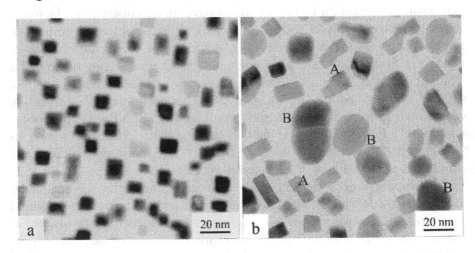

Figure 1. Overview image of silica supported Pt particles (a) as-grown and (b) after the reduction in H_2 at 873 K.

Fig. 1a shows an overview image of the as-grown Pt particles. Most of them exhibit square or rectangular shapes with size of 8–15 nm. Subsequent oxidation does not lead to significant changes, except for the appearance of higher-indexed facets. After the reductive treatment at 873K in hydrogen, particles consisting of two rectangular parts (denoted as A in Fig. 1b) and large particles with irregular forms (denoted as B in Fig. 1b) were observed.

The SAED patterns of the two samples are shown in Figs. 2a and 2b, respectively. Fig. 2a is identical to the diffraction pattern of a [001]-oriented Pt single crystal due to the epitaxial growth of particles on the (001) NaCl surface. The diffraction pattern of the sample after reduction shows rings together with some diffuse spots. The particles after the treatment are more randomly oriented than the as-grown ones. Calibrated by the electron diffraction pattern of the as-grown Pt, the interplanar spacings d ($\pm$ 1%) are calculated from diffractions in Fig. 2b and are listed in Table 1. The d values can be attributed to cubic Pt_3Si, monoclinic Pt_3Si, and tetragonal $Pt_{12}Si_5$. The seemingly forbidden diffractions corresponding to Pt 100 and 110 diffractions (arrows in Fig. 2b) probably arise from primary cubic Pt_3Si or monoclinic Pt_3Si, which is slightly distorted from the cubic one.

Table 1. Measured interplanar spacings d compared with those of Pt_3Si (cubic) Pt_3Si (monoclinic), $Pt_{12}Si_5$ and Pt

d (Å)	Pt_3Si (cubic)		Pt_3Si (monoclinic)		$Pt_{12}Si_5$ (tetragonal)		Pt	
	d (Å)	(*hkl*)	d (Å)	(hkl)	d (Å)	(hkl)	d (Å)	(hkl)
3.91	3.88	(100)	3.88	(002)				
3.45					3.48	(301)		
3.00					3.01	(420)		
2.76	2.75	(110)	2.78	(202)	2.76	(331)		
2.69			2.69	($\overline{2}$02)				
2.46								
2.36			2.36	(113)	2.36	(222)		
2.20			2.21	($\overline{2}$22)				
2.13					2.13	(620)		
1.96	1.94	(200)					1.96	(200)
1.81			1.80	(313)	1.82	(003)		
1.50			1.50	(115)				
1.38	1.37	(220)	1.39	(404)			1.39	(220)
1.31								
1.18	1.17	(311)					1.18	(331)

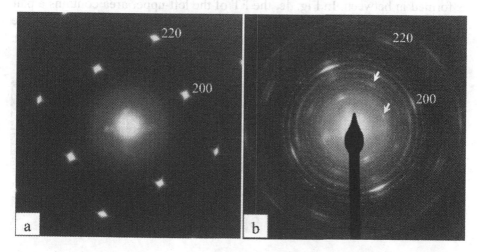

Figure 2. Electron diffraction patterns of silica supported Pt particles (a) as-grown and (b) after the reduction in H_2 at 873 K.

Microdiffraction patterns are taken from individual particles after the reduction treatment and are shown in Figs. 3a–d. Most particles with platelet shape and straight edges produce similar microdiffraction patterns, one of which is shown in Fig. 3a. It is indexed as Pt_3Si with Cu_3Au structure on [100] zone axis. Figs. 3b and 3c show the diffraction patterns from the not reacted Pt on [100] and [310] zone axes, respectively. Particles with irregular forms show various diffractions and a considerable amount of them can be attributed to $Pt_{12}Si_5$. One such pattern is shown in Fig. 3d, exhibiting $Pt_{12}Si_5$ on [152] zone axis.

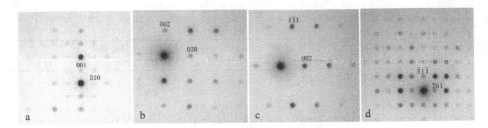

Figure 3. Microdiffraction patterns from four single particles after the reduction, indexed as
(a) [100] Pt_3Si, (b) [100] Pt, (c) [310] Pt and (d) [152] $Pt_{12}Si_5$.

Besides from the silicides, other influences on the microstructure after
the reduction at 873 K are observed in HRTEM images. Fig. 4a shows the
beginning stages of a coalescence process of three particles. The curved
edges in contacting area indicate the diffusion and the rearrangement of
atoms. Fig. 4b shows a coalescence process at an advanced stage. The
particle consists of two grains that may stem from two particles. An interface
is formed in between. In Fig. 4c, the FT of the left-upper area contains a pair
of strong reflections corresponding to the lattice fringes (indicated by the
arrow) as well as the reflections arising from Pt (indicated by the circles).
This indicates the overlapping of two phases.

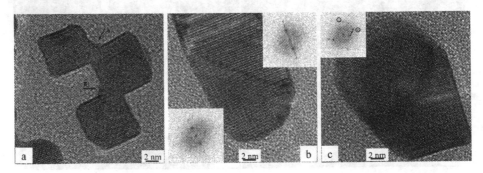

Figure 4. HRTEM images of particles after the reduction, showing (a) the coalescence of
three particles, (b) the coalescence of two crystallites with an interface formed in between and
(c) the overlapping of different phases.

ELNES on Si L ionisation edge can be used to reveal the change in
coordination of Si atoms due to the reductive treatment. After background
subtraction and removal of multiple scattering [49], the spectra from
different areas for both samples are plotted in Fig. 5. Before reduction, the Si
L ELNES from as-grown silica and from the area with Pt particles exhibit
the typical Si L ELNES of SiO_2, identical with the one measured from the
SiO_2 substrate after the treatment. Some new features appear in the Si L
ELNES spectrum obtained from particle after the reduction. This is a

signature of silicon in a changed chemical environment compared to SiO_4 tetragonal coordination, probably due to the reduction of silica and the formation of silicide proceeding through the interface between the Pt particle and the silica substrate.

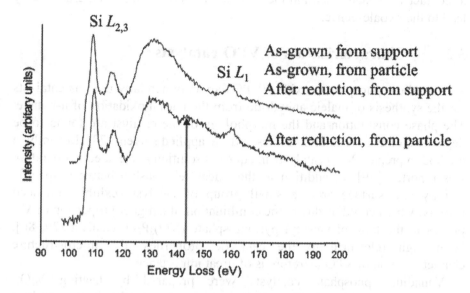

Figure 5. Si *L* ELNES spectra taken from the free silica substrate and from areas with particles for as-grown sample and that after heating in hydrogen at 873 K.

As suggested in [25], the formation of platinum silicides results from the reduction of SiO_2 by atomic hydrogen in the presence of platinum. This process involves the dissociative adsorption of hydrogen on platinum, the penetration of the metal support interface by atomic hydrogen and the reduction of SiO_2 accompanied by the migration of Si atoms into the Pt particles. The above-described observations support the mechanism. In addition, by growing dispersed Pt particles with a common crystalline orientation and regular shapes as the initial state, a topotactic growth of the Pt_3Si phase from the Pt particles is deduced from the azimuth orientation of Pt_3Si, which more or less coincides with that of the as-grown Pt. The formation of the Pt rich Pt_3Si phase was assumed as the first step of Pt/SiO_2 interaction in hydrogen at high temperature [28]. The regular (mostly rectangular platelet-like) shape of Pt_3Si particles is due to the surface reconstruction during reduction, leading the system to minimum surface energy [76].

Melting and recrystallisation must be taken into account to interpret the formation of the irregularly shaped $Pt_{12}Si_5$ particles oriented on various zone axes. The reason may be the decreased melting temperature of small particles caused by surface pre-melting and the surface tension of the solid-liquid interface [77]. The flattening of the particles increases the contacting area with the substrate and the silicide formation also increases the particle size. These effects may cause the as-grown particles with very close spacing to contact with each other and the surface diffusion at high temperature may lead to their coalescence.

3.2   Characterisation of VPO catalysts

Vanadium phosphorus oxides (VPO) are commercially used as catalysts for the synthesis of maleic anhydride from the partial oxidation of n-butane. The phase constitution and the morphology of the catalyst are found to be dependent on the preparation routes and the applied solvent [78]. Recently, a method to prepare VPO catalysts in aqueous solution at elevated temperature was reported [79]. In addition to the linear relationship between specific activity and surface area, a small group of catalysts exhibit enhanced activity, which could be due to the combination of a higher proportion of V^{5+} phases in the bulk of vanadyl pyrophosphate $(VO)_2P_2O_7$ catalyst [79, 80]. With high relevance to the catalytic properties, the microstructure characterisation of VPO therefore is of great importance.

Vanadium phosphate catalysts were prepared by heating V_2O_4, phosphorus acid, either H_3PO_4 or $H_4P_2O_7$, and water together in an autoclave at 145°C for 72 hours. Afterwards, the solid produced was recovered, washed with distilled water and dried in air at 120°C for 16 hours. Detailed preparation procedure is described in [79]. Such prepared precursors were activated in n-butane/air at 400°C to form the final catalysts. TEM and EELS are used to study the catalysts in Philips CM200 FEG microscope.

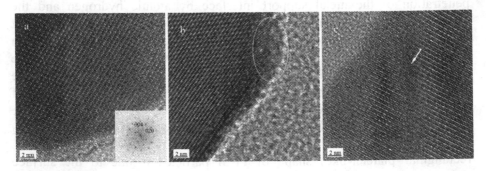

Figure 6. HRTEM images of the activated catalyst prepared from H_3PO_4, showing (a) [100] projected $(VO)_2P_2O_7$, (b) $VOPO_4$ at the edge of $(VO)_2P_2O_7$ crystallite and (c) amorphous region (indicated by the arrow) in $(VO)_2P_2O_7$ crystallite.

In the activated catalyst prepared from H_3PO_4, $(VO)_2P_2O_7$ is the dominant phase. Most particles provide HRTEM images with highly ordered structure, as shown by HRTEM image in Fig. 6a. From the FT inset, the image can be indexed as $(VO)_2P_2O_7$ on its [100] zone axis. Other structures are also observed in some microdomains. Fig. 6b shows different contrast from the bulk at the edge of a crystallite. Some regions with defect structures (e.g. in Fig. 6c) may indicate the existence of beam sensitive V^{5+} phases.

In the activated catalyst prepared from $H_4P_2O_7$, many crystallites exhibit $(VO)_2P_2O_7$ structure and one HRTEM image is shown in Fig. 7a. Another distinct phase present in the sample can be assigned to αII-$VOPO_4$, as revealed by HRTEM image in Fig. 7b and its FT inset. The αII-$VOPO_4$ crystallites are particularly beam sensitive and some disorder introduced by electron irradiation can be seen in Fig. 7b. Moreover, in the middle or at the edge of $(VO)_2P_2O_7$ crystallites, different features and different degrees of beam sensitivity are observed, which can also be attributed to V^{5+} phases.

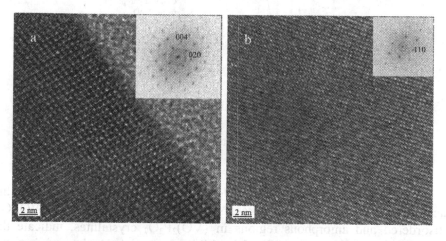

Figure 7. HRTEM images of the activated catalyst prepared from $H_4P_2O_7$, showing (a) [100] projected $(VO)_2P_2O_7$ and (b) [001] projected αII-$VOPO_4$.

ELNES on V $L_{2,3}$ edge and O K edge shows characteristic features reflecting the vanadium valence and the oxidation state. In Fig. 8, the EEL spectra from the two catalysts are compared with the spectra from the reference $VOPO_4$ samples. Since the EEL spectra of different $VOPO_4$ phases are identical, only one of them is shown. For the V^{5+} $VOPO_4$ phases, the maximum of V L_3 edge is at about 519.2 eV, while for the two catalysts, the maximum of V L_3 edge obviously shifts to a lower energy loss at about 517.9 eV. Such shift is associated with the reduction in vanadium valence, corresponding to V^{4+} in $(VO)_2P_2O_7$. Furthermore, an obvious change in relative heights of V L_3 and L_2 features is also observed. The present results are in agreement with the case of V_2O_5 thermal decomposition [81]. A similar shift takes place for O K edge and some changes in ELNES of O K edge can also be seen, indicating the different oxidation state in the catalysts and the $VOPO_4$ phases.

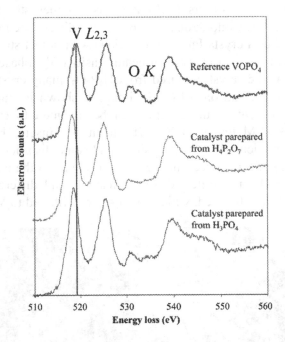

Figure 8. EEL spectra of the two catalysts and the reference VPO samples.

With HRTEM imaging, $(VO)_2P_2O_7$ is identified as the major phase in the activated catalysts. Other features, such as different phases of $VOPO_4$, disordered and amorphous regions in $(VO)_2P_2O_7$ crystallites, indicate the existence of V^{5+} species. EELS exhibits potential in determining the vanadium valence and the oxidation states and can be used for detecting the local electronic structures for VPO catalysts. The catalytic reaction and the present structural investigations suggest that the interaction between V^{4+} and V^{5+} phases is essential to the improvement of specific catalytic activities.

4.   SUMMARY

TEM and associated techniques such as EELS are powerful tools in investigations of heterogeneous catalysts. Electron diffraction can provide structural information in phase constitutions by electron crystallographic analysing methods. Microdiffraction offers the possibility in studying small particles down to several nanometers. In combination with electron diffraction and in aid of FT, HRTEM can be widely used in direct observations of features from perfect crystalline structure of bulk and surface to dislocations, planar defects and even amorphous regions in catalysts, which are often of great importance to catalytic properties. EELS, with high spatial resolution and high energy resolution, is a valuable analytic tool to access the local electronic structures of catalysts. ELNES on appropriate ionisation edges shows characteristic features that can be used to deduce the valence state of elements and coordination environment of atoms.

Nevertheless, one must be aware of the possible discrepancy between the observation in an electron microscope and the real structure of catalysts. The discrepancy may come from (1) using model catalyst, which is different from the real one, (2) electron beam induced effects and (3) conditions in an electron microscope that is very different from the ambient of the activated catalysts. For ex situ investigation, it is important to work with good designed model systems whose catalytic activities can be experimentally measured and care must be taken when relating the observed structure with reactivity of catalysts.

Acknowledgements

The author thanks Prof. R. Schlögl and Dr. D.S. Su for the advices on investigations in Fritz Haber Institute, Prof. K. Hayek and Mr. S. Penner (Leopold-Franzens University, Innsbruck, Austria) for the Pt/SiO_2 samples, Prof. G.J. Hatchings and his group (Cardiff University, UK) for the VPO samples, and Mr. A. Liskowski (Fritz Haber Institute) for the TEM work on VPO. The work is partly supported by the SFB564 of the Deutsche Forschungsgemeinschaft (DFG).

References

[1] R. Schlögl, *Cattech* 5 (2001) 146.
[2] P.L. Gai, E.D. Boyes, *Electron Microscopy in Heterogeneous Catalysis*, Institute of Physics Publishing, Bristol and Philadelphia (2003).
[3] A.K. Datye, *Catal. Rev.* 34 (1992) 129.
[4] J.V. Sanders, *J. Electron Micr. Tech.* 3 (1986) 67.
[5] A. Howie, L.D. Marks, S.J. Pennycock, *Ultramicroscopy* 8 (1982) 163.
[6] H. Poppa, K. Heinemann, *Optik* 56 (1980) 183.
[7] M.J. Yacaman, *Appl. Catal.* 13 (1984) 1.
[8] J.S. Anderson, *Chem. Scripta.* 14 (1978-79) 129.
[9] F. Delannay, in: *Characterization of Heterogeneous Catalysts*, F. Delannay (ed.), Marcel Dekker, New York and Basel (1984) p. 71.
[10] M. Valden, X. Lai, D.W. Goodman, *Science* 281 (1998) 1647.

[11] G. Rupprechter, H.-J. Freund, *Top. Catal.* 14 (2001) 3.

[12] S.A. Nepijko, M. Klimenkov, M. Adelt, H. Kuhlenbeck, R. Schlögl, H.-J. Freund, *Langmuir* 15 (1999) 5309.

[13] G.L. Haller, D.E. Resasco, *Adv. Catal.* 36 (1989) 173.

[14] G.J. den Otter, F.M. Dautzenberg, *J. Catal.* 53 (1978) 116.

[15] G. Rupprechter, G. Seeber, H. Goller, K. Hayek, *J. Catal.* 186 (1999) 201.

[16] T. Wang, C. Lee, L.D. Schmidt, *Surf. Sci.* 163 (1985) 181.

[17] R. Lamber, W. Romanowski, *J. Catal.* 105 (1987) 213.

[18] P.J.F. Harris, *Surf. Sci.* 185 (1987) L459.

[19] G. Rupprechter, K. Hayek, L. Rendon, M.J. Yacaman, *Thin Solid Films* 260 (1995) 148.

[20] W. Zhou, J.M. Thomas, *Curr. Opin. Solid St. M.* 5 (2001) 75.

[21] L.D. Marks, D.J. Smith, *Nature* 303 (1983) 316.

[22] O. Terasaki, J.M. Thomas, G.R. Willward, *Proc. Roy. Soc. Lond. A* 395 (1984) 153.

[23] D.A. Jefferson, J.M. Thomas, G.R. Willward, K. Tsuno, A. Harriman, R.D. Bryson, *Nature* 323 (1986) 428.

[24] K. Hayek, M. Fuchs, B. Klötzer, W. Reichl, G. Rupprechter, *Top. Catal.* 13 (2000) 55.

[25] R. Lamber, N. Jaeger, *J. Appl. Phys.* 70 (1991) 457.

[26] R. Lamber, N. Jaeger, G. Schulz-Ekloff, *J. Catal.* 123 (1990) 285.

[27] R. Lamber, N. Jaeger, G. Schulz-Ekloff, *Surf. Sci.* 227 (1990) 268.

[28] S. Penner, D. Wang, D. S. Su, G. Rupprechter, R. Podloucky, R. Schlögl, K. Hayek, *Surf. Sci.* 532-535 (2002) 276.

[29] D. Wang, S. Penner, D.S. Su, G. Rupprechter, K. Hayek, R. Schlögl, *J. Catal.* 219 (2003) 434.

[30] A. Magneli, *Acta. Cryst.* 4 (1951) 447.

[31] A.D. Wadsley, in: *Non-Stoichiometric Compounds*, L. Mendelcorn (ed.), Academic, New York (1964) p. 98.

[32] J. S. Anderson, B.G. Hyde, *J. Phys. Chem. Sol.* 28 (1967) 1393.

[33] L.A. Bursill, *Proc. Roy. Soc. A* 311 (1969) 267.

[34] P.L. Gai, *Phil. Mag. A* 43 (1981) 841.

[35] L.A. Bursill, *Acta Cryst. A* 28 (1972) 187.

[36] W. Sahle, M. Sundberg, *Chem. Scripta.* 16 (1980) 163.

[37] T. Ohno, Y. Nakamura, S. Nagakura, *J. Solid State Chem.* 56 (1985) 318.

[38] J. Van Landuyt, R. Vochten, S. Amelinckx, *Mater. Res. Bull.* 5 (1970) 275.

[39] O. Bertrand, N. Floquet, D. Jacquot, *Surf. Sci.* 164 (1985) 305.

[40] J.M. Thomas, O. Terasaki, P.L. Gai, W. Zhou, J. Gonzalez-Calbet, *Accounts Chem. Res.* 34 (2001) 583.

[41] L.A. Bursill, E.A. Lodge, J.M. Thomas, *Nature* 286 (1980) 111.

[42] S. Philippe, C. Massimiliano, K. Philippe, *Appl. Catal. A* 253 (2003) 337.

[43] H.W. Kroto, J.R. Heath, S.C. O'brien, R.F. Curl, R.E. Smalley, *Nature* 318 (1985) 162.

[44] D.S. Bethune, C.H. Kiang, M.S. de Vries, G. Gorman, R. Savoy, J. Vasquez, R. Bayers, *Nature* 363 (1993) 605.

[45] S.C. Tsang, J. Daniels, M.L.H. Green, H.A.O. Hill, Y.C. Leung, *J. Chem. Soc. Chem. Comm.* 17 (1995) 1803.

[46] B. Coq, J.M. Planeix, V. Brotons, *Appl. Catal. A* 173 (1998) 175.

[47] K.P. De Jong, *Curr. Opin. Solid St. M.* 4 (1999) 55.

[48] K.P. De Jong, J. W. Geus, *Catal. Rev.-Sci. Eng.* 42 (2000) 481.

[49] R.F. Egerton, *Electron Energy-Loss Spectroscopy in the Electron Microscope*, Plenum, New York (1986).

[50] P.J. Thomas, P.A. Midgley, *Top. Catal.* 21 (2002) 109.

[51] C. Hebert, M. Willinger, D.S. Su, P. Pongratz, P. Schattschneider, R. Schlögl, *Eur. Phys. J. B* 28 (2002) 407.

[52] D.A. Muller, D.J. Singh, J. Silcox, *Phys. Rev* B 57 (1998) 8181.

[53] H. Kurata, E. lefevre, C. Colliex, R. Brydson, *Phys. Rev. B* 47 (1993) 13764.

[54] C. Mitterbauer, G. Kothleitner, W. Grogger, H. Zandbergen, B. Freitag, P. Tiemeijer, F. Hofer, *Ultramicroscopy* 96 (2003) 469.

[55] L.A. Grunes, R.D. Leapman, C.N. Wilker, R. Hoffmann, *Phys. Rev. B* 25 (1982) 7157.

[56] S. Nakai, T. Mitsuishi, H. Sugawara, H. Maezawa, T. Matsukawa, S. Mitani, K. Yamasaki, T. Fujikawa, *Phys. Rev. B* 36 (1987) 9241.

[57] F.M.F. de Groot, M.Grioni, J.C. Fuggle, J. Ghijsen, G.A. Sawatzky, H. Petersen, *Phys. Rev. B* 40 (1989) 5715.

[58] P.L. Gai, *Curr. Opin. Solid St. M.* 5 (2001) 371.

[59] R.T.K. Baker, P.S. Harris, *J. Phys. E Sci. Instrum.* 5 (1972) 793.

[60] R.T.K. Baker, M.A. Barber, P.S. Harris, F.S. Feates, R. J. Waite, *J. Catal.* 26 (1972) 51.

[61] T.W. Hansen, J.B. Wagner, P.L. Hansen, S. Dahl, H. Topsoe, C.J.H. Jacobsen, *Science* 294 (2001) 1508.

[62] P.L. Hansen, J.B. Wagner, S. Helveg, J.R. Rostrup-Nielsen, B.S. Clausen, H. Topsoe, *Science* 295 (2002) 2053.

[63] J.M. Cowley, *Diffraction Physics* (2ed ed.), North-Holland, Amsterdam (1984).

[64] C. J. H. Spence, *Experimental High-Resolution Electron Microscopy* (2ed ed.), Oxford University Press, Oxford (1988).

[65] D.B. Williams, C.B. Carter, *Transmission Electron Microscopy*, Plenum Press, New York and London (1996).

[66] P.R. Buseck, J.M. Cowley, L Eyring (eds.), *High-resolution transmission electron microscopy and associated techniques*, Oxford University Press, New York (1988).

[67] R. Ludwig, *Transmission electron microscopy: physics of image formation and microanalysis* (3rd ed.), Springer, Berlin (1993).

[68] J.C.H. Spence, J.M. Zuo, *Electron Microdiffraction*, Plenum Press, New York (1992).

[69] W. Kossel, G. Möllenstedt, *Ann. Der Phys.* 36 (1939) 113.

[70] F.H. Li, D. Tang, *Acta Cryst. A* 41 (1985) 376.

[71] O. Scherzer, *J. Appl. Phys.* 20 (1949) 20.

[72] F. Hofer, *Microsc. Microanal. M.* 2 (1991) 215.

[73] H. Shuman, C.F. Chang, A.P. Somlyo, *Ultramicroscopy* 19 (1986) 121.

[74] C.R. Henry, *Surf. Sci. Rep.* 31 (1998) 235.

[75] M. Bäumer, H.J. Freund, *Prog. Surf. Sci.* 61 (1999) 127.

[76] A.-C. Shi and R.I. Masel, *J. Catal.* 120 (1989) 421.

[77] Z.L. Wang, J.M. Petroski, T.C. Green, M.A. El-Sayed, *Phys. Chem. B* 102 (1998) 6145.

[78] A. Liskowski, D.S. Su, R. Schlögl, J. Lopez-Sanchez, J.K. Bartley, G.J. Hutchings, *Microsc. Microanal.* 9 (Suppl. 3) (2003) 316.

[79] J. Lopez-Sanchez, L. Griesel, J.K. Bartley, R.P.K. Wells, A. Liskowski, D.S. Su, R. Schlögl, J.C. Volta, G.J. Hutchings, *Phys. Chem. Chem. Phys.* 5 (2003) 3525.

[80] C.J. Kiely, A. Burrows, G.J. Hutchings, K.E. Bere, J.C. Volta, A. Tuel, M. Abon, *Faraday Discuss.* 105 (1996) 103.

[81] D.S. Su, R. Schlögl, *Catal. Lett.* 83 (2002) 115.

Extended abstracts

STRUCTURAL INVESTIGATIONS OF COLD WORKED IRON BASED ALLOYS AFTER NITRIDING

L. Demchenko, V. Nadutov, D. Beke, L. Daroczi

1 National Technical University of Ukraine "KPI", 37 Peremogy prospect, Kyiv, Ukraine,
2 G.V. Kurdyumov Institute for Metal Physics of the NAS of Ukraine, 36 Vernadsky Blvd, Kyiv,
Ukraine, 3 University of Debrecen, 4010 Debrecen, P.O. Box 2, Hungary

Key words: electron diffraction, iron-based alloy, iron nitrides

1.   INTRODUCTION

At present the iron-based alloys diffusion saturation by nitrogen is widely used in industry for the increase of strength, hardness, corrosion resistance of metal production. Inexhaustible and unrealized potentialities of nitriding are opened when applying it in combination with cold working [1-3]. It is connected with one of important factors, which affects diffusion processes and phase formation and determines surface layer structure, mechanical and corrosion properties, like crystal defects and stresses [4, 5]. The topical question in this direction is clarification of mechanisms of interstitial atoms diffusion and phase formation in cold worked iron and iron-based alloys under nitriding.

Purpose of this work was to study the structure, phase composition and mechanical properties of surface diffusion layers formed in the preliminary cold deformed α-Fe and Fe-Ni alloys after nitriding.

491

T.E. Weirich et al. (eds.), Electron Crystallography, 491–495.
© 2006 *Springer. Printed in the Netherlands.*

2.   EXPERIMENTAL

The specimens of pure iron and Fe-Ni(0.5÷5.0 wt.%) alloys were made as thin foils of 20÷25 μm thickness for electron microscopy, Mössbauer spectroscopy and as plates of 0.8÷1.5 mm thickness for mechanical tests.

Preliminary plastic deformation (PPD) was realized by rolling with deformation degrees: ε = 3, 5, 8, 10, 15, 20, 25, 30, 35, 40, 50, 60, 70 %. The following gaseous saturation with nitrogen was performed in ammonia (NH_3) environment at T = 623÷853 K during τ = 0.5÷2.0 hrs for plates and τ = 5÷15 min for foils.

The diffusion layers obtained in specimens were investigated by means of metallography, electron microscopy, microhardness test, X-ray diffractometry and Mössbauer spectroscopy.

3.   RESULTS

As a result of the microstructure SEM investigations (fig. 1 and 2) it was established that the diffusion layer was a combination of surface layers of nitride phases and zone of internal saturation.

Thus, the diffusion layers in iron foils after nitriding consist of phases as follows:

ε-phase as a white homogeneous ~5 μm thickness acid-proof layer;

γ'-and ε-phases eutectic as a heterogeneous sub-layer;

α-solid solution of nitrogen in *bcc* iron lattice with disperse ε- and γ'-nitride precipitations representing a non-homogeneous core of foils.

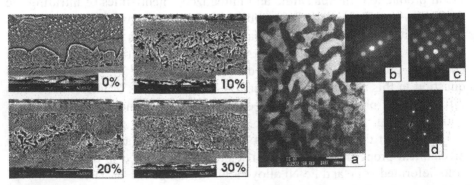

Figure 1. The electron-microscopic images of microstructure (on the left) and eutectic (a) of the phases formed in the Fe-foils after PPD and nitriding at T=853 K during 15 minutes, their electron-diffractions (b, c, d)

PPD effects on surface layers microstructure. The porosity is observed in the core of deformed specimens unlike non-deformed ones as a result of ε-and

γ'-phases' supersaturation by nitrogen and molecular N_2 precipitation and can be caused stresses relaxation nascent when nitrided phases grow. The thickness of ε-phase layer also depends on deformation degree and changes in the range of 25-43 μm.

The TEM data (fig. 1) showed that the dark component of the eutectic has the *fcc* structure with lattice parameter a=0.378±0.004 nm (γ'-phase) and the light component has the *hcp* structure (ε-phase). The lattice constant of the α-solid solution of N in Fe is a=0.286 ±0.004 nm.

The nitrided layers formed in Fe-Ni alloys some differ from those in α-Fe. These differences consist in the presence of additional layer of ξ-phase on the surface (fig. 2). Due to ξ-phase high hardness and brittleness this layer can be destroyed during mechanical treatment.

Moreover, PPD effects on phase formation. The γ'-phase precipitations (fig. 2) form in the zone of internal saturation, in the core of only non-deformed (0% PPD) Fe-Ni foils as thin dark needles of 1-2 μm length. The γ'-phase needle orientation depends on the grains orientation, they oriented along slip planes.

The ε-phase precipitations form as white non-etching precipitations of 0.3-0.5 μm size both in non-deformed and deformed Fe-Ni foils (fig. 2).

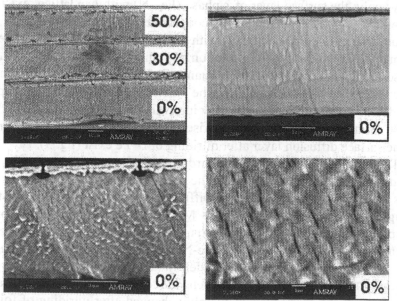

Figure 2. The electron-microscopic images of microstructure formed in the Fe-0.5 % Ni *foils after deformation and nitriding at T=623 K during 5 minutes*

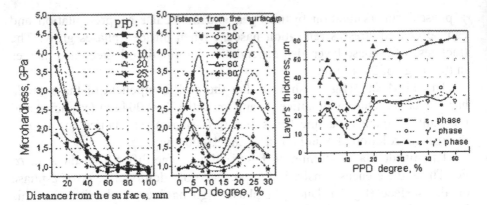

Figure 3. The microhardness distributions in depth of the nitrided layers in Fe after different degrees of deformation at T=853 K during 1 hr.

Figure 4. The thickness of ε- and γ'- nitrides in Fe after deformation and nitriding at T=853 K during 1 hr.

X-ray diffractometry confirmed that the surface diffusion layer depending on PPD degree consists of: the surface layer of ξ-phase (Fe$_2$N) with orthorhombic lattice, contenting 11.0-11.35 wt.% N, which forms in Fe-Ni alloys only; the surface layer of ε-phase (Fe$_{2-3}$N) with *hcp* lattice, existing in the wide 8,15-11,0 wt.% range of N concentration, forms both Fe and Fe-Ni; the underlayer of γ'- phase (Fe$_4$N) with *fcc* structure, forms in α-Fe; the zone of internal saturation is a solid solution of N in *bcc* α-phase with ε-or/and γ'-phase precipitations. The relative quantity of nitrides in the diffusion zone depends on the degree of PPD. The microhardness test in the depth of nitrided layers discovered the narrow intervals of deformations of 3-8 % and 20-30 % in which the considerable rise (in about 2 times) of microhardness of the surface diffusion layer after nitriding of α–Fe exist (fig. 3).

The correlation between microhardness (fig.3) and thickness (fig. 4) of the diffusion layers was revealed. The distribution of nitrogen atoms in the foils of Fe after deformation and diffusion saturation with N was studied using Mössbauer spectroscopy. The Mössbauer spectra obtained from the samples in initial state have typical shape for such absorbers and they consist of one sextet (Fe) and two sextets (Fe-Ni, Fe-Cr). The shape of the spectra is not changed in general after plastic deformation of samples. The line width increases that is attributed to broaden due to high density of dislocations. The shape of the spectra is considerably changed after nitriding of foils. An analysis of the hyperfine parameters has shown that the NGR spectra consist of several components relating to iron atoms in α-Fe and iron nitrides (ε-, γ'-phase).

4. CONCLUSION

The preliminary plastic deformation considerably effects on the phase formation, structure, microhardness and thickness of nitrided layers in α-Fe and Fe-Ni alloys. The *high microhardness* of the diffusion layers results from the formation of the ε- and γ'- nitrides. Iron doping with Ni leads to changing of the ε-, γ'-phases composition. The existence of narrow intervals of deformations of 3-8 % and 20-30 %, in which the considerable (about 2 times) rise of microhardness of surface nitrided layers due to accelerated formation of ε- and γ'-phases, was found.

The *non-monotonous* dependence of surface layer microhardness on deformation degree results from different mechanisms of nitrogen diffusion in deformed material. In our point of view, under the deformations of 3-8 and 20-30 % the greatest number of *mobile dislocations,* capable to provide the additional transfer of nitrogen interstitial atoms with *Cottrell's* atmospheres by *the dislocation-dynamic mechanism* [6-8], can be formed.

The maximum rise of number of mobile dislocations in the deformed materials occurs in the range of 10-20 % [9, 10]. Such processes influence on kinetic of phase formation that results in the accelerated growth of ε- and γ'- nitrides and in increase of microhardness of the surface diffusion layers.

Acknowledgements

The work was carried out with financial support of INTAS (fellowship grant N° YSF 2001/2-77

References

[1] S.I. Sidorenko, et al. Met.Phys.Adv.Tech. Vol.18 (2000) p. 753-762
[2] S.I. Sidorenko, et al. Met.Phys.Adv.Tech.Vol.18 (2000) p. 1383-1392
[3] L.D.Demchenko, et al. Met.Phys.Adv. Tech. Vol.22 (2000) p. 72-79
[4] L.D.Demchenko, et al. Met.Phys.Adv.Tech.Vol.19(2001) p. 711-718
[5] L.D.Demchenko, et al. Bulletin Cherkasy State University. Phys. Vol.37-38 (2001-2002) p. 229-233.
[6] O.V. Klyavin, et al. Fiz. Tverd. Tela Vol. 20 (1978) p. 3100-3106
[7] A.N. Orlov: Fiz. Tverd. Tela Vol. 22 (1980) p. 3580-3585
[8] V.B. Brik Ukr. Fiz. Zh. Vol. 29 (1984) p. 1059-1061
[9] L.D. Demchenko. Acta Cryst. Vol.A58 (2002) (Suppl) p. 146
[10] L.Demchenko, et al. Def.Diff.Forum. Vol.216-217 (2003) p. 87-92

STRUCTURAL REFINEMENT OF NANOCRYSTALLINE TiO$_2$ SAMPLES

I. Djerdj, A. M. Tonejc and A. Tonejc
Dept. of Physics, Faculty of Science, University of Zagreb, Bijenička 32
P.O. Box 331, 10002 Zagreb, Croatia

Key words: electron diffraction, structure determination, titanium dioxide

1. INTRODUCTION

TiO$_2$ is used in many applications such as photovoltaic devices, integrated wave guides, gas and humidity sensors, inorganic membranes, catalyst supports and electrochemical displays [1, 2, 3]. Functional properties of TiO$_2$-based ceramics can be improved by a small amount of doping. The substitutional doping of TiO$_2$ with iron Fe (III) has a profound effect on the charge carrier recombination time in this colloidal semiconductor [1, 4, 5, 6]. In the present work, we used two methods for structural characterization, X-ray diffraction (XRD) and selected area electron diffraction (SAED), to analyse the structural properties of pure and iron doped TiO$_2$.

2. EXPERIMENTAL PROCEDURE

Nanocrystalline iron-doped TiO$_2$ samples (as-prepared S2 and annealed at 500^{0}C S4 [7]) and undoped samples (S$_1$ and S$_3$ [8]) were synthesised by a modified sol-gel method. The details of preparation were reported earlier [7, 8]. The X-ray diffraction of the samples was carried out at room temperature using a Philips powder diffractometer (PW 1820) with monochromatized CuKα radiation. Transmission electron microscopy (TEM) and SAED investigations were carried out by using a JEOL JEM 2010 200 kV microscope, Cs=0.5 mm, point resolution 0.19 nm.

T.E. Weirich et al. (eds.), Electron Crystallography, 497–501.

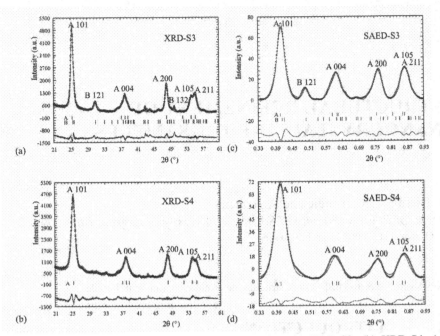

Figure 1. Results of Rietveld refinement of the samples: (a) XRD-S3; (b) XRD-S4; (c) SAED-S3; (d) SAED-S4. The positions of the Bragg reflections of anatase, A, and brookite, B, are indicated by vertical bars (|). The difference curves between the experimental and the calculated intensities from the refined model are shown in the lower part of the diagrams.

In the Rietveld method, the structure and the profile refinements were carried out using the program FULLPROF [9]. The crystal structure was refined in the space group of anatase $I4_1/amd$ (141) and in the space group of brookite *Pbca* (61). The percentages (in wt%) of the phases in the undoped samples were obtained according to the procedure of Hill and Howard [10]. In order to prepare the SAED images for the Rietveld refinement, a digitisation of the film negatives was performed. Then, after a correction due to the non-linearity in the blackening curve with the help of the ELD program [11], we extracted integrated intensities as a function of the 2θ angle and obtained XRD-like patterns. In this case atomic scattering factors for electrons replaced the ones for XRD [12].

3.   RESULTS AND DISCUSSION

Figs. 1(a) and 1(b) show the best fits between calculated (solid line) and observed (dotted line) X-ray diffraction patterns for samples S3 and S4.

Inspecting the difference between the experimental and calculated curves one notices a good agreement between observed and calculated data. The fitted electron diffraction profiles of S3 and S4, together with the difference curve, are shown in Figs. 1(c) and 1(d). Details and results of the refinement are shown in Table 1 for the XRD patterns and in Table 2 for the SAED patterns. The refinement of the XRD data yielded lattice parameters of the anatase phase given in the Table 1. It was found that the lattice parameter c increased in the Fe doped samples in comparison to the undoped ones, the percentages increase is 0.14 % and 0.69 % (XRD), and 0.42 % and 0.53 % (SAED). The lattice parameter a was found to increase for 0.03 % and 0.08 % (XRD), and 0.8 % and 1.6 % (SAED). According to our analysis the increase of lattice parameters in the as-prepared iron doped sample was mainly due to the iron cations action as well as due to the PEG presence [7].

Comparing the values of lattice parameter c and a obtained by XRD with those obtained by SAED (Table 2), one can notice a slight disagreement between their values for the both kinds of samples. This came from the nature of the SAED method itself. The possible cause, which introduces slight difference in the Rietveld refined parameters of the SAED data related to the XRD ones, is the limited area of sample from which the SAED record is taken. This area can contain, for example, more localized defects than the bulk as a whole and can contribute to different values of unit cell parameters. Contrary to the SAED data, the results obtained from the refinement of XRD data are volume averaged. The second source of disagreement could be systematic errors introduced during transformation of Debye-Scherrer rings into Debye-Scherrer XRD-like pattern. The first introduced error is the zero shift, originated from precise determination of SAED centre. Since the program ELD refines SAED centre automatically, its precision determines zero shift value. Inelastic scattering also distorts intensities over the examined range of scattering angles. Dynamical scattering also alters intensities considerably. Similarly, as in the case of inelastic scattering the dynamic effects can be minimised taking SAED pattern from the thin samples. To justify applicability of Rietveld method onto SAED pattern in this work, the absence of dynamical scattering is confirmed with a finding that experimental intensities extracted from SAED pattern are unaltered by dynamical scattering. This was achieved by comparison with X-ray diffracted intensities, shown in Figs. 1(a) and 1(b). The best way to compare the XRD and SAED intensities is to look at the phase composition calculated from XRD and SAED pattern. The phase composition is actually calculated from integrated intensities by refining scale factors of presented phases. The results are also included in Table 1 (XRD) and Table 2 (SAED). One should note that the amount of anatase in sample S1 is 73.3 wt.% calculated from XRD, and 74.5 wt.% calculated from SAED. Clearly, this fairly good agreement of phase composition calculated by refinement of XRD and SAED patterns, indicates the absence of dynamical scattering of electrons and supports the validity of the use of SAED data in structural refinement.

Table 1. Crystallographic data of TiO_2 (anatase) from Rietveld refinement with XRD data.

Sample	S1	S2	S3	S4
Lattice const. (nm)	a=0.3796(1) c=0.9444(3)	a=0.3799(1) c=0.9509(3)	a=0.3785(1) c=0.9482(3)	a=0.3786(2) c=0.9495(1)
Ti-Position (x, y, z)	(0, 0.25, 0.375)	(0, 0.25, 0.375)	(0, 0.25, 0.375)	(0, 0.25, 0.375)
O-Position (x, y, z)	(0, 0.25, z) z=0.168(1)	(0, 0.25, z) z=0.1706(4)	(0, 0.25, z) z=0.166(1)	(0, 0.25, z) z=0.1713(4)
Anatase (%)	73.3	100	73.8	100
Brookite (%)	26.7	0	26.2	0
R_{wp} (%)	9.3	8.1	10.3	7.0
GoF- index	1.1	1.5	1.9	1.8

Table 2. Crystallographic data of TiO_2 (anatase) from Rietveld refinement with SAED data.

Sample	S1	S2	S3	S4
Lattice const. (nm)	a=0.379(1) c=0.941(1)	a=0.385(1) c=0.946(3)	a=0.377(1) c=0.942(2)	a=0.380(1) c=0.946(2)
Ti-Position (x, y, z)	(0, 0.25, 0.375)	(0, 0.25, 0.375)	(0, 0.25, 0.375)	(0, 0.25, 0.375)
O-Position (x, y, z)	(0, 0.25, z) z=0.158(2)	(0, 0.25, z) z=0.169(3)	(0, 0.25, z) z=0.151(2)	(0, 0.25, z) z=0.169(2)
Anatase (%)	74.5	100	77.4	100
Brookite (%)	25.5	0	22.6	0
R_{wp} (%)	2.94	17.3	12.5	14.0
GoF- index	0.9	0.7	0.9	0.8

4. CONCLUSIONS

Iron doping causes the increase of lattice parameters of anatase a and c and unit-cell volume in comparison to the corresponding undoped TiO_2 samples. The Rietveld refinement of SAED patterns gave the values of structural parameters (lattice parameters, atomic coordinates, bond length and angles) close to those refined from XRD patterns. The comparison of the phase composition derived from both diffraction techniques confirmed the absence of dynamical scattering in electron diffraction. In this way, we have shown that the application of the Rietveld method originally developed for determining structures from X-ray powder data is possible and useful in electron powder diffraction data analysis of nanocrystalline materials.

References

[1] J. Moser, M. Grätzel, R. Gallay, Helv. Chim. Acta 70 (1987) 1596.
[2] A. Turković, A. M. Tonejc, S. Popović, P. Dubček, M. Ivanda, S.   Musić, M. Gotić, Fizika A6 (1997) 77.
[3] D. S. Bae, K. S. Han, S. H. Choi, Solid State Ionics 109 (1998) 239-245.
[4] J. Moser, M. Grätzel, Helvetica Chimica Acta, 65 (1982) 1436.
[5] A. R. Bally, E. N. Korobeinikova, P. E. Schmid, F. Lévy, F. Bussy, J. Phys. D: Appl. Phys. 31 (1998) 1149.
[6] F. C. Gennari, D. M. Pasquevich, J. Mat. Sci., 33 (1998) 1571.
[7] A. M. Tonejc, I. Djerdj, A. Tonejc, Mat. Sci. Eng. B85 (2001) 55-63.
[8] A. M. Tonejc, I. Djerdj, A. Tonejc, Mat. Sci. Eng. C19 (2002) 85-89.
[9] J. Rodriguez-Carvajal, FULLPROF- A program for Rietveld   Refinement, Laboratorie Leon Brillouin, CEA-Saclay, France, 2000.
[10] R. J. Hill, C. J. Howard, J. Appl. Cryst. 20 (1987) 467-476.
[11] X. D. Zou, Y. Sukharev, S. Hovmöller, Ultramicroscopy 52 (1993) 436-444.
[12] J. S. Jiang, F. H. Li, Acta Phys. Sin. 33 (1984) 845-849.

References

RELATION BETWEEN MAGNETIC PROPERTIES AND THE STRUCTURE OF IRON-BASED AMORPHOUS ALLOYS DETERMINED BY ELECTRON DIFFRACTION

O.V. Hryhoryeva

National Technical University of Ukraine "Kyiv Polytechnic Institute"37, Peremohy Ave., 03056, Kyiv, Ukraine

Key words: electron diffraction, structure determination, iron-based alloy

The method of complex study of magnetic effects and the changes of electron diffraction patterns during the heating of amorphous alloys is developed. The study was carried out on the alloys Fe-Si-B, that were the bands in amorphous state. The phase composition that correspond to registrated diffraction patterns and to magnetic effects is established.

The equations that describe the magnetic effects and the changes of electron diffraction patterns are got in consequence with the data of X-ray investigation of amorphous alloys and the products of crystallization.

The method of the differential thermomagnetic analysis with use of two etalon of armko-iron was applied [1]. The method of so-called internal etalon was applied also, in which a graduation of registered changes of magnetization was carried out using the well-known magnetic effect in the Curie point of carbide of iron θ-Fe_3C (fig. 1). We hoped that the information about the magnetization of amorphous foils $Fe_{82}Si_2B_{16}$ and about their Curie points might be obtained by using the differential thermomagnetic analysis in the combination with X-ray study. The X-ray researches were carried out using standard diffractometry (Fe-anode). The thermomagnetic curve of heating and cooling of amorphous alloy $Fe_{82}Si_2B_{16}$ is shown on fig. 2. The principal feature of the thermomagnetic curve is that this curve has the reversible character in the temperature interval from room temperature up to approximately 395°C (668 K). This reversible character of thermomagnetic

T.E. Weirich et al. (eds.), Electron Crystallography, 503–506.
© 2006 *Springer. Printed in the Netherlands.*

curve says on our opinion that the changes of magnetization, which are observed, are caused by the nearing to the Curie temperature of the amorphous alloy $Fe_{82}Si_2B_{16}$.

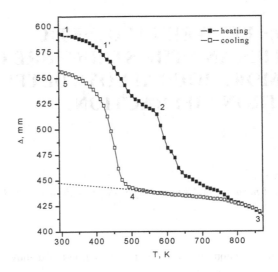

Figure 1. Magnetic effect in the Curie point of the θ-Fe3C carbide.

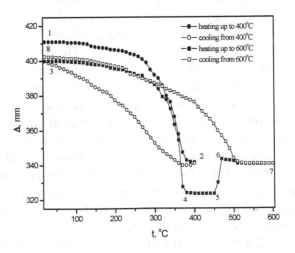

Figure 2. The thermomagnetic curve of cyclic heating of alloy Fe82Si2B16.

The well-known relations of the thermodynamical theory of the temperature dependence of the spontaneous magnetization of a ferromagnetic near Curie point permit to estimate approximately the value of Curie temperature of amorphous alloy $Fe_{82}Si_2B_{16}$. In this case we to use the concrete value of the corresponding thermodynamic coefficient. This value is unknown for the amorphous alloys. We calculated the principal thermodynamic coefficient for pure ferromagnetics Fe, Ni, co and for phases θ-Fe_3C, Fe_3B, Fe_4N. The middle value this coefficient was used for estimating Curie temperature of amorphous alloy $Fe_{82}Si_2B_{16}$. The value we got (390°C, 663 K) is according to data published [2] very reasonable.

The second typical feature of all thermomagnetic curves we got is the fact that the curves have the irreversible character in the temperature interval 425–490°C (698–763 K) during the heating the irreversible increase of magnetization is observed. This process is accompanied by the appearing of X-ray patterns the sharp lines of α-Fe and boride phase Fe_3B and FeB (fig. 3). Also weak interferentions of phases Fe_3Si were observed. These data permit to content that the increase of magnetization saying about is the result of the crystallization of the amorphous alloy:

$$(Fe_{82}Si_2B_{16})_{am} \rightarrow (Fe+FeB)+Fe_3Si+Fe$$

We used the common principles of quantitative analysis of additive properties of alloys developed in work [3].The following equation that describes the magnetic effect during the crystallization of the amorphous alloy $Fe_{82}Si_2B_{16}$ was got:

$$\frac{\Delta_{cr}}{\Delta_\theta} = \frac{\sigma_{Fe}^{(773)}}{\sigma_\theta^{(293)}} \cdot \frac{M'(1-0,24-0,36)}{15C \cdot 10^{-2} M_0}$$

Δ_{cr} and Δ_θ are the corresponding magnetic effects in units of the scale of magnetometer.

In the same way the equation for the determination of the specific magnetization of boride Fe_3B was obtained. This phase is formed during the crystallization amorphous alloy $Fe_{82}Si_2B_{16}$. As we found the specific magnetization of boride Fe_3B is 192 $Tl \cdot 10^{-4}$ cm^3/m. This value is in 1,5 times greater then the specific magnetization of cementite θ-Fe_3C and 1,1 times lower then the specific magnetization of α-iron.

According to experimental data, which we obtained the specific magnetization of alloy $Fe_{82}Si_2B_{16}$ after it's crystallization measured at room temperature, does not differ of the magnetization of this alloy in amorphous state. The calculation shows that the numerical value of the specific magnetization σ ($Fe_{82}Si_2B_{16}$) is 159 Tl·10^{-4} cm^3/g. This value is close to the values in references [4].

The calculation of measured in our experiments changes of magnetization in the temperature interval of phase transformation permit to determine the specific magnetization of the amorphous alloy $Fe_{82}Si_2B_{16}$ at 470°C (743 K, this is the middle temperature of narrow interval of crystallization).

The method of study of magnetic properties that is described above could be used in the investigation of diffusion layers, coatings of different nature; zones are obtained by laser treatment etc.

References

[1] Belous M.V., Hryhoryeva O.V., Lakhnik A.M., Moskalenko Yu.N., Sidorenko S.I., "Thermomagnetic analysis of amorphous alloys Fe-B", Met. Phys. Adv. Tech., Vol. 23(12), (2001) p.1639-1650.

[2] Luborsky F.E., et al. " Magnetic properties of metals and alloys", Eds. Kahn, R.U. and Haasen, P., Metallurgy, Moscow, Vol. 3(29), (1987) p. 624.

[3] Belous M.V., Cherepin V.T., Vasiliev M.A., "Transformations during the tempering of the steels", Metallurgy, Moscow, (1973) p. 232.

[4] Zolotukhin I.V., "Physical properties of amorphous metallic alloys", Metallurgy, Moscow, (1986) p. 176.

SELECTED ISSUES IN QUANTITATIVE STRUCTURE ANALYSIS OF NANOCRYSTALLINE ALLOYS

D. M. Kepaptsoglou*, M. Deanko**, D. Janičkovič**,
E. Hristoforou* and P. Švec**

* National Technical University of Athens, Athens, Greece; ** Institute of Physics, Slovak Academy of Sciences, Bratislava, Slovakia; **Center of Excellence NANOSMART; Corresponding author: P. Švec, e-mail: fyzisvec@savba.skn

Key words: amorphous state, metallic glas, nanocrystalline material, TEM, XRD

1. INTRODUCTION – CRYSTALLIZATION FROM CLUSTERED AMORPHOUS STATE

Quenching alloy melts by rates exceeding 10^6K/sec in certain cases leads to formation of amorphous systems known as metallic glasses. Their local structure and atomic ordering reflects the ordering established in the liquid state and preserved by rapid quenching – atoms are arranged in nonequilibrium positions with a certain degree of short range ordering which is superimposed over complete disorder and structural and chemical randomness [1]. The existence of non-random local ordering in form of clusters containing several atoms with interatomic bonding stronger than that of the surrounding matrix, discussed on the basis of ab-initio simulations [2] and certain experimental evidence [3, 4] is reflected in certain specific features of metallic glasses. These features become manifest especially with relation to their stability and behavior during transformations from amorphous state. As suggested in [3], the crystallization process in clustered media is energetically more advantageous if, instead of disruption of clusters into single atoms diffusing over certain distances, entire clusters as quasi-

507

T.E. Weirich et al. (eds.), Electron Crystallography, 507–511.
© 2006 Springer. Printed in the Netherlands.

stable entities take part in the formation of nuclei and in their subsequent growth. Motion of single atoms over shorter or longer distances becomes thermodynamically more feasible only after exhaustion of the clusters containing higher number of constituent atoms [1] or of the cluster types with lower thermodynamic stability. This phenomenon thus easily explains the generally observed two-stage transformation behavior of metallic glasses. As a rule the first stage is characterized by formation of mainly metallic crystalline grains, often metastable and exhibiting a certain degree of supersaturation with atoms of glass-forming species (*e.g.* metalloids), however, with very low degree of impingement. Further transformation of these metastable phases takes place in form of crystal-crystal transition. The second stage, frequently separated from the first one in temperature and/or time, leads to complete crystallization involving the remaining amorphous matrix, where phases with higher content of glass-forming atoms are formed.

In all investigated cases rapidly quenched amorphous ribbons ~20 μm thick and 6-10 mm wide were prepared by the method of planar-flow casting. Temperatures of subsequent annealing were chosen according to the intervals of crystallization determined from temperature dependencies of electrical resistivity R(T) during linear heating, yielding samples in different transformation stages or with varying (nano)crystalline content. X-ray diffraction (XRD) was performed in a (Bragg–Brentano) $\theta/2\theta$ geometry using Cu–Kα radiation. Microstructure was analysed by transmission electron microscopy (TEM) and electron diffraction using JEM-1200EX microscope. TEM samples were prepared by ion beam milling using a Gatan PIPS. An example of classical amorphous-to-crystalline transformation is shown in Fig. 1 on the case of Fe-Co-Si-B. Crystallization monitored by R(T) exhibits two decreasing steps with apparent classical Arrhenius behavior. Polycrystalline grains ~200nm large containing bcc-Fe(Co) start to appear at 800K. The grain morphology resembles dendritic structure; however, a detailed TEM analysis, confirmed by XRD, shows that the grains are composed of particles ~60nm large (Fig. 1, left). The particles do not grow to significantly larger sizes by the end of the first stage (Fig. 1 middle) or even after the end of the second stage (Fig. 1, right), where formation of tetragonal $(FeCo)_3B$ and additional bcc-Fe(Co) from the amorphous remains can be seen. Low level of impingement of bcc particles formed during the first stage is well noticeable.

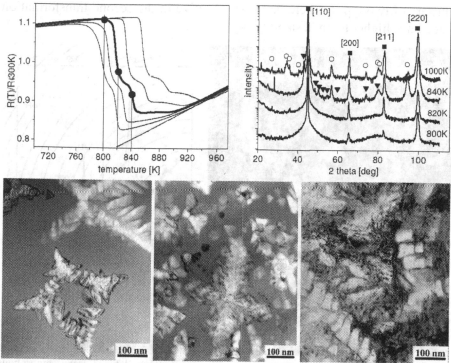

Figure 1. Top left - evolution of electrical resistivity R(T) during two-stage crystallization of amorphous $Fe_{61}Co_{19}Si_5B_{15}$ under different linear heating regimes (2.5, 5, 10, 20 and 40K/min from left to right, respectively). Top right – X-ray diffraction patterns of samples annealed at the temperatures indicated by bullets on R(T) for 60 min showing the presence of bcc-Fe (■), t-Fe$_3$B (▼) and t-Fe$_2$B (○). Bottom – the corresponding TEM pictures for samples annealed at 800, 820 and 840K (from left to right, respectively).

2.   NANOCRYSTALLIZATION IN RAPIDLY QUENCHED ALLOYS

Nanocrystalline alloys can be formed by annealing of certain amorphous precursors, where after the end of the first transformation stage usually at most 50 vol.% are transformed into particles not exceeding few tens of nanometers in size and not in contact with each other. The particles, in most cases spherulitic or nearly equiaxed polyhedric grains, are embedded in amorphous matrix acting as suitable intergranular coupling for enhancement of properties. Specific chemical composition of these systems is given by the formal need for high nucleation and low growth rates, *e. g.* by addition of small amounts of Cu, Au, Pt and Zr, Nb, Mo, respectively. Again, cluster structure of the amorphous precursor is expected, with clusters of two types, where the clusters active in the second transformation stage have higher thermal stability [4].

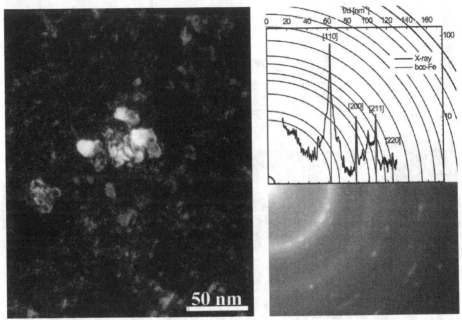

Figure 2. Formation of nanocrystals with bcc-Fe(Co) lattice from amorphous (Fe3Co2)73Nb7Si5B15 annealed at 840K/800min with the corresponding XRD and electron diffraction patterns (right). Vertical lines and indices indicate the positions of the bcc-Fe(Co) peaks; higher order reflections not measured by XRD are easily identifiable on the electron diffraction pattern with sufficient intensity.

Addition of Nb into the Fe-Co-Si-B system from the previous case leads to transformation of ~40 vol.% into grains of bcc-Fe(Co) with dimensions ~30nm (Fig. 2). Even smaller nanograins of bcc-Fe(Mo), not exceeding 8nm are obtained by crystallization of Fe-Mo-Cu-B (Fig. 3), where the stability of the clustered amorphous remains keeps the content of nanocrystalline phase lower than 25 vol.% till almost 1000K. The reasons for this behavior can be traced to drastically enhanced nucleation rate via heterogeneous or instantaneous nucleation, which can decrease the amount of nanocrystallized volume in the first transformation stage even below 20 vol.% [5].

3. CONCLUSION

Three selected cases of crystallization from amorphous state were presented. Two-stage transformation process was observed where the first transformation stage leads to formation of small crystals with sizes and amounts depending on the chemical composition. The content of (nano)crystalline phase was briefly related to the character of nucleation processes and subsequently to the specific quenched-in structure of the amorphous state clustered on atomic scales.

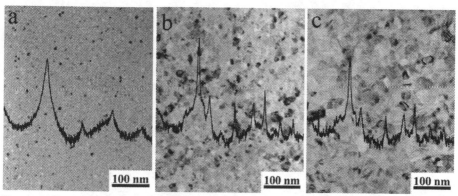

Figure 3. Formation of tiny nanocrystals of bcc –Fe(Mo) from the amorphous matrix in $Fe_{76}Mo_8Cu_1B_{15}$ alloy (a – sample annealed at 823K/1 hour), followed by formation of additional irregular phase with fcc lattice (b – sample annealed at 923K/1 hour) and (c) recrystallization after annealing at 973K/1 hour into polyhedral grains with the same lattice parameters. The insets show the corresponding XRD patterns.

Support of the Slovak Grant Agencies for Science (VEGA 2/2038, APVT-51-052702 and SO 51/03R8 06 03), the SAS Center of Excellence "Nanosmart" and the Greek-Slovak bilateral project No. R&D 01/2001 is acknowledged.

References

1. K. Kristiakova, K., Svec, P. (2001) Continuous distribution of thermodynamic microprocesses in complex metastable systems, Phys. Rev. **B 64**, 184202.
2. Krajci, M., Hafner, J. (2002) Covalent bonding and bandgap formation in intermetallic compounds: a case study for Al3V, J. Phys. – Condens. Matter **14**, 1865.
3. Svec P, Duhaj, P (1990) Growth of crystalline phase in amorphous alloys, Mat. Sci. Eng. **B 6**, 265.
4. Kristiakova, K., Svec, P., Janickovic, D. (2001) Short range order and micromechanism controlling nanocrystallization of Iron-Cobalt based metallic glasses, Mater. Transaction JIM **42**, 1523.
5. Kristiakova, K., Svec, P. Deanko, M. (2004) Cluster structure and thermodynamics of formation of (nano)crystalline phases in disordered metastable metallic systems, Mat. Sci. Eng. A **375-377**, 136.

CHARGE ORDERING AND TILT MODULATION IN MULTIFERROIC FLUORIDES WITH TTB STRUCTURE BY ELECTRON DIFFRACTION AND SINGLE CRYSTAL XRD

E.Montanari [1)], S.Fabbrici [1)], G.Calestani [1)] and A. Migliori [2)]

[1] *Dip. di Chimica GIAF, Università di Parma, Parco Area delle Scienze 17A, I-43100 Parma, Italy, [2]C.N.R.-IMM, Area della Ricerca di Bologna, Via Gobetti 101, I-40126 Bologna, Italy*

Key words: tetragonal tungsten bronze, charge ordering, electron diffraction

Multiferroic materials are characterised by the coexistence of two or three degrees of order (among ferroelectricity, ferroelasticity or magnetic order) and their feature opens interesting fields either for practical application and theoretical investigation. In fact, in recent years, such kind of compounds has gained a larger and larger attention and new compounds are always required.

The term "tetragonal tungsten bronze" (TTB) describes compounds with general formula $(A_1)_x(A_2)_yC_zB_{10}X_{30}$ ($0<x,z<4$ and $0<y<2$) having a layered structure ideally derived from the perovskite structure by rotation of some $[BX_6]$ octahedra and by corner sharing of those units. Linking of rotated blocks produces different types of channels containing A_1, A_2, C ions and the possibility of a partial filling of that tunnels gives rise to a very flexible system either for cation compositions and valence state.

Nowadays a large number of functional crystals belongs to the TTB family and presents electro-optic, ferro- piezo- pyroelectric properties (e.g. the well-known barium sodium niobate $Ba_2NaNb_5O_{15}$ (BNN)). In this work iron and manganese based $K_x(M^{+2})_x(M^{+3})_{1-x}F_3$ fluorides ($0.4<x<0.6$) with TTB structure has been synthesised by solid state reaction and characterised by different diffraction techniques in order to investigate their crystalline structure.

T.E. Weirich et al. (eds.), Electron Crystallography, 513–516.
© 2006 *Springer. Printed in the Netherlands.*

Previous characterisations of similar compounds ($K_{0.6}FeF_3$[1], $K_{0.54}(Mn,Fe)F_3$[2] and $K_{0.38}Mg_{0.38}In_{0.62}F_3$[3]) were mostly performed by powder X-ray diffraction, or sometimes by single crystal X-ray diffraction, assigning the typical TTB structure or related variants. Otherwise, up to date, a unified description of the whole system was not presented. In fact two distinct modulations coexist in the structure[4]: charge ordering of M ions and cooperative tilting involving the MF_6 octahedra of the bronze framework. Each modulation is characterized by its own symmetry and periodicity and, in first approximation, can be considered independent from the other one. General characteristics of those modulations remain unaltered not only on the whole compositional range between $K_{0.4}MF_3$ and $K_{0.6}MF_3$, but also by changing the nature of the M cations (Mn, Fe, Co, Cr, and In). The real structure is a superposition of the two independent modulations as shown in fig.1. The charge order is peculiar of the mixed valence fluoride, while the nature of the tilt modulation is clearly associable to the structural features of TTB niobates, suggesting also for the fluoride bronzes the existence of a generalized ferroelectric-ferroelastic behaviour at room temperature. Since that kind of fluoride bronzes has already been studied for their magnetic properties (frustrated ferrimagnetism[5] at low temperatures), they can be considered as a promising class of multiferroic materials in which coexistence of ferroelectric and ferrimagnetic order is expected in the low temperature region, coupled to ferroelasticity, giving rise to interesting magnetoelectric phenomena.

The structural characterization of TTB fluorides represents a plain example of the wide possibilities offered by the synergic use of different diffraction techniques in solving complex structural problems.

In fact, a preliminary investigation was carried out by powder x-ray diffraction: all the compounds $K_x(M^{+2})_x(M^{+3})_{1-x}F_3$ where $0.4<x<0.6$ and M belongs to metal transition, show a tungsten bronze structure with tetragonal symmetry. Only the composition $K_{0.6}FeF_3$ presents a slight orthorhombic distortion. The existence of the modulation related to a charge ordering is detectable by the presence of weak modulation peaks, which leads to a larger tetragonal cell ($a_{tetr} = b_{tetr} = a_{TTB}$, $c_{tetr} = 2c_{TTB}$) as illustrated in the insertion of fig.2. On the contrary, it is impossible to detect the tilt modulation, because of the to weak intensity of the modulation satellites, which are recordable just by electron diffraction technique or by single crystal X-ray diffraction in proper samples (i.e. single ferroelastic domain crystal). Regarding the approach by single crystal x-ray diffraction, it was easy to realize that it is impossible to solve the structure without preliminary information on the nature of the modulations, which, otherwise, can be obtained by systematic analysis of different samples by selected area electron diffraction technique.

A careful study, performed on SAED patterns taken along different zone axis for different compositions, revealed the existence of two different types of modulation spots (fig.3): the former is strictly related to tilt wave propagation usually observed in TTB oxide, while the latter is clearly compositional depend and associated to the metal transition type. The *a priori* information obtained by SAED, indicating a fairly independent character of the two modulations, suggested a step approach to the single crystal data refinement. The two modulations were refined independently by coupling the fundamental reflections with the correspondent satellites and by taking into account the proper symmetry and periodicity. This multi step approach to the structure solution was successful, revealing the effective independence of the two modulations and therefore the correctness of the SAED indications.

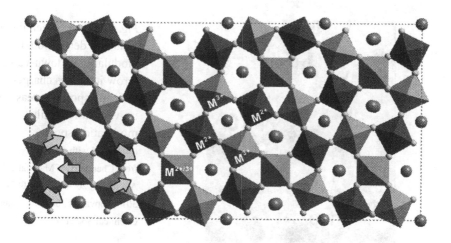

Figure 1. View of the general structure of metal transition TTB fluorides: different colours mean different valence state while the arrows in the low right corner give an idea of the tilt wave propagation. (Reprinted with the permission from Chemistry of Materials, August 10, 2004 16(16) 3007-3019 Copyright 2004 American Chemical Society)

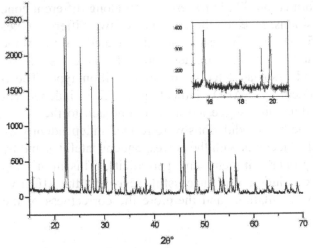

Figure 2. A typical TTB powder diffraction pattern. Insertion: part of the diffraction pattern of $K_{0.4}(Fe,Mn)F_3$ where the presence of slight modulation peaks is pointed out.

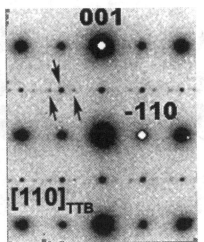

Figure 3. SAED technique is able to detect both the modulations: down-verse arrow indicates charge-ordering spot, whereas up-verse arrows tilt modulation spots.

(Reprinted with the permission from Chemistry of Materials, August 10, 2004 16(16) 3007-3019 Copyright 2004 American Chemical Society)

References:

1. A.M. Hardy, A. Hardy, G. Ferey "Structure Cristalline du Bronze pseudo-Quadratique $K_{0.6}FeF_3$; Transition Pyrochlore-Quadratique pour les Composés $KMM'X_6$" *Acta Cryst.* **1973** B 29, 1654
2. E. Banks, S. Nakajima, G.J.B. Williams "The Crystal Structure of $K_{0.54}(Mn,Fe)F_3$ at Room Temperature" *Acta Cryst.* **1979**, B 35, 46
3. A. Caramanian, N. Dupont, P. Gredin, A. de Kozak "The Crystal Structure of $K_x(Mg_xIn_{1-x})F_3$ (x = 0.38): a new Magnéli-Bronze Type Fluorides" *Z. Anorg. Allg. Chem.* **1999**, 625, 933
4. S.Fabbrici, E.Montanari, G.Calestani and A. Migliori "Charge ordering and tilt modulation in multiferroic $K_xM^{II}_xM^{III}_{1-x}F_3$ (0.4<x<0.6) transition metal fluorides with tetragonal tungsten bronze structure" *Chem. Mater.* **2004** 16(16) 3007-3019
5. P. Lacorre, J. Pannetier, G. Ferey, "The magnetic structure of the tetragonal bronze $KMnFeF_6$" *J. Magn. Magn. Mat.* **1991**, 94, 331

ELECTRON DIFFRACTION BY DISTORTED NANOCRYSTALS
Application of the 'Eikonal Representation'

M.B. Shevchenko, O.V. Pobydaylo

Institute for Metal Physics, 36 Vernadsky St., 03680 Kiev, Ukraine

Abstract: The original approach called the 'Eikonal representation' is used to study the dynamical diffraction of electrons in distorted nanocrystals. The new equations are derived from the Takagi-Howie-Whelan equations and have been used to calculate the diffracted and transmitted waves for a crystal containing a screw dislocation.

Key words: electron diffraction, nanocrystal, dynamical diffraction

1. INTRODUCTION

Up to now, the dynamical approach developed by Takagi is applied in electron diffraction structure analysis for accurate diagnosis of a thick distorted nanocrystals. Within this approach, the amplitudes of the transmitted and diffracted waves $\Psi_{0,g}$ are given by the following Takagi-Howie-Whelan (THW) equations [1]:

$$\begin{cases} \dfrac{d\Psi_0(z)}{dz} = \dfrac{i\pi}{\xi_g}\Psi_g(z) & (1) \\[2ex] \dfrac{d\Psi_g(z)}{dz} = \dfrac{i\pi}{\xi_g}\Psi_0(z) + 2i\pi\big(s + \vec{g}\vec{u}'(z)\big)\Psi_g(z) & (2) \end{cases}$$

where $\vec{u}(z)$, s, ξ_g stand for a displacement field at the depth in the crystal z, the deviation from the Bragg's low, the extinction length, respectively. These equations have the hyperbolic form and may be solved

517

T.E. Weirich et al. (eds.), Electron Crystallography, 517–521.

rigorously by the method of Riemann's [2]. In doing so, one can obtain the amplitudes $\Psi_{0,g}$ in the terms of the Riemann's functions. However, these functions have a very complicated form and, therefore, the great difficulties may happen, when they are employed for structure determination. This problem is the most serious in the case of a strong deformations and is due to intensification of the interbranch transitions. As a result of interbranch processes, electrons are redistributed between the Bloch's waves considerably and interpretation of electron diffraction data runs into great difficulties. In the given paper we develop the new dynamical approach, which is based on the 'Eikonal representation' for amplitudes of the transmitted and diffracted waves. In such way, we could separate contributions of interbranch and into-a-branch processes to amplitudes of these waves and simplify study of interbranch transitions. Using this approach, we examined electron diffraction by distorted crystal with screw dislocation and calculated the asymptotic expressions for intensities of the diffracted and transmitted waves.

2.   RESULTS

With the help of the substitutions $\Psi_{0,g} = \widetilde{\Psi}_{0,g}(z)\exp\left\{i\int q(z)dz\right\}$ where $q(z) = \pi\left(s + \vec{g}\vec{u}'(z)\right)$, it is convenient to rearrange the THW equations (1), (2) to the form:

$$\frac{d^2\widetilde{\Psi}_{0,g}(z)}{dz^2} + \left\{\left(\frac{\pi^2}{\xi_g^2} + q^2(z)\right) \pm i\pi\vec{g}\vec{u}''(z)\right\}\widetilde{\Psi}_{0,g}(z) = 0 \quad (3)$$

In the case of a slightly deformed crystal, the Eikonal approach can be applied for solution of Eqs. (3) approximately. This approach for electrons and X-Rays scattering in distorted crystalline media was developed by Kato [3], who used the wave-optical foundation of geometrical optics in inhomogeneous media. Moreover, the 'modified Bloch-waves' associated with the Eikonal solutions were also introduced by Kato so that to study Pendellosung fringes in distorted crystal. However, the 'modified Bloch-waves' describe into-a-branch dynamical process only, which is specified by moving of the tiepoints along the branches of the dispersion surface. This fact may be used to separate contributions of interbranch and into-a-branch processes. Then, passing from the Bloch-waves representation to Darvin approximation of the dynamical theory for amplitudes of the transmitted and diffracted waves, we write the amplitudes $\widetilde{\Psi}_{0,g}(z)$ as follows:

$$\tilde{\Psi}_{0,g}(z) = \sum_{j=1}^{2} C_{0,g}^{j}(z) \exp\left\{i \int Q_{0,g}^{j}(z) dz\right\} / \sqrt{p(z)} \qquad (4)$$

where $p(z) = \sqrt{q^2(z) + \pi^2/\xi_g^2}$; $Q_{0,g}^1(z) = p(z) + i\pi\vec{g}\vec{u}''(z)/(2p(z))$ and $Q_{0,g}^1(z) = -Q_{0,g}^2(z)$. In the expression (4), the exponential factors divided by $p^{-1/2}(z)$ are the Eikonal solutions of (3). Expression (4) introduces a new representation of the dynamical diffraction, which we call by the 'Eikonal representation'. Within this representation, the amplitudes $\tilde{\Psi}_{0,g}(z)$ may be expressed by sum of the modulated Eikonal solutions of the THW equations. This means that interbranch process may be treated as a quantum beat effect [4], such that the coefficients $C_{0,g}^{j}(z)$ should be considered as a modulated amplitudes, which describe interbranch transitions only. Substituting expression (4) into equations (1) and (2), we obtain the following set of the fundamental equations for $C_{0,g}^{j}(z)$

$$\frac{dC_{0,g}^1(z)}{dz} = -\frac{\pi\varepsilon(z)}{\xi_g} \frac{\exp\left\{\pm i\pi\vec{g}\int \vec{u}''(z)dz/p(z)\right\}}{p^2(z)\left(\omega \pm \sqrt{1+\omega^2}\right)} C_{0,g}^2(z) \qquad (5)$$

$$\frac{dC_{0,g}^2(z)}{dz} = -\frac{\pi\varepsilon(z)}{\xi_g} \frac{\exp\left\{\mp i\pi\vec{g}\int \vec{u}''(z)dz/p(z)\right\}}{p^2(z)\left(\omega \mp \sqrt{1+\omega^2}\right)} C_{0,g}^1(z) \qquad (6)$$

where $\varepsilon(z) = \xi_g^2 \vec{g}\vec{u}''(z)/(2\pi)$, and $\omega = s\xi_g^{-1}$. In the present work, we study electron diffraction by crystal with a screw dislocation parallel to the exit surface. For this purpose we exploit the idea about resonance interbranch scattering by a distorted crystal, which has been given in [5]. Following this idea, we introduce into the consideration the resonance extremal notation. By this one, we will understand the curve along which the local Bragg's conditions are satisfied exactly. Obviously, this curve can be determined by means of the relation:

$$s + \vec{g}\, d\vec{u}/dz = 0 \qquad (7)$$

where according to [6] the displacement field $\vec{u} = \vec{b}\, arctg\{z - y/x\}/(2\pi)$, such that $\vec{b}$ is Burgers vector and x, y stand for distance from dislocation in direction parallel to the surface, the depth of its location, respectively. Then, one can obtain from (7) that the resonance extremal for a screw dislocation is given by the expression:

$$z_{ext} = y + \sqrt{nx/(2\pi|s|) - x^2} \qquad (8)$$

Here $n = \vec{g}\vec{b}$ and we consider $s < 0$. Due to the resonance character of interbranch process near the resonance extremal, one can establish by using (5) and (6) that the coefficients $C_0^{1,2}(z_{ext})$ have to contribute mainly to the diffracted intensity. Bearing this in mind, we can get the following expression for the asymptote of the diffracted intensity $R = |\Psi_g|^2$ when $z \gg y$ and ω is sufficiently large:

$$R = \frac{1}{2}\left(1 - \frac{\omega}{\sqrt{1+\omega^2}}\right)\frac{1}{1+|\varepsilon(z_{ext})|^2/4} \tag{9}$$

Using expression (8), we can find the value $\varepsilon(z_{ext})$ so that to calculate the intensity R with the help of (9). At the same time, because we neglect inelastic scattering, asymptotic intensity of the transmitted wave $T = |\Psi_0|^2 = 1 - R$. In the Fig. 1 (a) and (b) R intensities are plotted as versus x/ξ_g for $n = 1$ and $n = 2$, respectively.

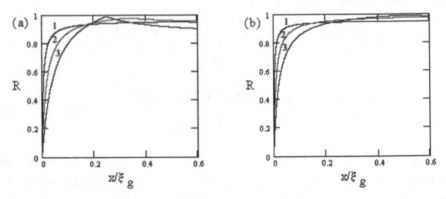

Figure 1. R intensity as a function of x/ξ_g for (a) $n = 1$ and (b) $n = 2$.
The curves 1,2,3 correspond to $\omega = -4\pi, -6\pi, -8\pi$, respectively

As is seen from these figures, the diffracted intensity becomes vanishingly small near the dislocation. Such decrease of the value R is associated with the intensification of the interbranch transitions of the transmitted electrons. Due to this fact, intensity of the transmitted wave increases considerably in this region. It should be noted, that the larger the angle variable ω the stronger this effect is. Indeed, the steepness of the R intensity curve decreases with increasing ω. On the other hand, appreciable weakening of interbranch transitions and domination of the into-a-branch process takes place far from dislocation [7]. Therefore, in accordance with the well-known prediction of the Eikonal theory, R intensity tends to its maximal value 1 as x/ξ_g increases.

It should be noted that R intensity depends on the diffraction number too. Comparing the appropriate curves in the Fig. 1 (a) and (b), one can establish that in the case of $n = 2$, interbranch process extinguishes in more fast way. Consequently, the energy of incident beam may be used as characteristic parameter to vary interbranch transitions.

3.   CONCLUSION

The original approach called the 'Eikonal representation' was presented to investigate the electron dynamical diffraction by deformed nanocrystals.

The new fundamental equations of the dynamical theory were written within the 'Eikonal representation'. Such equations are valid for any kind and strength of a regular deformation and, in opposite to the THW equations, they describe only interbranch transitions of electrons.

The asymptotic expressions for intensities of the diffracted and transmitted waves were obtained for crystal with a screw dislocation, which are also correct in the case of intensive interbranch transitions near the dislocation.

References

1. Takagi S. (1962). Acta Cryst. **15**, 1311.
2. Takagi S. (1969). J. Phys. Soc. Jap. **26**, 1239.
3. Kato N. (1963). J. Phys. Soc. Jap. **18**, 1785.
4. Shevchenko M. & Pobydaylo O. (2003). Acta Cryst. **A59**, 45.
5. Shevchenko M. (2003). Acta Cryst. **A59**, 481.
6. Friedel J. (1964). *Dislocations*. Pergamon Press.
7. Authier A. (2001). *Dynamical Theory of X-Ray Diffraction*. Oxford University Press.

ELECTRON MICROSCOPY INVESTIGATIONS OF GLASSY-LIKE CARBON

Shumilova T.G.*, Akai J.**, Golubev Ye.A.*
*Institute of Geology, Komi SC, UB, RAS, Syktyvkar, Russia, **Niigata University, Niigata, Japan, E-mail: shumilova@geo.komisc.ru

Key words: HRTEM, glassy-like carbon, AFM, SEM

High-resolution methods of substances investigations allow understand not only structure of those or that material but may be used more much widely in application fields. One of them may be mineralogical-geological reconstructions which get possibility to discover new mineral deposits and new natural raw materials [1].

Especially methods of electron microscopy are important at study of X-ray amorphous substances and polyphase nanomixtures which are distributed very widely in the nature such as agate, bauxite, bitumen, coal, natural glasses etc.. as X-ray diffraction is almost useless at analyzing such mostly disordered materials.

By this work we would like to demonstrate the complex analysis of different high-resolution methods data and genetic nature of materials demonstrating possibility of various mechanisms of their forming. At our experimental work we used X-ray almost amorphous glassy-like carbon of different genesis – glassy-carbon (SU-2000 standard) produced by the Novocherkassk electrode factory (Russia) and natural glassy-like carbon – shungite (Shunga deposit, Russia).

During investigations we were analyzing samples by methods of X-ray diffraction, electron scanning microscopy, microprobe analysis, atomic force microscopy, high-resolution transmission electron microscopy with preliminary attracting of the another methods including optical microscopy, Raman spectroscopy, thermal analysis and some of others.

T.E. Weirich et al. (eds.), Electron Crystallography, 523–526.
© 2006 *Springer. Printed in the Netherlands.*

By means of scanning electron microscopy and microprobe analysis the alike character of fresh crushed surfaces for both analyzed substances was established (Fig. 1, 2). It demonstrates the large smooth areas with the special form of edge parts with ribs of different levels (Fig. 3, 4). The ribs have subparallel orientation and may be restructured by crossing ribs of smaller level. Natural shungite is characterized by the thicker ribs, more roughness and presence of more quantity of mineral inclusions such as quarts, micas, feldspatoids et.al. Glassy-carbon includes some of geteroelements – Al, S, Ca, Ti. By the other hand both are quite pure carbon substances with carbon content 98-99 %.

By the method of atomic force microscopy it was established the both substances have globule-like structure with the globules size about 50 nm in diameter (Fig. 5, 6). According to detail analysis of globule-like forms it is clear that shungite consists of a bit outstretched globules that provides lightly anisotropic structure. By this reason optical anisotropy may be watched at reflection with crossed nikols. At the same time glassy-carbon is absolutely optically isotropic.

According to the data of high-resolution transmission microscopy glassy-like carbon is presented as really amorphous substance mainly with groups of carbon atoms collected to curved subparallel surfaces laying among nonstructured atoms (Fig. 7). Such structure feature belongs as to glassy-like carbon [1] as to shungite [2].

Analysis of the all received data shows the both substances have similar images which demonstrate onion-like structure with tight composition of globule-like units and alike physical and chemical properties. Just small difference between them exists – natural glassy-like carbon is a little crumbly that explains light variations of their other features.

Summarizing the structural data described above and conditions of substances forming we conclude the same carbon structures may be formed at extremely various P-T-conditions by different mechanisms as glassy-carbon is produced at temperatures 2000-3000^{0}C, but natural shungite at the conditions of sedimentary rocks (no more than 300^{0}C).

Glassy-carbon is an industrially important carbon material. In addition the received data allow demonstrate shungite is the analogue of synthetic glassy-carbon. It let us preview shungite may be used at the same industrial needs like glassy-carbon.

The investigations were supported by the Russian Science Support Foundation (2004 year project).

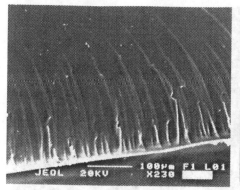

Figure 1. SEM image of glassy-carbon.

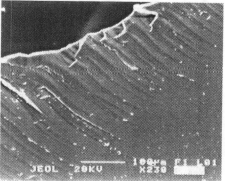

Figure 2. SEM image of shungite.

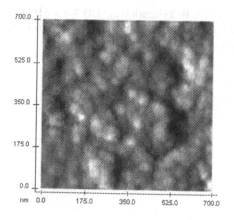

Figure 3. SEM image of glassy-carbon ribs.

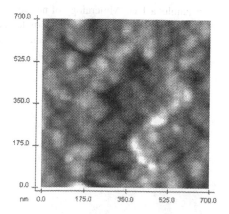

Figure 4. SEM image of shungite ribs.

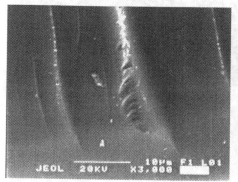

Figure 5. AFM image of glassy-carbon.

Figure 6. AFM image of shungite.

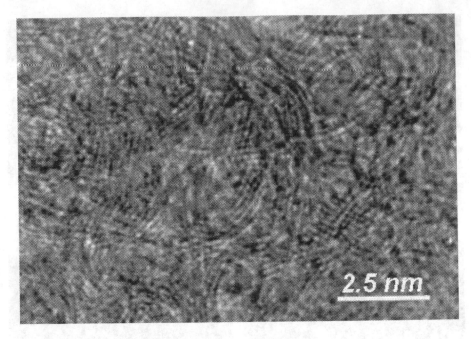

Figure 7. HRTEM image of glassy-like carbon.

References

1. Shumilova T.G. Mineralogy of native carbon (2003). Yekaterinburg: UD RAS, 316 p.
2. Kovalevski V.V., Buseck P.R., Cowley J.M. Comparison of carbon in shungite rocks to other natural carbons: An X–ray and TEM study (2001). Carbon, 39, 243-256 pp.

IMAGE DECONVOLUTION ON CRYSTALS WITH INTERFACES

C.Y. Tang and F.H. Li*
Beijing National Laboratory for Condensed Matter Physics, Institute of Physics, Chinese Academy of Sciences, Beijing 100080, China

Key words: HRTEM, image deconvolution, interface structure

1. INTRODUCTION

Interfaces in crystals are of great importance on the mechanical, physical and chemical properties of materials. It has long been recognized that our understanding of interface properties depends strongly on the availability of information about the atomic configuration of the interface. Many efforts have been made to improve the resolution of electron microscope images aiming at reaching to 0.1 nm or better. Image resolution of 0.089 to 0.11 nm was attained with the high-voltage high-resolution microscope operated at 1.25MV [1, 2]. The point resolution of amplitude-contrast images taken with the spherical aberration-corrected microscope can extend to the information limit [3-5]. Pairs of Cd and Te columns separated by 0.093 nm were resolved in high-angle annular dark-field images taken with the scanning transmission electron microscope [6]. The three means mentioned above provide a perspective of crystal structure at atomic scale directly. On the other hand, the resolution of images taken with the most popular medium-voltage field-emission microscopes can be improved by posterior image processing, for instance, the exit wave reconstruction [7, 8] and image deconvolution [9, 10]. In both cases, the resolution of images taken with the medium voltage field-emission microscope can be improved down to the microscope information limit together with the crystal structure restoration. The present

527

T.E. Weirich et al. (eds.), Electron Crystallography, 527–531.
© 2006 Springer. Printed in the Netherlands.

paper aims at developing the image deconvolution technique to study the planar defects following to Refs. [10-12]. The focus is on the peculiarity of the Fourier filtering and dynamical scattering effect correction for this purpose. A test for simulated field-emission electron microscope images of Si {111} twin boundary has been carried out.

2.   IMAGE PROCESSING

Images of Si [110] consisting of 1024×1024 pixels were simulated by multislice method. The simulation parameters were set by referring to field-emission electron microscopes with accelerating voltage 200 kV, spherical aberration coefficient 0.5 mm, defocus spread due to the chromatic aberration 3.8 nm and defocus values -120 nm. The corresponding CTF curve is shown in Fig. 1, in which the four independent reflections 111, 220, 113 and 004 of Si up to the resolution 0.136 nm are indicated as line segments with the length proportional to the amplitude. All the four reflections are far from zero crosses of CTF curves. It indicates that images obtained at this defocus value are appropriate for image deconvolution [10]. Figs. 2a-2d are the center parts of simulated images with crystal thickness 0.384 nm, 3.84nm, 7.68 nm and 9.60 nm, respectively. The vertical {111} twin boundary lies in the middle in each image.

At present the initial images are obtained by simulation and the defocus value is known. For the case of experimental images, the defocus value can be determined roughly by a Thon diffractogram [13] for the amorphous region at the edge of the crystal and then refined by matching the image contrast of the deconvoluted image for the perfect structure region [12]. The deconvoluted diffractograms shown in Figs. 2e-2h are obtained after deducting the modulation of CTF from the Fourier transforms of the simulated images given in Figs. 2a-2d, respectively, with the transmitted beam omitted. In Fig. 2e, four main independent reflections $\bar{1}11$, $\bar{2}20$, $\bar{1}13$ and 004 are pointed. It can be seen by comparing these diffractograms with one another carefully that the intensity of weak reflection increases with the thickness increasing and vice versa. Some forbidden reflections such as 002 and $\bar{2}22$ (pointed in Figs. 2f-2h) appear. The intensity of forbidden reflections for big thickness becomes even stronger than that of the fundamental reflections. Such phenomena are in accordance with the behavior of dynamical scattering effect. Therefore, to obtain the right structure map, it is necessary to correct the dynamical scattering effect.

According to Ref. [11], the dynamical scattering effect was reduced by forcing the integral amplitudes of reflections to be equal to the corresponding structure factor amplitudes of the corresponding perfect crystal. Like electron diffraction patterns for crystals with planar defects, the reflections in the deconvoluted diffractogram are elongated. Therefore, different from the circle windows utilized in Ref. [11], here elliptical windows were applied to every reflection in the deconvoluted diffractograms to perform the Fourier filtering and amplitude integration, of which the short axis is 0.4 nm^{-1} and long axis 0.8 nm^{-1}. The length of axes is designed carefully so that most contribution of boundary area of interface could be included in while the overlapping of adjacent reflections be avoided. After the Fourier filtering and amplitude correction, the resultant patterns named as corrected deconvoluted diffractograms are shown in Figs. 2i-2l, in which all forbidden reflections were omitted and no obvious change of fundamental reflections is seen with the increase of thickness. This implies the success of dynamical scattering effect correction.

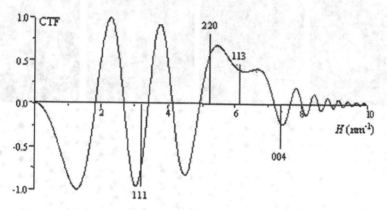

Figure 1. CTF curve for defocus value -120 nm.

The inverse FT of the corrected diffractogram yields to the corrected deconvoluted images, which is corresponding to the pseudo projected potential map (PPM), revealing the projected structure with correct atomic positions, but the contrast of heavy atoms lower than the actual contrast and vice versa. The inverse FTs of Figs. 2i-2l are shown in Figs. 2m-2p, respectively. All the four corrected deconvoluted images show the twin structure at the interface with atom columns revealed clearly as black dots. The measured distances between two adjacent black dots of the atom pairs in the perfect area are exactly the same as those in the PPM and slightly larger for the pairs at the interface. In the latter case, the average deviation of measured distances from the ones in the PPM is about 0.007 nm. Atomic columns at the interface were revealed correctly with rather accurate atomic positions in the final images.

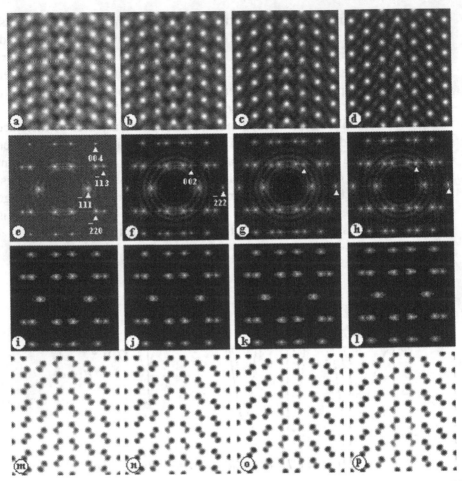

Figure 2. (a)-(d) Simulated images of Si [110] with {111} twin boundary, the defocus is -120 nm and crystal thickness is 0.384 nm, 3.84 nm, 7.62 nm and 9.60 nm, respectively. (e)-(h) Deconvoluted diffractograms and (i)-(l) corrected deconvoluted diffractograms corresponding to (a)-(d), respectively. (m)-(p) Corrected deconvoluted images obtained from inverse FTs of (i)-(l), respectively

3.   CONCLUSION

It has been demonstrated that the dynamical scattering effect correction technique is also effective in image deconvolution for restoring the atomic configuration for crystals with interface when the elliptical windows are applied. The technique is essential in improving the quality of deconvoluted images so that the available crystal thickness extends to 10 nm or even bigger for Si.

This project is supported by the National Natural Science Foundation of China (grant 50072043). C.Y. Tang acknowledges Dr. D. Wang for the helpful discussion.

References

[1] E. Takuma, H. Ichinose and F.R. Chen, J. Electron Microsc. 51 (2002) 297.

[2] H. Ichinose, in: Proc. International Kunming Symposium on Microscopy (2000) 11.

[3] M. Haider, S. Uhlemann, E. Schwan, H. Rose, B. Kabius and K. Urban, Nature 392 (1998) 768.

[4] M. Lentzen, B. Jahnen, C.L. Jia, A. Thust, K. Tillmann and K. Urban, Ultramicroscopy 92 (2002) 233.

[5] C.L. Jia, M. Lentzen and K. Urban, Science 299 (2003) 870.

[6] P.D. Nelist and S.J. Pennycook, Phys. Rev. Lett. 81 (1998) 4156.

[7] P. Schiske (1968), in: Proceedings of the 4th European Conference on Electron Microscopy, Rome 145.

[8] D. Van Dyck, H. Lichte and D. Van der Mast, Ultramicroscopy. 64 (1996) 1.

[9] F.H. Li and H.F. Fan, Acta Phys. Sinica 28 (1979) 267.

[10] W.Z. He, F.H. Li, H. Chen, K. Kawasaki and T. Oikawa, Ultramicroscopy 70 (1997) 1.

[11] F.H. Li, D. Wang, W.Z. He and H. Jiang, J. Electron Microsc. 49 (2000) 17.

[12] D. Wang, J. Zou, W.Z. He, H. Chen, F.H. Li, K. Kawasaki and T. Oikawa, Ultramicroscopy 98 (2004) 259.

[13] F. Thon, in: Electron Microscopy in Materials Science, ed. Valdre U (1971) pp. 570, (Academic Press, New York and London).

CRYSTAL STRUCTURE DETERMINATION BY MAXIMUM ENTROPY IMAGE DECONVOLUTION IN COMBINATION WITH IMAGE SIMULATION

Y.M. Wang, H.B. Wang and F. H. Li

Beijing National Laboratory for Condensed Matter Physics, Institute of Physics, Chinese Academy of Sciences, P.O. Box 603, Beijing 100080, People's Republic of China

Key words: HRTEM, image deconvolution, maximum entropy method

In the present paper, the crystal structure of $Nd_{1.85}Ce_{0.15}CuO_{4-\delta}$ has been studied by maximum entropy image deconvolution in combination with image simulation. This work is focused on demonstrating the effectiveness of image deconvolution technique rather than determining the structure itself.

Specimens of single crystal $Nd_{1.85}Ce_{0.15}CuO_{4-\delta}$ were observed by JEM-2010 high-resolution electron microscope operated at accelerating voltage 200 kV. Electron diffraction study indicates that the crystal structure of $Nd_{1.85}Ce_{0.15}CuO_{4-\delta}$ belongs to the tetragonal system with $a = 0.394$ and $c =1.209$ nm. Fig.1a and b show respectively the [001] and [100] electron diffraction patterns (EDPs) which indicate that the possible [100] plane groups of the crystal are *cm* and *c2mm*. Because the phase residual for symmetry *c2mm* is smaller than that for *cm*, the correct plane group is *c2mm*. One of the images taken near the Scherzer focus condition is shown in Fig. 1c together with the inset corresponding diffractogram. The Fourier filtering and symmetry averaging were performed with VEC program [1] on the circled thin area in Fig. 1c and the symmetry-averaged image is shown in Fig. 2a. Image deconvolution is carried out based on the principle of maximum entropy, which has been described in detail in Refs. [2, 3]. It is necessary to select the correct deconvoluted image carefully among several possible solutions because of the Fourier image effect [4, 5]. Fig. 2b shows

T.E. Weirich et al. (eds.), Electron Crystallography, 533–536.

the correct deconvoluted image where all metallic atoms are clearly resolved as black dots, while oxygen atoms are not visible owing to the insufficient resolution of electron microscope. Referring to the deconvoluted image and the crystallographic chemistry information, the structure model was constructed (see Fig.2c). Images were simulated with different defocus values and crystal thickness. The simulated image for thickness 4.334 nm and defocus value −36 nm (Fig. 2d) fits symmetry averaged image (Fig. 2a).

Other images from the through-focus series were processed one by one with the same procedure. The corresponding symmetry averaged and deconvoluted images are shown in top and middle rows of Fig. 3. It can be seen that all deconvoluted images are in agreement with one another in contrast. The simulated images of thickness 4.334nm are in accordance with the experimental images when the defocus values are equal or close to the ones utilized for obtaining the corresponding deconvoluted images. They are shown in the bottom row of Fig. 3.

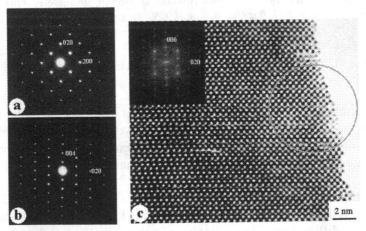

Fig. 1. (a) [001], (b) [100] EDPs of $Nd_{1.85}Ce_{0.15}CuO_{4-\delta}$ and (c) [100] experimental image with the corresponding diffractogram inserted on the top left.

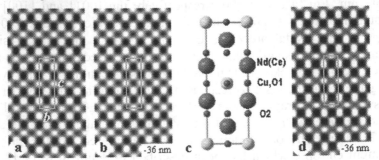

Fig. 2 (a) symmetry averaged image, (b) deconvoluted image corresponding to figure 1(c), (c) constructed model and (d) simulated image.

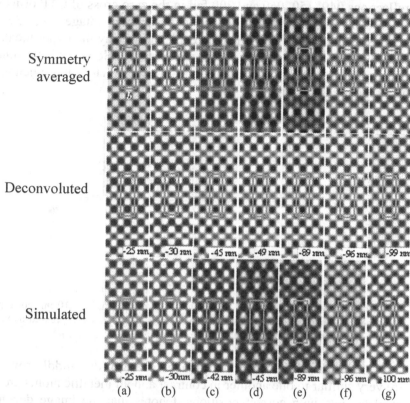

Symmetry averaged

Deconvoluted

Simulated

Fig. 3. Experimental average images (top row), deconvoluted images (middle row) and simulated images (bottom row) with $U = 200$ kV, $C_s = 0.5$ mm, $D = 10$ nm, and crystal thickness 4.334 nm. The defocus value is given in every image on the bottom right.

It can be seen that all simulated images fit the corresponding experimental ones shown in the top row of Fig. 3. It is reasonable to say that the proposed model is correct.

To avoid the infinity or big error of the calculated structure factor set, in most of previous works dealing with the image deconvolution [2, 6, 7, 8], a threshold value of CTF was set up to exclude those reflections which fall in the zero cross of CTF or its vicinity. Later a modification was made to include the number of reflections forming the deconvoluted image as large as possible [9]. In the modified method when the absolute value of CTF corresponding to a reflection is smaller than the threshold value, it is replaced by the threshold value so that this reflection remains included. Such modification was employed in the present paper. The CTF curve shown in Fig. 4 corresponds to the image given in the top of Fig. 3e. There are altogether seven independent reflections contributing to the image within the present resolution. In the CTF curve each reflection is expressed as a vertical bar with the length proportional to the modules. It can be seen that reflections 040, 150, 060 and 200 fall in the zero cross of CTF or its vicinity. Hence, their contributions to the experimental image Fig. 3e top are seriously attenuated. This offers an explanation why the experimental image (see Fig.3e) has a rather dark appearance. However, the quality of deconvoluted image is as good as that of others. The same phenomenon is also seen in Fig. 3c and d.

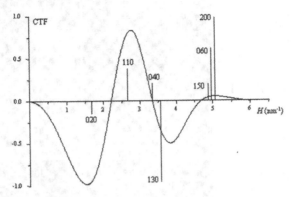

Fig. 4. CTF curve calculated with $U = 200$ kV $C_s = 0.5$ mm, $D = 10$ nm corresponding to experimental images figure 4e. Seven independent reflections are expressed as vertical bars with the length proportional to the amplitude values.

The fact that all deconvoluted images given in the middle row of Fig. 3 are very similar to one another in contrast and all metallic atoms are resolved as black dots with correct positions denotes that the image deconvolution technique is powerful to transform the image taken at an arbitrary defocus into the structure image. It is for the first time to clarify that the deconvoluted images still reveal the projected structure even if some reflections fall in the vicinity of zero cross of CTF.

References

[1] Z.H.Wan, Y.D. Liu, Z.Q. Fu, Y. Li, T.Z. Cheng, F.H. Li, and H.F. Fan, Z. Kristallogr. 218 (2003) 308.

[2] J.J. Hu and F.H. Li, Ultramicroscopy 35 (1991) 339.

[3] H.B.Wang, Y.M. Wang and F.H. Li, Ultramicroscopy 99 (2004) 165.

[4] J.M. Cowley and A.F. Moodie, Proc. Phys. Soc. B 70 (1957) 486.

[5] J.M. Cowley and A.F. Moodie, Proc. Phys. Soc. 76 (1960) 378-384.

[6] D.X. Huang, W.Z. He and F.H. Li, Ultramicroscopy 62 (1996) 141.

[7] F.S. Han, H.F. Fan and F.H. Li, Acta Crystallogr. Sect. A 42 (1986) 353.

[8] J.J. Hu, F.H. Li and H.F. Fan, Ultramicroscopy 41 (1992) 387.

[9] S.X. Yang and F.H. Li, Ultra microscopy 85 (2000) 51.